Substituent Effects in
Organic Polarography

Substituent Effects in
Organic Polarography

Petr Zuman
Heyrovský Institute of Polarography
Czechoslovak Academy of Sciences
Prague, Czechoslovakia

With a Foreword by
J. Heyrovský

℗ PLENUM PRESS · NEW YORK · 1967

ISBN 978-1-4684-8663-6   ISBN 978-1-4684-8661-2 (eBook)
DOI 10.1007/978-1-4684-8661-2

Library of Congress Catalog Card Number 63-17643

©1967 Plenum Press
Softcover reprint of the hardcover 1st edition 1967

A Division of Plenum Publishing Corporation
227 West 17 Street, New York, N. Y. 10011
All rights reserved

To the memory of
Professor J. Heyrovský,
my esteemed teacher

Foreword

During the forty years which have passed since Masuzo Shikata published his paper on the reduction of nitrobenzene at a dropping mercury electrode, the number of polarographic studies of organic compounds in the literature has risen to several thousands. The ever-increasing amount of experimental data was in need of some unified method of classification which would yield unambiguous and possibly complete information on the polarographic behavior of organic substances.

Dr. Zuman's book presents an original attempt to meet this need by providing a system based on correlations between the polarographic half-wave potentials of organic depolarizers and their Hammett constants. I consider this a very happy conception, for, more than any other book yet written, it brings polarography nearer to the organic chemist; and it will undoubtedly convince him that, in its application to his subject, the method is more than a mere analytical tool.

The author hardly needs any introduction. During many years of research in the field of organic polarography, he has published numerous papers on a variety of problems; his latest interest is the application of the Hammett–Taft equation to polarographic measurements, in which he has done pioneering work.

It remains for me to hope that this book, which opens up new prospects for the fruitful application of polarography, may inspire some reader with useful ideas in his search for new paths in his research problems.

Prague   J. HEYROVSKÝ
November, 1966   **1890–1967**

Preface

"Habent sua fata libelli"

Though the above quotation usually is made about books which have already been published, the fates of manuscripts which have not yet appeared in print may sometimes be even more devious and uncertain. Even among scientific books, there are some for which the road between the first draft and the final text is short and smooth, but for many it is unfortunately long and tedious. These are the cases in which the "finish" in Michael Faraday's "Work, finish, publish!" is definitely the most difficult of the three, and it is to this category that the present manuscript belongs.

Its story started as early as 1955, when Plenum Press asked me for permission to prepare and publish an English translation of my review article entitled "Effects of Constitution on the Polarographic Behavior of Organic Substances," which appeared in Czech in *Chemicky Listy* in 1954. I agreed, but at the same time I made a decision which—young authors note and be warned!—was much easier to make than to implement, namely, to rewrite the article and bring it up to date. As the years passed, this project merged with one associated with my research interests, and it culminated in a D.Sc. thesis entitled "General Equation for Substituent Effects on Half-Wave Potentials of Organic Substances." The collection of material for the projected book was complete in 1959, but the matter was not in the form of a final manuscript, and also the question of a translator remained open. In the meantime I tried to improve my English, and during a short but pleasant visit to the New York offices of Plenum Press (where it was realized with some surprise that I am indeed a human being, and not just a ghost appearing only in the correspondence) it was agreed that I, myself, should prepare an English manuscript. However, during the rewriting of the text, which started in 1963, I found that the appearance of new material since 1959 made it necessary to make a very extensive revision of most of the chapters. These changes were sometimes so substantial that, for example, the last chapter (Alicyclic Compounds), which was some 12 pages long in the 1959 version, grew to about 60 in 1965. May I, again, interpose a word to younger authors based on my experience with two other monographs? It is much more pleasant and interesting to make a new dress than to remake an old one! Because this revision was time-consuming, the coverage of the literature varies; whereas, the polarographic literature was considered up to at least 1963 in the first chapters and up to 1965 in the last ones, the treatment of the theory of linear free energy relations corresponds to its state in 1961–62. The important contributions published in the 1963–65 period could not be considered because of the technical problems involved.

I was extremely pleased when a reviewer described one of my books as "readable," but I am afraid that no such epithet can be applied to the present monograph. This book has two principal aims. The first is to show organic polarographers the types of structural problems which can be treated quantitatively, to stress the experimental conditions which must be observed for such treatments to be valid, and, above all, to show what kinds of experiments it would be important to carry out in the future and what criteria should be used in selecting organic compounds for polarographic studies. Furthermore, the present treatment reveals the possibility of generalizing interpretations of the effects of substituents on polarographic behavior, and it presents an advantageous organization of experimental data. From the electrochemical point of view, the present treatment is important in revealing the fundamental fact that electrode processes, which are essentially heterogeneous, are, in principle, affected by structure in the same way as homogeneous reactions.

The second aim is to show physical organic chemists the possibilities that polarography offers for the solution of fundamental, and also applied, problems in their field. The greatest obstacle, in this respect, is the limited number of compounds in most of the reaction series for which half-wave potentials have been measured so far.

My thanks must be expressed to Drs. O. Exner and V. Hanuš, who read the whole of the Czech manuscript, and to Dr. M. Charton (chapter on Quinones), and Drs. W. Klyne and V. Delaroff (chapter on Alicyclic Compounds), who read the manuscripts of some of the chapters in an uncomplete form.

My special thanks are due to Dr. A. E. Stubbs, who not only polished by English and style with a subtle feeling for my ideas and the subject matter, but also unified the scientific terminology and organic nomenclature and drew my attention to several inconsistencies. When I consider my own timetable, I am put to shame by the speed with which this thorough and careful revision work was done.

Last, but definitely not least, my thanks are directed to the publisher and all his associates for their kind patience and understanding. One of the editors recently wrote "I look forward to the publication with considerable satisfaction; that a project of so many years' standing has finally become reality affirms one's belief that not all human planning is in vain," and this best expresses their kind and courteous attitude to all the troubles which they have had with me and my manuscript. I thank them.

Prague   P. ZUMAN
October, 1966

Contents

Chapter VI

Effects of Substituents in Condensed Polycyclic Hydrocarbons . 219

Chapter VII

Effects of Substituents in Polycyclic Heterocyclic Compounds . 247

Chapter VIII

Chapter IX

The Most Important Equations

$$\Delta E_{1/2} = P + M_\pi + S \quad (20) \qquad \Delta E_{1/2} = \rho_{\pi,R}^* \sigma_X^* + M_\pi \qquad\qquad (33)$$

$$\Delta E_{1/2} = \rho_{\pi,R}^I \sigma_X^I \qquad\qquad (25) \qquad \Delta E_{1/2} = \rho_{\pi,R}\sigma_X + \delta_{\pi,R}(E_S)_X \qquad (36)$$

$$\Delta E_{1/2} = \rho_{\pi,R}\sigma_X \qquad\qquad (27) \qquad \Delta E_{1/2} = \rho_{\pi,R}\sigma_{o-X} + \delta_{\pi,R}(E_S)_{o-X} \quad (37)$$

$$\Delta E_{1/2} = \rho_{\pi,R}^* \sigma_X^* \qquad\qquad (29) \qquad \Delta E_{1/2} = \rho_{\pi,R}\Sigma\sigma_X \qquad\qquad (38)$$

$$\Delta E_{1/2} = \rho_{\pi,R}\sigma_X^- \qquad\qquad (30) \qquad \Delta E_{1/2} = \Sigma\rho_{\pi,R}^I \sigma_X \qquad\qquad (39a)$$

$$\Delta E_{1/2} = \rho_{\pi,R}\sigma_X^+ \qquad\qquad (31) \qquad E_{1/2} = \rho_{\pi,R}\sigma_X + \varkappa\rho_{\pi,R} + K \qquad (44)$$

$$(E_{1/2})_{1,X} - (E_{1/2})_{1,H} = \rho_\pi'[(E_{1/2})_{2,X} - (E_{1/2})_{2,H}] \qquad (49)$$

I: Introduction

1. Development of the Study of Structural Relations

The discovery of the reducibility or oxidizability of certain organic compounds at a dropping mercury electrode was followed very soon by a search for relations between the polarographic behavior of organic substances and their structure. Heyrovský[1,2] was the first to correlate these properties for a representative number of compounds on the basis of the available experimental evidence. He stressed especially the significant role that conjugated double and triple bonds and aromatic rings play in determining the reducibility of a given compound. From the available material Heyrovský deduced that polarographic reduction becomes easier as the number of conjugated bonds in the organic molecule increases. The polarographic reduction wave is observed, therefore, at more positive potentials, the more extended the conjugated system in the molecule. This principle remains one of the important guiding rules for organic polarographers, even though various other factors have been recognized recently as affecting polarographic oxidizability or reducibility. It plays an especially important role in predicting whether a compound (which has not yet been studied polarographically and whose polarographic behavior has not been described in the literature) will undergo polarographic reduction in the available potential range.

The first attempt to describe the effects of substituents on half-wave potentials in the polarographic reduction of organic compounds was made by Shikata and Tachi.[3] They postulated the "electronegativity rule," which states that the more electronegative the substituent, the more positive the half-wave potential. They defined electronegativity in terms of the effect of the introduction of the substituent into the *para* position of benzoic acid on the dissociation of the acid. This rule satisfactorily described the direction and sequence of the observed shifts for a number of benzene compounds substituted in the ring and with the reducible group in the side chain. But it could not explain the shifts in the half-wave potentials of some straight-chain compounds (e.g., aliphatic nitro compounds in acid media, chloroacetic esters) for which shifts in a direction opposite to the expected were observed.

For a long time these two rules—the influence of conjugation and the electronegative rule—were the only guidelines for organic polarographers. Many review articles discussing the polarography of organic substances[3–20] exhibited original viewpoints and introduced different approaches to the subject, but hardly any of them made generalizations concerning the relations between polarographic data and the structure of organic compounds.

The first generalizations of our existing knowledge were possible[4,21] only after systematic attempts had been made to explain the structural effects in organic polarography in terms of theoretical organic chemistry and after the accumulation of sufficient experimental data. Efforts were made to explain the changes in the polarographic waves in terms of concepts developed to describe the different types of electron displacement occurring in organic molecules under the influence of substituents. These include the concepts of inductive I_s and inductomeric I_d effects, which express the displacement of electrons in one direction in the molecule. These effects are caused by substituents, or by external electric fields (such as those resulting from the presence of another molecule). For systems involving π-π or π-p conjugation, mesomeric M and electromeric E effects, which express the effect of a shift of an electron pair under the influence of a substituent or of an external field, also have been considered.

Inductive I_s and mesomeric M effects are considered to be static effects, i.e., effects which exist in the molecule after the introduction of the substituent even when the molecule is isolated and not undergoing chemical changes. Inductomeric I_d and electromeric E effects, on the other hand, are dynamic effects, exhibited by molecules during chemical reactions when the reacting molecule is influenced by the electric field of the reagent.

The static effects primarily affect equilibrium potentials which are measured in the polarography of reversible systems. In irreversible systems, for which the half-wave potential is a function of an activation energy, the predominant role is attributed to dynamic effects. Unfortunately, it is not possible to estimate the respective contributions of the static and dynamic effects quantitatively. It has proved useful, therefore, to denote the two first-mentioned effects simply as inductive effects I and to lump the latter two as tautomeric effects T, though the term "tautomeric" is somewhat misleading.

The effect of a substituent is expressed relative to that of hydrogen. If the substituent is electrophilic, i.e., more electronegative than the hydrogen atom, the effect is taken as negative $(-I, -T)$. If the group is nucleophilic, i.e., less electronegative than the hydrogen atom, and repels electrons, the resulting effect is taken as positive $(+I, +T)$. The effects of some common substituents can be summarized as follows:

NO_2	$-I$	$-T$	Cl	$-I$	$+T$
SO_3H	$-I$		CH_3	$+I$	$+T$
COOH	$-I$	$-T$	OCH_3	$-I$	$+T$
I	$-I$	$+T$	OH	$-I$	$+T$
Br	$-I$	$+T$	NH_2	$-I$	$+T$

In the application of this classification of substituent effects to polarographic data,[21] the following two rules have been recognized for reduction processes for which the potential is determined by a nucleophilic process:

1. Groups with $-I$ and $-T$ effects shift the half-wave potentials toward more positive values. Groups with $+I$ and $+T$ effects shift the half-wave potentials toward more negative values.

2. In the absence of other effects, the reduction of benzene derivatives with a reducible group in the side chain and a $-T$ substituent in the *ortho* or *para* position occurs at more positive potentials than that of the *meta* isomer. On the other hand, in the presence of a $+T$ group, the reduction of the *meta* derivatives takes place at a more positive potential than that of the *ortho* or *para* isomer.

The validity of these rules has been verified for the reduction of numerous quinonoid and aromatic carbocyclic and heterocyclic compounds, but they are often found inapplicable in the discussion of substituent effects in the polarographic reduction of aliphatic and alicyclic compounds and in some polarographic oxidations.

Even in those cases where the aforementioned rules prove applicable, they allow us to determine only whether the introduction of the given substituent will result in a shift to more positive or to more negative potentials. For some groups of substituents it is possible to predict what the sequence of the substituent effects is likely to be, i.e., to estimate which of the substituents will cause a greater, and which a smaller effect. But even where this is possible, it will not provide us with even an approximately quantitative measure of the value of the predicted shift.

Numerous examples of successful qualitative discussion of the shifts of half-wave potentials can be cited. In the early 1950's the application of I and T effects in the explanation of shifts of half-wave potentials caused by substituents was quite common in polarographic literature. Because most of these effects can be treated quantitatively, using the relations discussed later in this monograph, such examples of qualitative discussion are not cited here.

Further progress of our understanding of the relations between polarographic data and organic structure has been made possible by the considerable progress in the theory of organic chemistry. During the last three or four decades the evolution of this branch of chemistry has enabled us to move from a qualitative characterization of substituent effects to attempts to express these effects quantitatively. It has been possible to ascribe numerical coefficients to each particular substituent in a given type of molecular skeleton (the substitution being carried out in a given position relative to the reacting group) for a given type of reaction. These coefficients characterize the effect of the substituents on the values of the rate and equilibrium constants. The coefficients have been correlated in equations of the Hammett[22-24] and Taft[25] types (generally called "linear free energy relations," even though it has been stressed by Taft[25] that not only simple free energies may be involved) with rate and equilibrium constants for numerous homogeneous reactions.

During the last decade the theory of irreversible polarographic waves has been elaborated in considerable detail.[26-30] It has been deduced that for irreversible electrode processes there exists a simple relation between the half-wave potentials and the logarithm of the rate constant of the electrode reaction. On the basis of this and the progress in theoretical organic chemistry, it has been possible to show[31-34] that in electrode processes, which are largely heterogeneous, quantitative relations hold which are analogous to those applicable to homogeneous reactions.

In this monograph, the experimental evidence has been summarized, and it has been shown that polarographic data can be treated by the use of coefficients which characterize the effects of substituents quantitatively. Such a treatment enables us to construct a cohesive and uniform picture of structural effects on half-wave potentials in the light of the present knowledge of physical organic chemistry.

For this purpose, first a general equation for the shift in half-wave potential due to substituent effects is derived. The possibility of the application of special forms of this equation in various groups of organic compounds then is demonstrated. The importance of such relations for polarographic theory and practice is discussed at some length. Their general validity is substantiated by the fact that, so far, they have been shown to apply to the half-wave potentials of over 600 benzene derivatives, over 100 monocyclic heterocyclic compounds, and over 100 polycyclic substances, as well as to over 300 values for substances in which alkyl and aryl groups are replaced by one another, and to 100 oxidation-reduction potentials of quinonoid substances, either measured by us or given in the literature.

The success of this treatment of polarographic data has shown that the customary classification of compounds in organic polarography and the usual division of data are not the most advantageous. Organic compounds are usually first divided into classes in accordance with the polarographically active grouping present. Within each chapter devoted to a particular grouping the discussion may be subdivided into subclasses, e.g., aliphatic, aromatic, etc., compounds, each containing the same electroactive grouping. With such a division, the common features of substances bearing various polarographically active groups on the same molecular frame are usually lost, because these substances are scattered over several chapters. Moreover, such a division often leads to a search for analogies within a particular chapter, i.e., among substances bearing the same electroactive group on different types of molecular frames. Such a search is frequently unsuccessful since the reaction mechanism of the electrode process can be profoundly affected by the type of molecular frame that carries the polarographically active grouping. It will be demonstrated later that, for example, the reduction mechanism of aliphatic nitro compounds differs substantially from that of nitrobenzene derivatives. Similar differences also have been found for the reduction of haloalkanoic acids and aryl halides. Thus, it can be readily seen that it is not possible to find a general explanation of structural effects for all nitro compounds or all halogen derivatives.

It has been proved that a classification similar to the classical division of organic chemistry into aliphatic, alicyclic, and aromatic compounds also is useful as the first classification principle in organic polarography. Experimental evidence has shown that it is useful to include aryl derivatives which can be derived by exchanging aryl for some of the alkyl groups in the molecule with the straight-chain compounds of the first-named group, and that aromatic heterocyclic compounds may be included with benzene derivatives in the last-named group.

If this scheme is followed, it is found that in most of the subclasses the mechanism responsible for the electrode reaction of the polarographically active group is the same for most of the members. The suggested classification has clarified some factors that had remained unexplained, e.g., the role of the phenyl substituent, or the effect of the exchange of alkyl for hydrogen which results in a shift in the half-wave potentials sometimes toward more negative and sometimes toward more positive values. It can, thus, help us to a better understanding of structural effects within these homogeneous groups of substances and in organic polarography in general.

2. Classification of Structural Effects

Structural changes affect polarographic curves in various ways. They may cause shifts in the half-wave potentials, or may affect the diffusion coefficient as reflected in changes in the limiting current. Changes in the structure of an organic molecule can result in changes in the mechanism, reversibility, or rate of the electrode process. Moreover, polarography makes possible the study of structural effects on the rate of fast antecedent or consecutive reactions associated with the electrode process proper. Finally, it is possible to use the polarographic method as an analytical tool in the study of structural effects on the kinetics of certain homogeneous reactions of organic compounds in solution.

In this monograph, the discussion of effects of substituents on polarographic curves will be limited to the influence on half-wave potentials. However, at least it should be mentioned here that changes in an organic molecule also can be reflected on polarographic curves as changes in the dependence of the wave height, potential, or slope, respectively, on pH. Structural changes can cause changes in the number of waves observed, in the wave height (in the number of electrons exchanged), and in the character of the waves (e.g., diffusion, kinetic, catalytic, adsorption current, etc.). Sometimes a relatively small change in an organic molecule can result in a complete change in the character of the observed currents. For example, a relatively small structural change in a sugar molecule can transform the substance from one showing kinetic current into one showing diffusion-controlled current. Furthermore, depending on the structure of the depolarizer, the shape, height, and potential of the wave can depend in various ways on the composition of the supporting electrolyte and on the presence of surface-active substances. These factors must be kept in mind when optimum conditions for the comparison and discussion of structural relations are to be determined.

When the half-wave potentials of irreversible systems are compared, it should always be remembered that for such systems (which are very common in organic polarography) the shifts in the half-wave potentials are simple functions of the logarithm of the rate constant, or of the activation energy of the electrode process only when the systems under comparison do not differ in the value of the transfer coefficient α. The magnitude of this coefficient can be determined from the slope of the polarographic wave and from the slope of the curve for the pH dependence of the half-wave potential.

For a group of substances that fulfill all the necessary conditions (i.e., one for which the half-wave potentials can be quantitatively correlated), the shifts of half-wave potentials can be caused either by an interchange of the electroactive group, or by the effect of substitution of various groupings in parts of the molecule other than the electroactive group itself. The first case is analogous to an exchange of chromophores, and the second to the effect of auxochromes in the discussion of spectra. Comparison of substances in which changes occur simultaneously in several parts of the molecule is usually avoided.

The first category comprises exchanges that involve the kind of atoms or the number of atoms in the electroactive group (e.g., the comparison of chloro, bromo, and iodo derivatives, and the comparison of nitro and nitroso compounds). Also, groups of substances that differ in the extent of the conjugated system may be included in this category, since the entire conjugated system can be considered as a single electroactive group, and changes in its extent can be considered analogous to an interchange of electroactive groups (i.e., the effect of chain length in polyenic aldehydes, or of the role of the magnitude and type of the conjugated system in condensed aromatic hydrocarbons is a pertinent subject for discussion). Some questions concerning the role of the extent of the conjugated system have recently been treated by quantum-chemical methods.

The extension of the conjugated system is not considered to be a simple effect of substitution, because such molecules contain a common system of highly mobile π-electrons. Under such conditions, it seems that a discussion of the part of such a molecule to which electrons are transported from the electrode is meaningless. Also, it should not be forgotten that the endproducts of a constant-potential electrolysis are not necessarily identical with the products formed at the surface of the dropping electrode.

The two types of comparisons included in the first category (i.e., of various electroactive groups and of various conjugated systems) are usually not very accurate and, thus, not very reliable. This is because two conditions must be fulfilled if a series of half-wave potentials are to be compared as simple functions of the energy changes during the electrode process, namely, the condition of identity of the mechanisms of the electrode reactions to be compared and the condition of approximate constancy of the value of the transfer coefficient for the irreversible processes.

When the polarographic behaviors of substances containing different electroactive groups are compared, the condition of identity of reaction mechanisms rarely is fulfilled. Only in exceptional circumstances and for small groups of substances, (e.g., in the comparison of the reduction waves of alkyl chlorides, bromides, and iodides), can be operating mechanisms be expected to be identical. This is not necessarily true when, for example, compounds bearing a nitro group are compared with compounds bearing a nitroso group on the same molecular frame. In some carefully selected groups of conjugated systems an analogous reduction mechanism can be expected. Satisfactory experimental evidence for the role of the transfer coefficient has been given in scarcely any of the comparisons of this type described in the literature.

It seems the only case in which the reactivities of various electro-active groups can be compared quantitatively is that of reversible system. In all other instances the best that can be achieved is an approximate qualitative comparison or a crude estimate of the reactivities.

The second category comprises comparisons that can be described briefly as substituent effects. Substituent effects on half-wave potentials, which form the subject of this monograph, are treated in a way that has been used for the quantitative treatment of substituent effects on reaction rates and equilibria, where it has been proved useful and advantageous to compare the substituent effects in groups of compounds called reaction series. These reaction series are groups of compounds of the type X-Y-R, where the reactive group R undergoes a certain reaction (e.g., electroreduction) under specified reaction conditions. The compounds compared in a given reaction series differ only in the kind of the substituent X (or in the molecular grouping X-Y-), which is not directly involved in the reaction. It has been demonstrated that in such groups of compounds, polar, mesomeric, and steric effects can often be distinguished and sometimes separated.

Accurate comparison of polarographic half-wave potentials is made possible because in a properly chosen reaction series the mechanism of the electrode process is the same for all members of the series and, moreover, that the values of the transfer coefficient derived from the waves of various members of the reaction series usually do not differ very much.

It can be seen readily that in the classification of structural effects according to their nature, either as exchanges of the polarographically active group or as substituent effects, it is often difficult to draw a sharp line between these two categories. To make the division less arbitrary, we stipulate that the introduction of a substituent does not change the number or kind of atoms in the polarographically active group, but only affects the electron distribution in the electroactive part of the molecule. Changes in the course of polarographic electrolysis act as an indicator of change in the polarographically active group. Hence, in a given series of compounds the electroactive group is assumed to be the same if the course of the polarographic curves (and hence also the electrode mechanism) remains the same. On the other hand, change in the electroactive group is marked by a substantial change in the polarographic behavior.

Thus, introduction of a group into an aromatic ring with an electroactive group in a side chain, extension or branching of an aliphatic chain containing a reactive group, or changes in the size of rings attached to polarographically active groups are regarded as substituent effects. On the other hand, the changes in the sequences

$$CHO \longrightarrow COOH \longrightarrow COOR$$
$$CHO \longrightarrow COCH_3$$
$$COOR \longrightarrow COCH_3$$

etc.

are not considered substituent effects. The treatment of comparisons of compounds of the type C_6H_5COR (where $R = CH_3$, COO^-, $COOH$, $COOCH_3$, or $COCH_3$) as substituent effects[13] cannot be recommended, because the substances involved are reduced by completely different mechanisms. Similarly, comparisons of half-wave potentials of substances,[13] such as

$$HCHO, CH_3CHO, \text{ and } CH_3COCH_3$$
$$C_6H_5COCH_3 \text{ and } (CH_3)_2C{=}CHCOCH_3$$
$$C_6H_5CHO \text{ and } CH_2{=}CHCHO$$
$$C_6H_5COCH{=}CHCH_3 \text{ and } C_6H_5COC_6H_5$$

that undergo polarographic reduction by mechanisms that are different in principle are not very meaningful.

While the effect of the extent of the conjugated system has been included in the same category as the exchange of electroactive groups, it has proved useful to include the introduction of groups where $\pi\text{-}p$ conjugation takes place (e.g., substitution by -OH or -NH_2 in aromatic systems), and even some groups (like phenyl) where $\pi\text{-}\pi$ conjugation is expected, with the substituent effects.

It should be stressed here that results for as many substances as possible should be considered in the discussion of the structural effects. The larger the group of substances, the smaller the possibility that the discussion will be substantially affected by values for one or two compounds that show anomalous behavior. Specific suggestions for the selection of the most suitable and important members of the reaction series are made for each type of reaction series separately.

3. Techniques in the Polarographic Study of Structural Effects

When we wish to compare the polarographic properties of various compounds, usually it is not sufficient to measure the half-wave potentials of the substances in one or a few buffers with deliberately chosen pH values and to compare these potentials. Of course, it is possible that by luck the experimental conditions chosen will be just right, but the chances are that values measured under conditions selected at random will not be strictly comparable. Even less suitable is the use of unbuffered solutions as supporting electrolytes in the measurement of half-wave potentials of organic substances whose reduction or oxidation is accompanied by a proton transfer. Comparison of such half-wave potentials cannot possibly give a clear picture.

As mentioned in Section 2 of this Introduction, the necessary condition for a discussion of structural effects can be stated as follows: The mechanism of the electrode processes must be the same for all substances involved in the comparison, and the value of the transfer coefficient computed for these substances must be approximately constant in the entire group. Strictly speaking, the mechanism of the electrode process should be studied for each of the compounds in question. But this would be very time consuming, and some other acceptable practice must be found.

Before the general practice suggested can be discussed, we must turn our attention to the concept of "mechanism" and to its limitations as applied to electrode processes. The detailed mechanism of an electrode reaction would have to describe the transport of particles to the electrode surface (probably into an inner double layer), including chemical processes that precede the electrode reaction proper, the orientation of the particle at the moment of reaction, the structure—i.e., the composition and stereochemistry—of the transition state, the reactions that follow the electrode reaction proper, and the transport from the surface of the electrode. Such complete information about electrode reactions is not available for even a single electrode process. While some information is available on the transport problem and the accompanying chemical reactions, the approaches of physical chemistry and stereochemistry to the electrode process proper are merely speculative. Most frequently, only the course of the reaction, i.e., the overall reaction, the structure or composition of the electroactive species, the number of electrons transferred, the type of the electrode process (reversible or irreversible), and the products of the electrode reaction are established and verified. Since this sort of information is often sufficient for structural studies, it is accepted as the "mechanism of the electrode process" for purposes of this monograph.

Usually, when a group of compounds or a reaction series is to be studied, it is necessary to determine the mechanism of the electrode process (in the above sense) for one of a few typical members of the series. Sometimes sufficient information can be found in the literature. The mechanism then is assumed to be identical for the entire reaction series if,

1. The number of the electrons transferred in the reaction, as indicated by the wave height, is the same.
2. The courses of plots of $E_{1/2}$ versus pH and i_{lim} versus pH are analogous for all substances studied; the slope of the $E_{1/2}$ versus pH graph remains practically constant* in the entire reaction series.
3. The character of all the compared limiting currents is the same (i.e., if in all cases the limiting currents are diffusion-controlled; when kinetic currents are involved a correction of the measured values of the half-wave potential is necessary).
4. The degree of irreversibility of all compared systems is the same, i.e., either all the systems involved are reversible, or for irreversible systems the slope of the wave determined, for example, from a logarithmic analysis, and expressed in terms of the transfer coefficient does not vary substantially* for the substances in question.
5. When the composition of the supporting electrolyte (in particular, the presence of nonaqueous solvents and the type and concentration of cations or anions) or the presence of surface-active substances has a pronounced effect on the polarographic behavior of any of the substances, it has the same effect on all members of that particular reaction series.

We assume that the electrode reaction proceeds by the same mechanism for any group of compounds for which the above conditions are fulfilled. For such compounds, and only for such compounds, can the structural effects on shifts of the half-wave potential be treated quantitatively.

From this point of view, it will be understood why it is possible to make quantitative comparisons of the half-wave potentials of pyridinecarboxaldehydes[35] or pyridinecarboxylic acids[36,37] in which the side chains are in positions 2 or 4, but not in the case of derivatives with the aldehydic or carboxylic group in position 3. Whereas the

---

* Or is a linear function of structural parameters; see p. 51.

first two types of compounds show similar polarographic behavior, the 3-derivatives differ from the other two isomers in the number of waves, in their change with pH, and in the shape of the $E_{1/2}$ versus pH curve. A comparison of the half-wave potentials of all three isomers, at best, can be only a rough estimate. Similarly, it is not possible to compare the polarographic reduction of p-nitrophenol, reduced with the consumption of six electrons to p-aminophenol, with the reduction waves of other substituted nitrobenzenes, which yield the corresponding phenylhydroxylamine derivatives as reduction products at the dropping mercury electrode after the uptake of four electrons in the first wave. Another example that may be cited is the reduction of substituted acetophenones in acid media. Most of the ring-substituted aceto-phenones are reduced in the protonated form in two one-electron steps. Hence the half-wave potentials of the two waves, the first due to the reduction of the protonated form and second to the reduction of the radical (formed as an intermediate), can be compared, and structural effects in both these species can be discussed; p-aminoacetophenone cannot be included in this group because it is reduced in the entire available pH region in one single two-electron step.[38,39] The energy necessary for the reduction of the radical does not differ in this particu-lar case sufficiently from that for the reduction of protonated p-aminoacetophenone. The two steps in the reduction process, thus, are not separated, and hence, the mechanism of the electrode reaction is different and the shift of the half-wave potentials of acetophenone due to the introduction of the amino group cannot be treated quantitatively.

If the course of the electrode process for the reaction series studied has not been investigated earlier, than, in addition to ascertaining that conditions (1) to (5) are fulfilled, one must determine the actual number of electrons transferred for one or a few members of the series—not merely prove that this number is the same for all the substances—and determine the structure of the product (or products) of the electrode reaction. The techniques used for this purpose are discussed else-where.[40]

Only when it has been established that the half-wave potentials of the studied group of substances are really comparable can the experi-mental conditions for the polarographic electrolysis best suited for the study of structural effects be discussed. As mentioned above, polarographic data are usually dependent on the composition of the supporting electrolyte, in particular the pH value, the kind and con-centration of cations and/or anions of strong electrolytes, the presence

of complex-forming reagents, and the presence of surface-active substances. Hence, the most important condition for measurements
intended for structural correlations is that all values to be compared
be obtained under exactly identical experimental conditions, with
particular emphasis on identical supporting electrolytes. Since the half-
wave potentials of irreversible processes are to a certain degree dependent on the capillary constants, temperature, and depolarizer concentration, it is preferable to keep even these factors constant, although these
effects are often negligible compared with structural effects. The concentration of the depolarizer should preferably be kept between
1×10^{-4} and 5×10^{-4} molar. At lower concentrations more accurate
corrections for capacity current are necessary, and such operations
are time-consuming. At depolarizer concentrations of greater than
about 1×10^{-3} molar complications sometimes arise which are non-
existent or less marked at lower concentrations. These may be caused
either by adsorption phenomena or by reactions of higher order that
accompany the electrode process as side reactions or secondary
reactions.

The measurements of half-wave potentials should be very precise,
but the three-electrode system of measurement[41] and manual point-
by-point recording of the curves need be used only when a reproducibility of better than about ± 0.002 V is required. To achieve the usually
satisfactory reproducibility of ± 0.002 to 0.005 V, it has proved useful
to record the polarographic curve on an extended voltage scale (e.g.,
having exactly* 50 mV per abscissa division) using a photographically
recording polarograph. Two curves are recorded: one from positive
to negative potentials, and the other in the opposite direction. The half-
wave potentials of both curves are measured and the mean value is
calculated. In this way systematic errors, such as those caused by galvanometer hysteresis, can be eliminated. To obtain well formed waves
that can be measured exactly (i.e., for which especially the mean current
in the rising part of the curve can be determined accurately), a slower
rate of voltage scanning (e.g., 100 mV/min) is recommended.

For those good pen-recording polarographs for which the current
recording system possesses the characteristics of a critically damped
galvanometer with a period of swing of about 10 sec, the recording of
two curves and the determination of the average value of the half-wave
potentials can be carried out in similar fashion. However, many of the

---

* Accurate control of the applied voltage is essential.

commercially available pen-recording instruments do not fulfill these conditions, especially with respect to the properties of their damping circuit and the slow response of the recorder. Curves recorded automatically with such instruments are distorted and unsuitable for exact measurements of half-wave potentials. With such instruments currents must be measured at selected voltages, the current-voltage curve plotted manually, and the half-wave potentials measured from this curve.

An important problem in the measurement of half-wave potentials is the choice and preparation of the reference electrode. First of all, this electrode should be unpolarizable. This means that during the recording of a polarographic curve, when currents of a few microamperes are flowing, the potentials of the electrode should remain constant. If possible, the potentials of the reference electrode should be reproducible, i.e., differences between the potentials of similarly prepared electrodes should be small. However, it follows from the description of the potential measurements given below that this condition is not essential. On the other hand, it is important that the reference electrode be combined with the dropping (indicator) electrode in such a way that the diffusion potential and the iR drop are either negligibly small or constant and measurable. The diffusion potential depends on the electrolyte surrounding the dropping and reference electrodes, and the iR drop on the conductivity of the medium and the distance between the two electrodes.

To eliminate the influence of diffusion potentials and to decrease the effect of resistance, the reference electrode can be placed directly into the solution being measured. In aqueous solutions containing $0.1\text{-}M$ chloride a silver chloride electrode has proved efficacious. The potential of this electrode is practically unaffected by buffer components in buffered solutions of pH 2–12. Only in the presence of substances that form stable silver complexes, such as ammonia or barbital, is it impossible to use this type of reference electrode. It is not necessary to measure the absolute value of the potential of this reference electrode: it is preferable to measure potentials against an internal standard (pilot). For this purpose curves are recorded for the standard substance, using the same technique and equipment as for the substance under study, and the half-wave potential of the standard substance is measured. The differences between the half-wave potentials of the substances studied and the standard are then determined. When absolute potential values—usually against a saturated calomel electrode (SCE), the use of the normal hydrogen electrode (NHE) being less frequent in practical measurements—are looked for, the thallous ion is used as the internal standard for aqueous solutions. The half-wave potential of the thallous ion is practically independent of the composition of the supporting electrolyte in most commonly used base solutions, and is equal to -0.45_5 V (SCE). Hence, values of the half-wave potentials of substances measured against the half-wave potential of the thallous ion can be easily converted into SCE values. The reproducibility of half-wave potentials measured in this way is about ± 0.002 V, as demonstrated on some few, randomly selected, examples of repeated measurements (always with a fresh solution; the measurements were carried out over a period of two months) for substances **A, B,** and **C**:

A:	-0.45_0 V	B:	-0.46_2 V	C:	-0.47_3 V
	-0.45_1 V		-0.46_4 V		-0.47_4 V
	-0.45_1 V		-0.46_0 V		-0.47_3 V
	-0.45_8 V				

Moreover, in a reaction series the absolute values of the half-wave potentials (versus SCE) are usually far less important than the relative values, i.e., the shifts resulting from the introduction of a substituent. It is, thus, generally sufficient in a reaction series to measure half-wave potentials relative to that of one particular compound selected from the series. Usually the unsubstituted, parent compound is chosen for such comparisons.

A mercury pool electrode at the bottom of the electrolysis vessel is sometimes used as the reference electrode. The application of this type of electrode for exact measurements of half-wave potentials cannot be recommended. Polarization phenomena cause the potential of this electrode to show a tendency to change during the recording of a current-voltage curve even under normal polarographic conditions. Due to this effect, which plays an important role even in solutions containing 0.1 M chloride, the reproducibility of half-wave potential values determined with a mercury pool reference electrode is rarely better than ± 0.01 V.

An unsolved problem is the choice of directly immersed reference electrodes for nonaqueous solvents. Silver chloride electrodes behave erratically in both glacial acetic acid and dimethylformamide, in the latter probably because of the solubility of the silver chloride and the presence of amines. In glacial acetic acid satisfactorily reproducible results were obtained with a graphite reference electrode, but in dimethylformamide the same electrode gave less reproducible results. Even less information is available about other types of electrodes.

The main advantages of separate reference electrodes when used for measurements in aqueous solutions are convenience in handling, and independence of the composition of the supporting electrolyte. The saturated calomel electrode and the mercurous sulfate electrode are the most frequently used types of separate reference electrodes. The reproducibility of the potential of the former is better, but the latter is less polarizable and, hence, generally is more useful. For the connection between the compartment containing the reference electrode and the compartment containing the solution under study and the dropping electrode, liquid junctions are preferred to sintered glass, porous ceramics, salt bridges, etc. To avoid unnecessary resistance, the connecting part should be as short and wide as practically possible. When a separation is necessary (e.g., when an agar bridge is used), a cellophane membrane is preferable to sintered glass. Little or no mercurous salt is placed on the surface of the reference electrode which is large to keep the resistance low.

To avoid estimating the diffusion potentials the use of an internal standard (such as the thallium ion) is again recommended.

The use of separate reference electrodes for measurements in nonaqueous solutions presents a rather involved problem. No exact definition of the potential scale in such solvents exists. Data on the diffusion potentials between various solutions in the same nonaqueous solvent, e.g., $N(C_4H_9)_4I$ in dimethylformamide and potassium chloride in dimethylformamide, or between aqueous and nonaqueous phases usually are not available. Two types of approach are usually recommended. In the first, a nonpolarizable electrode is devised for use in the nonaqueous solvent (such as a calomel or a mercury acetate electrode in glacial acid, or a mercurous bromide electrode in dimethylformamide). The plausible assumption is then made that the liquid junction potential at the boundary of two solutions in the same solvent will be negligible. In the second approach, a well defined aqueous reference electrode, such as a calomel electrode, is used. Data on the diffusion potential at the solvent—water phase boundary usually are not available. With the present state of limited knowledge about the electrochemistry of nonaqueous solutions, it seems possible to use either of these types of reference electrodes in conjunction with an internal standard. For relative measurements a member of the reaction series again can be used. No simple correlation with an "absolute" potential scale (such as the aqueous SCE scale) exists. The potential of the thallous ion changes substantially with change of solvent and cannot be recommended. The reversibility of the standard system seems to be advantageous. For example, in solutions of sulfuric acid in glacial acetic acid, the use of 2-methylnaphthoquinone[42] gave good results.

After this brief discussion of the importance of the reference electrode, we can now consider the supporting electrolyte. With the exception of systems and conditions in which the electrode process is not a function of pH, well buffered solutions are always used. Universal buffer solutions (such as Britton–Robinson, McIlvaine, or Clark–Lubs buffers) are well suited for a general survey and for obtaining initial information about the polarographic behavior of an organic substance. For exact measurements of half-wave potentials, the use of simple buffer solutions is recommended. For the comparison of half-wave potentials measured at different pH values, it is preferable that the compared solutions have the same ionic strength. In some cases (e.g., with some halogen derivatives) not only the total concentration of ions (cations and anions) is of importance, but also the kind of the ion. Thus, in the preparation of buffers of different pH, Li^+ ions should not be changed for K^+, and the ionic strength should be kept constant by addition of a salt containing the cation present in the buffer.

The rules for the choice of the most suitable pH value (or values) for the measurement of half-wave potentials for structural correlation purposes now can be discussed. It is preferable to compare the half-wave potentials under conditions under which they are all independent of pH. Sometimes it is experimentally impossible to achieve the very low (or very high) acidities at which the half-wave potentials are no longer pH-dependent. For this reason, aprotic solvents, such as dimethylformamide or acetonitrile, sometimes are suggested.[43] It should always be established in such experiments that during the electrode process the composition of the solution in the neighborhood of the electrode does not change, i.e., that complications typical of aqueous unbuffered solutions do not take place. For example, the effect of traces of water[44] on the course of the electrode process can be sometimes very marked.

When conditions, under which an acidity-independent half-wave potential can be measured, are inaccessible experimentally, a pH range is chosen where the $E_{1/2}$ versus pH plots are parallel, i.e., where the slopes ($d\,E_{1/2}/d$ pH) are approximately the same for all the derivatives compared.

The choice of an arbitrary pH value for the comparison, such as values extrapolated to pH 0 or 7, is unfounded and sometimes leads to erroneous conclusions about the influence of the substituent. When values of half-wave potentials are compared for compounds that give slopes, $d\,E_{1/2}/d$ pH, varying over a wide range, it can be shown that according to the value of pH chosen as reference state, not only quantitatively different results but also sometimes even a different sequence of substituent effects, may be obtained. Nevertheless, comparisons of this type appeared recently in the literature (e.g., see Ref. 13).

When it is not possible to find a pH region in which the plots of $E_{1/2}$ versus pH are parallel, a quantitative discussion of the structural effects must be abandoned. For example, Sasaki[45] carefully measured the half-wave potentials of 13 nitrofuran derivatives, especially semicarbazones, over a wide range of pH, and Del Marco[46] and co-workers measured the half-wave potentials of 12 nitroguanidine derivatives. But in neither of these groups of compounds can a pH region be found in which conditions are comparable. It is thus impossible to use these data for a quantitative treatment.

In cases where structural deductions are to be based on values of half-wave potentials given in the literature, usually it is necessary to restrict the comparison to results obtained by one single author under defined conditions. Only rarely are the experimental conditions published in sufficient detail, and the parameters needed for re-evaluation usually are not known with sufficient accuracy, so that it is impossible to make an accurate comparison of values obtained by various authors. The number of useful data can sometimes be extended when several values overlap for two groups of compounds belonging to the same reaction series measured under similar conditions by two different authors (i.e., when the two groups contain several compounds in common).

Having defined the scope of problems to be discussed in this monograph and stated conditions for the choice of the substances to be compared and of the experimental conditions, we proceed now in the next chapter to a discussion of the general form of the equation for the shifts in half-wave potential caused by substituent effects.

In subsequent chapters special forms of this equation are illustrated in their application to benzene derivatives, aromatic heterocyclic compounds, aliphatic compounds, polycyclic hydrocarbon derivatives, and polynuclear heterocyclic compounds. Quinonoid and alicyclic compounds are discussed separately. This division of the subject matter, rather unusual in general organic chemistry, has been chosen because our knowledge of the behavior of aromatic carbocyclic and heterocyclic compounds is most advanced.

In addition to the application of special forms of the general equation, in several chapters correlations with some other physical quantities are mentioned. In cases in which the quality or quantity of the experimental material does not permit a quantitative treatment, the experimental facts are interpreted qualitatively.

References

[1] J. Heyrovský, "A Polarographic Study of the Electrokinetic Phenomena of Adsorption, Electroreduction and Overpotential Displayed at the Dropping Mercury Cathode," *Actualites Scientifiques at Industrielles*, No. 90, Paris, 1934.

[2] J. Heyrovský, "Polarographie," 185, Springer Verlag, Wien, 1941.

[3] M. Shikata and I. Tachi, *Collection Czech. Chem. Commun.* **10**: 368 (1938).

[4] R. Brdička, *Časopis Lékařů Českých* **61**: 53 (1948).

[5] V. Hanuš, Proceedings of the First International Polarography Congress, Prague, 1951, Part III, p. 103, Přírodověd. nakl., Prague, 1962.

[6] J. E. Page, *Quart. Rev.* **6**: 262 (1953).

[7] S. Wawzonek, *Anal. Chem.* **21**: 60, (1949); **22**: 30 (1950); **24**: 32 (1952); **26**: 65 (1954); **28**: 635 (1956); **30**: 66 (1958); **32**: 114 (1960); **34**: 182 R (1962); **36**: 220 R (1964).

[8] S. Wawzonek, "Organic Polarography," in *Polarography*, Second Ed., edited by I. M. Kolthoff and J. J. Lingane, Interscience Publishers Inc., New York, 1952.

[9] M. V. Stackelberg, "Polarographie organischer Stoffe," in *Methoden der organischen Chemie*, Part III, Vol. 2, Hauben-Weyl, Stuttgart, 1955.

[10] F. Ender, *Z. Elektrochem.* **54**: 219 (1950).

[11] H. J. Gardner and L. E. Lyons, *Rev. Pure Appl. Chem.* **3**: 134 (1953).

[12] P. J. Elving, "Application of Polarography to Organic Analysis," in *Organic Analysis*, Vol. II, p. 195, edited by J. Mitchell Jr., Interscience Publishers Inc., New York, 1954.

[13] C. Prévost and P. Souchay, *Chim. Anal.* **37**: 3 (1955).

[14] I. A. Korshunov, *Zavod. Lab.* **24**: 543 (1958).

[15] J. Tirouflet and E. Laviron, *Ricerca Sci.* **29**: Suppl. 189 (1959) (Contributi teor. sper. polarografia, Vol. IV).

[16] J. Tirouflet and R. Dabard, *Ricerca Sci.* **29**: Suppl. 211 (1959) (Contributi teor. sper. polarografia, Vol. IV).

[17] P. J. Elving, *Ricerca Sci.* **30**: Suppl. 205 (1960) (Contributi teor. sper. polarografia, Vol. V).

[18] H. W. Nürnberger, *Angew. Chem.* **72**: 433 (1960).

[19] P. J. Elving, "Polarography in Organic Analysis," in *Progress in Polarography*, Vol. II, p. 625, edited by P. Zuman and I. M. Kolthoff, Interscience Publishers Inc., New York, 1961.

[20] Ju. P. Kitajev and G. K. Budnikov, *Uspechi Khim.* **31**: 670 (1962).

[21] P. Zuman, *Chem. Listy* **48**: 94 (1954).

[22] L. P. Hammett, *Chem. Rev.* **17**: 125 (1935).

[23] L. P. Hammett, "Physical Organic Chemistry," p. 184; McGraw-Hill Book Co., New York, 1940.

[24] H. H. Jaffé, *Chem. Rev.* **53**: 192 (1953).

[25] R. W. Taft Jr., "Separation of Polar, Steric, and Resonance Effects in Reactivity," in *Steric Effects in Organic Chemistry*, edited by M. S. Newman, John Wiley & Sons Inc., New York, 1956.

[26] H. Mejman, *Zh. Fiz. Chem.* **22**: 1454 (1948).

[27] J. Koutecký, *Chem. Listy* **47**: 323 (1953); *Collection Czech. Chem. Commun.* **18**: 597 (1953).

[28] P. Delahay, New Instrumental Methods in Electrochemistry, Interscience Publishers, New York, 1954, str. 82; *J. Am. Chem. Soc.* **75**: 1430 (1953); **76**: 6417 (1954).

[29] R. Brdička, *Collection Czech. Chem. Commun.* **19**: 541 (1954).

[30] A. I. Lopushanskaya and A. V. Pamfilov, *Usp. Khim.* **30**: 386 (1961).

[31] P. Zuman, D.Sc. Thesis, Czechoslovak Academy of Science, Prague, 1959.

[32] P. Zuman, *Collection Czech. Chem. Commun.* **25**: 3225 (1960).

[33] P. Zuman, *Chem. Listy* **54**: 1244 (1960); *J. Polarograph. Soc.* **7**: 66 (1961).

[34] P. Zuman, *Ricerca Sci.* **30**: Suppl. 229 (1960) (Contributi teor. sper. polarografia, Vol. V).

[35] J. Volke, *Chem. Listy* **52**: 16 (1958).

[36] F. Šorm, *Chem. Obzor* **18**: 213 (1943).

[37] J. Volke, *Chem. Listy* **51**: 414 (1957).

[38] G. Giacometti and A. Del Marco, *Atti. Accad. Lincei* [8] **14**: 511 (1953).

[39] E. Knobloch and E. Svátek, *Chem. Listy* **49**: 37 (1955); *Collection Czech. Chem. Commun.* **20**: 1113 (1955).

[40] P. Zuman, "Some Techniques in Organic Polarography," in *Advances in Analytical Chemistry and Instrumentation*, Vol. 2, p. 219, edited by C. N. Reilley, Interscience Publishers Inc., New York, 1963.

[41] A. A. Vlček, *Chem. Listy* **48**: 189 (1954); *Collection Czech. Chem. Commun.* **19**: 862 (1954).

[42] L. Stárka, A. Vystrčil, and B. Stárková, *Chem. Listy* **51**: 1440 (1957), *Collection Czech. Chem. Commun.* **23**: 206 (1958).

[43] P. H. Given and M. E. Peover, Advances in Polarography (Proceedings of the Second International Congress Polarography, Cambridge, 1959), Vol. III, p. 948, Pergamon Press, Oxford, 1960.

[44] L. Holleck and P. Becher, *J. Electroanalyt. Chem.* **4**: 321 (1962).

[45] T. Sasaki, *Pharm. Bull.* **2**: 104 (1954).

[46] P. Lanza, A. Delmarco, A. F. McKay, and G. Semerano, *Ricerca Sci.* **26**: 116, 129, 148 (1956).

II: General Equation for the Relation Between the Polarographic Half-Wave Potential and Substituent Effects

1. General

For a reversible oxidation–reduction process, the half-wave potential (corrected for the quotient of the diffusion coefficients of the oxidized and reduced forms) is proportional to the equilibrium constant K of the reaction

$$Ox + ne \rightleftharpoons Red$$

$$K = \frac{[Red]}{[Ox]}$$

namely,

$$-E^0_{1/2} = \frac{RT}{nF} \ln K \tag{1}$$

Since

$$RT \ln K = \Delta F^\circ \tag{2}$$

where ΔF° is the change in the standard free energy, it follows that

$$-nF E^0_{1/2} = \Delta F^\circ \tag{3}$$

For half-wave potentials of irreversible systems, the relationship (4) has been derived:[1-5]

$$E_{1/2} = \frac{RT}{\alpha n F} \ln 0.87 k^0_f \sqrt{\frac{t_1}{D}} \tag{4}$$

23

where R is the gas constant, T is the absolute temperature, α is the transfer coefficient, n is the number of electrons transferred in the electrode process, $\mathbf{F}$ is the faraday (96,500 coulombs), k_f^0 is the rate constant of the heterogeneous electrode process, t_1 is the drop time, and D is the diffusion coefficient.

According to the theory of absolute reaction rate, the rate constant can be correlated with the change of free activation energy $\Delta F^{\pm}$ as follows:

$$RT \ln k = \Delta F^{\pm} \tag{5}$$

Combining (4) and (5) results in the following:

$$E_{1/2} = -\frac{\Delta F^{\pm}}{\alpha n \mathbf{F}} \ln 0.87 \sqrt{\frac{t_1}{D}} \tag{6}$$

If we compare the half-wave potential of the unsubstituted substance (denoted by E_H if the standard substance contains hydrogen instead of the substituent, or by E_{CH_3} if it contains methyl) with the half-wave potential of the compound containing the substituent X, denoted by $(E_{1/2})_X$, then (insofar as α can be treated as constant) we can characterize their difference as $\Delta E_{1/2}$ for irreversible and $\Delta E_{1/2}^0$ for reversible processes:

$$(E_{1/2})_X - E_H = \Delta E_{1/2} \tag{7}$$

$$\Delta E_{1/2} \sim \Delta \ln k_f^0 \tag{8}$$

where

$$\Delta \ln k_f^0 = \ln \frac{(k_f^0)_X}{(k_f^0)_H} \tag{9}$$

and

$$(E_{1/2}^0)_X - E_H^0 = \Delta E_{1/2}^0 \tag{10}$$

$$\Delta E_{1/2}^0 \sim \Delta \ln K \tag{11}$$

where

$$\Delta \ln K = \ln \frac{K_X}{K_H} \tag{12}$$

Whereas, for reversible processes the relation (11) involves no assumptions additional to those mentioned in connection with equation (1), for irreversible processes the transition from equation (4) to equation (8) requires that the value of the transfer coefficient α be approximately the same for all compounds compared, or that it change linearly with $\Delta E_{1/2}$. Fortunately enough, experimental results show that in groups of structurally related substances, like those grouped into reaction series, values of the transfer coefficient α either remain reasonably constant, or are a linear function of $\Delta E_{1/2}$.* Only for certain electrode reactions, seemingly those where structural changes in the vicinity of the center can affect steric conditions in the transition state or the adsorbability, have substantial changes in the values of α been observed. The other factors, i.e., t_1 and D, remain unchanged when the polarographic electrolysis is carried out under identical experimental conditions for all the substances compared (i.e., in the same supporting electrolyte and for a given drop time t_1), and when these substances do not differ substantially in molecular shape and weight (i.e., when the diffusion coefficient does not change substantially).

When the above conditions are fulfilled, a measure of the influence of a substituent on the half-wave potential of an irreversible electrode process is provided by the change in the free activation energy increment, i.e., the free energy change in the conversion of the reactant state into the transition state (13). The effect of a substituent, thus, is not influenced by factors which change the free energy of the original state and simultaneously the free energy of the transition state by the same amount:

$$\Delta E_{1/2} \sim \Delta\Delta F^{\pm} \tag{13}$$

A measure of the influence of a substituent on the redox potential of a reversible electrode process is provided by the change in the standard free energy change for the interconversion of the oxidized and reduced forms.†

$$\Delta E_{1/2} \sim \Delta\Delta F^{0} \tag{14}$$

It would be of interest to separate $\Delta F^{\pm}$ into effects due to changes in potential energy and those in kinetic energy. Such separation, however, cannot be carried out in practice. In the study of structural effects

---

* The possibility cannot be excluded that even in series in which α seems to be constant this relation exists, but is obscured by experimental error.
† This is because in polarography irreversible systems are much more frequent than reversible, the subsequent treatment will be restricted to irreversible systems. All the deductions can be applied to reversible systems as well.

on reactivity in homogeneous chemical reactions it has proved useful to discuss correlation sometimes with $\Delta F^{\pm}$ and sometimes with $\Delta H^{\pm}$. Since data on the temperature coefficients of half-wave potentials are usually not available, we shall restrict the subsequent discussion to effects of substituents as measured by $\Delta\Delta F^{\pm}$.

By the use of thermodynamic or equivalent approaches, the effects of substituents, as measured by $\Delta\Delta F^{\pm}$, cannot be further separated into kinetic and potential energy contributions or into other components. Such a separation is at present possible only on the basis of empirical experience, which has accumulated during the past century in the course of the development of theoretical organic chemistry. When the behaviors of appropriately selected model compounds were compared, it was possible to deduce that reactivity (and hence $\Delta\Delta F^{\pm}$) is affected by the polar, steric, and mesomeric effects of substituents.

In our treatment here, definitions of the particular effects, as suggested by Taft,[6] are given first, and then it is shown how $\Delta\Delta F^{\pm}$ can be separated into energy contributions corresponding to particular effects in the molecule X-R, in which R is the polarographically active group and X is the substituent.

Polar effects can affect both the change in the potential energy increment and the contribution of the kinetic energy term. The former contribution results in coulombic interaction between the substituent X and the polarographically active group R.

Polar interaction can be propagated along single bonds (inner inductive effect and effect of bond polarization), or directly through space (field effect). The charge separation can be the consequence of difference in the electrophilic characters of the atoms involved, of mesomeric interaction inside the group which is regarded as the substituent, of mesomeric interaction inside the polarographically active group, and, finally, of the presence of a unit charge on the substituent X or on the electroactive group R.

When this convention is used, even effects that result in mesomeric interaction inside the substituent (e.g., in groups like NO_2, COO^-, etc.) are regarded as polar effects, when the resulting change in electron density is transmitted toward the reactive center by an inductive mechanism.

Furthermore, even effects in which mesomeric interaction occurs between the substituent and a part of the molecule other than the electroactive group are classified as polar effects. Polar effects can operate also in cases in which the polarographically active group shows a mesomeric interaction with a part of the molecule other than the substituent, i.e., with the skeleton or the unchanged part of the molecule. An example of the former type is the mesomeric effect of a substituent in a benzene ring in the *meta* or *para* position relative to the electroactive group R, which involves only the substituent X (e.g., COO^-, CN, or CHO) and the aromatic ring but not the electroactive group R (e.g., I). Here again, the resulting change in electron density is propagated to the electroactive group by an induction mechanism. This type of effect is sometimes called a polar mesomeric effect.

Also, all contributions corresponding to a kinetic energy change that depend directly on the polar effect of the potential energy term, are considered as polar effects. Hence, in cases in which linear relations between $\Delta\Delta F^{\pm}$ and $\Delta\Delta S^{\pm}$ or $\Delta\Delta H^{\pm}$ and $\Delta\Delta S^{\pm}$ have been found,[7] the change in $\Delta\Delta S^{\pm}$ is regarded as a polar effect.

Steric interactions may also affect the change in potential energy increment as well as the steric restriction of motion expressed in the kinetic energy term.

The change in the strain energy increment ($\Delta\Delta E_R^{\pm}$), which is the component of the steric interaction that affects the potential energy, arises from van der Waals attractive and repulsive forces acting between parts of the molecule and resulting in changes in the valence angles in the ground and transition states (to different extents) of the molecule. These differences in the orientation of valences cause changes in the energy of the X-R bond.

The influence of a substituent on steric hindrance to intramolecular motion, resulting in a change in the kinetic energy term, is a typical expression of the steric effect. As has been demonstrated by Taft,[6] steric effects can cause a hindrance to motion without any effect on the strain energy. On the other hand, no substituent increases the strain energy without simultaneously affecting the steric hindrance to motion.

Every change in $\Delta\Delta S^{\pm}$ that cannot be explained by the earlier-mentioned polar effect should be described as a steric effect.

Mesomeric interaction between the substituent X and the polarographically active group R can take place, when π-π, π-p, or π-σ conjugation occurs between the two groups. This kind of interaction can be visualized by using at least two contributing structures; an exact description requires a quantum-mechanical treatment. Mesomeric effects are assumed to contribute predominantly to the potential energy term.

Thus, it can be concluded that the change in the potential energy increment can be influenced by changes in forces of electrical origin, described as polar, steric, and mesomeric effects. The kinetic energy term can be affected by steric and sometimes even by polar effects. It should be mentioned here that in polarography the reactions studied occur in solutions. Thus, in addition to effects concerning the reacting molecules, effects (which, again, can be classified as polar, steric, and mesomeric) influencing the potential energy of solvation should be considered as well. In its relation to the effect of solvation the kinetic energy term is concerned not only with the intrinsic molecular motions, but also with changes in the position of the molecule (or its parts) relative to solvent molecules in the solvated particle.

Because of the overlapping of the components of the total change in the activation free energy increment by the various factors of the substituent effects, and because it is impossible, as stated above, to separate the changes in potential and kinetic energy, we shall abandon the thermodynamic interpretation of $\Delta\Delta F^{\pm}$ and adopt the division suggested by Taft[6] for homogeneous reactions. The change in the total potential energy increment $\Delta\Delta E_p^{\pm}$ is assumed to consist of contributions from the change in the activation energy increment connected with the polar substituent effect $\Delta\Delta E_\sigma^{\pm}$ (which will be denoted as polar activation energy), from the change in mesomeric energy increment $\Delta\Delta E_M^{\pm}$, and finally the above-mentioned change in the strain energy increment $\Delta\Delta E_R^{\pm}$. It should be stressed that for all the effects the particular energies correspond to the changes in the polar, mesomeric, or strain forces which occur in the transition from the initial to the transition state. Hence,

$$\Delta\Delta E_p^{\pm} = \Delta\Delta E_\sigma^{\pm} + \Delta\Delta E_M^{\pm} + \Delta\Delta E_R^{\pm} \tag{15}$$

Since

$$\Delta\Delta F^{\pm} = \Delta\Delta E_p^{\pm} - RT\ln(\Pi Q^{\pm}) \tag{16}$$

(where $\Pi Q^{\pm}$ is the partition function corresponding to the change in the kinetic energy increment accompanying the activation process), it follows that

$$\Delta\Delta F^{\pm} = \Delta\Delta E_\sigma^{\pm} + \Delta\Delta E_M^{\pm} + \Delta\Delta E_R^{\pm} - RT\ln\Pi Q^{\pm} \tag{17}$$

When the kinetic contributions of polar effects are included in $\Delta\Delta E_\sigma^{\pm}$ and when the definition of relative total steric change (18) is applied:

$$\Delta\Delta E_S^{\pm} = \Delta\Delta E_R^{\pm} - RT\ln\Pi Q^{\pm} \tag{18}$$

we obtain

$$\Delta\Delta F^{\pm} = \Delta\Delta E_{\sigma}^{\pm} + \Delta\Delta E_{M}^{\pm} + \Delta\Delta E_{S}^{\pm} \tag{19}$$

Equation (19) covers most substituent effects on half-wave potentials. It is transformed,[8,9] for applications in polarography, by combination with equation (13) into

$$\Delta E_{1/2} = P + M_{\pi} + S \tag{20}$$

where P represents the shift in the half-wave potential resulting from the change in the polar activation energy increment, M_{π} is the shift in the half-wave potential due to the change in the mesomeric (resonance) energy increment, and, finally, S is the shift in the half-wave potential due to the change in the total steric energy increment.

The polar P, steric S, and mesomeric M_{π} effects are often independent variables. Sometimes it is possible to discuss these contributions on the basis of experience and of estimates of their relative magnitudes. But in some of the cases in which we are interested here, the functional relations between these contributions and the structural parameters can be formulated.

The single value of a half-wave potential does not allow us to distinguish the various contributions of the substituent. In order to differentiate these contributions, a proper choice of the selection rules for a given reaction series and also a choice of suitable substituents should be made. First, we choose substances for which both the steric and mesomeric effects are negligible or practically constant for the whole group of substances under consideration. For such a group of substances the substituent effect can be assumed to be polar only. Equation (20) then reduces to (21). Having established the rules that govern the polar effect of substituents, we can consider substituents that cause both polar and steric effects [equation (22)], or both polar and mesomeric effects [equation (23)]. Having established correlations of this kind for reaction series in which for certain substituents the polar effects can be treated separately, we can find the appropriate forms of equations (22) and (23) for handling reaction series in which both polar and steric or both polar and mesomeric effects are operating throughout:

$$\Delta E_{1/2} = P \tag{21}$$

$$\Delta E_{1/2} = P + S \tag{22}$$

$$\Delta E_{1/2} = P + M_{\pi} \tag{23}$$

Finally, in some reaction series the effect of substituents with (a) only polar effects, (b) both polar and steric effects, and (c) polar, steric, and mesomeric effects can be studied by applying equation (20) in its full form. So far, no reaction series has been found in which only the steric effect, only the mesomeric effect, or only a combination of the steric and mesomeric effects influences the shifts in half-wave potential. The special forms of equation (20) will be discussed in the following sections.

2. Influence of Polar Effects

In most of the reaction series studied so far there is no conjugation interaction between the polarographically active group and the substituent, and steric interactions are negligible, or can be assumed to be constant throughout the reaction series. In these cases the shift in the half-wave potential is due entirely to the polar effect of the substituent, and equation (20) can be used in the simplified form of equation (21).

First, we shall deal with rigid, usually cyclic, systems in which the reactive group is fairly distant from the substituent and only polar effects are operating. In such molecules the relative positions of the polarographically active group and the substituent are fixed, and any change in the interaction of the substituent with the electroactive group is prevented.

Those rigid systems in which steric effects do not occur belong to two classes: saturated compounds, and certain types of aromatic systems. In saturated-ring compounds mesomeric interactions between the ring and the substituent, as well as between the ring and the reactive center, do not occur. Hence, only polar effects can operate. It is not certain, however, whether these consist solely of so-called internal inductive effects, which are transmitted along interatomic bonds, or include also the field effect, which can be propagated even through the space outside the molecule.

As suitable model compounds, in which the spatial configuration is similar to that of the benzene ring, Roberts and Moreland[10] have chosen substituted bicyclo[2.2.2]octane-1-carboxylic acids (I) (R = COOH) and the corresponding esters (R = $COOC_2H_5$), each with the substituent X in the 4-position.

$$R - \langle \underline{\quad} \rangle - X \qquad \qquad (I)$$

The changes observed in the rate and equilibrium constant as the substituent X is varied reflect the exclusively polar effects of the substituents X. The authors[10] expected that, due to the similarity in spatial arrangement, the distance between groups R and X and also their mutual interaction would be of the same order of magnitude as in benzene.

It has been shown that the polar effect P can be expressed as*

$$P = \rho_R^I \sigma_X^I \qquad (24)$$

and, by combination with equation (21)

$$\Delta E_{1/2} = \rho_{\pi,R}^I \sigma_X^I \qquad (25)$$

In these equations σ^I is the polar inductive substituent constant. Its value depends on the nature of the substituent X and expresses its inductive effect on the reactive group R. It depends further on the type of rigid system and on the relative positions of the groups X and R. It is independent of the type of reaction that takes place at the group R and of the reaction conditions. It is defined as $\sigma_X = \log K_X/K_H$, where K_X and K_H are the dissociation constants of the acids (I; X = a substituent and X = H).

The term $\rho_{\pi,R}^I$ is called the reaction constant—it expresses the susceptibility of the reactivity of the group R to the effects of the group X. The value of $\rho_{\pi,R}^I$ depends on the type of rigid system, on the kind of the reactive group R, and on reaction conditions. It is independent of the kind of substituent X.

Cyclic compounds of type (I) have not been studied polarographically, but the validity of equation (25) has been demonstrated for some straight-chain and heterocyclic compounds.

The second class of rigid systems in which the participation of steric effects can be excluded consists of compounds in which the reactive group and the substituent are separated by an aromatic ring. For steric effects to be neglected it is necessary to exclude *ortho*-substituted derivatives (or in general compounds in which the substituent and reactive groups are in vicinal positions).

First, only polar interactions between the substituent and the reactive group should be considered. This does not exclude cases in which the reactive group and the aromatic ring, or the substituent

* The form of the equation showing the dependence of the change in reactivity on the empirical substituent constant σ has been chosen on analogy with the Hammett and Taft equations discussed later.

and the nucleus, are in mesomeric interaction. It is only necessary that the change in electron density on the atom of the benzene ring, bearing the substituent, be transmitted to the reactive group by an inductive mechanism. In this part of our discussion, however, cases in which mesomeric interaction between the reactive group and the substituent takes place via the aromatic ring are excluded.

When all the conditions and limitations stated above are fulfilled, the polar effect P can be expressed by the Hammett equation:[11–13]

$$P = \rho_R \sigma_X \tag{26}$$

Hence, for the shift in the half-wave potential we may use the modified form of the Hammett equation:

$$\Delta E_{1/2} = \rho_{\pi,R} \sigma_X \tag{27}$$

In this equation σ_X stands for the total polar substituent constant, which is dependent on the kind and the position of the substituent and to a certain degree on the kind of aromatic ring, but is independent of the type of reactive group R, of the reaction involved, and of the reaction conditions. It is defined[11,12] as the logarithm of the ratio of the dissociation constants of the substituted and the unsubstituted benzoic acid. Its values, based on extensive experimental evidence, have been tabulated.[13]

The proportionality constant $\rho_{\pi,R}$ (measured in volts) is called the reaction constant. It expresses the susceptibility of the electrode reaction to the total polar effects of substituents. Its value depends on the kind of the electroactive group R, on the composition of the supporting electrolyte, and on the temperature, but it is independent of the kind and the position of the substituents. The value of the reaction constant is influenced by the mesomeric interaction between the polarographically active group and the aromatic nucleus. Moreover, this value is affected by those changes in the kinetic energy term that are directly dependent on changes in the potential energy due to the polar effect.

In virtue of its derivation, equation (27) is designed for application to benzene derivatives, and such applications are discussed in detail in Chapter III. In the following chapters it is shown that the application of equation (27), and perhaps of other forms of the relation (21), can be extended even to polycyclic hydrocarbons and to mono- and poly-cyclic heterocyclic compounds in which the polarographically active group is in the side chain. The reducible bond can be either directly between the aromatic ring and the active group (as in halobenzenes), in the α-position relative to the aromatic ring (as in carbonyl compounds or nitrobenzenes), or even in the β- or γ-position (as in phenylsydnones). The reactive center can be a single bond (as in halogen derivatives), a group of atoms with multiple bonds (as in the nitro, azo, azomethine, or carbonyl group), or even a whole heterocyclic ring (as in phenylsydnones or isoflavones).

It is evident that the application of equation (27) to heterocyclic and polycyclic systems will give only approximate results. When more experimental data are available in these reaction series, other forms of equation (21) can be expected to be derived. Such relations will enable us to determine more appropriate substituent constants for particular series, better expressing the effects of substituents on reactivity in heterocyclic and polycyclic rings than it is at present possible with the use of σ_X constants.

The second type of reaction series in which polar effects play a predominant role comprises groups of compounds in which the influence of the steric effect is practically constant throughout the group. A typical example is provided by reaction series containing groups of aliphatic compounds that do not differ considerably among themselves in the geometry of the transition state, which in the positions of the bonds must not differ considerably from the state of the compound before the reaction. Such reaction series consist mainly of compounds in which the introduction of a substituent does not change the steric conditions in the neighborhood of the electroactive group. Conjugated systems are excluded from such reaction series, so that mesomeric effects are not involved.

For reaction series in which the above conditions are fulfilled the following equations can be used:

$$P = \rho_R^* \sigma_X^* \qquad (28)$$

and hence,

$$\Delta E_{1/2} = \rho_{\pi,R}^* \sigma_X^* \qquad (29)$$

This is a modified form of the Taft equation.[6] Here σ_X^* is the polar substituent constant, in most cases proportional to the polar inductive constant σ_X^I; in fact, $\sigma_{CH_2X}^*$ is linearly proportional to σ_X^I. The value of the constant σ_X^* is dependent on the kind of substituent, but not on the nature of reactive group R, the type of reaction, and the reaction conditions. The value of constant σ_X^* usually does not include the mesomeric interaction between the reactive group and the substituent, which here is attached directly to the electroactive group or separated only by a methylene group. On the other hand, the value of constant σ_X^* expresses the direct polar interaction between the substituent and the polarographically active group.

The substituent constants were defined by Taft[6] from the difference of the logarithms of the base-and acid-catalyzed ester hydrolysis rate constants, and their applicability was checked for more than 1000 experimental values. As standards, substances for which $X = CH_3$ were chosen.

The term $\rho_{\pi,R}^*$ is called the reaction constant, expressed in volts. It is a measure of the susceptibility of the reduction process to the polar effects of substituents. The value of this constant depends on the nature of the electroactive group R, on the composition of the supporting electrolyte, and on the temperature. It is independent of the kind of substituents. The value of the reaction constant $\rho_{\pi,R}^*$ does not involve the mesomeric interaction of the reactive group with the molecular frame.

Proof of the validity of equation (29) is provided by the additivity of σ_X^* values, the fact that the equation holds for various types of reactions, and the parallelism that has been observed between σ_X values and polar inductive constants σ_X^I, which have been obtained by completely independent methods.

Since substitution in the immediate vicinity of the reducible group often can result in steric or mesomeric effects, the applicability of equation (29) is more restricted than that of equation (27). Before equation (29) can be applied in the treatment of shifts in half-wave potentials in a reaction series the limits of the validity of the equation must be established. It may happen that for a given substituent X the value of σ_X^* describes the shifts in half-wave potential sufficiently exactly for several polarographically active groups, but for other electroactive groups deviations may occur due to the participation of steric and mesomeric effects.

3. Influence of Polar and Resonance Effects in Combination

Up to now, we have discussed the treatment of reaction series which involve only polar effects. We now turn to systems in which polar and resonance effects are operating simultaneously.

In aromatic compounds containing a polarographically active group which is in considerable mesomeric interaction with the aromatic ring and simultaneously a *para* substituent which is also in considerable mesomeric interaction with the ring (e.g., in p-HO-C_6H_4-NO_2), a considerable mesomeric interaction between the reactive group and the substituent can occur via the nucleus. Mesomeric interaction can involve quinonoid contributing forms, like

In these instances it has proved useful to replace the substituent constant σ_X (e.g., $\sigma_{p\text{-}NO_2}$) in equation (27) by some other substituent constants denoted by σ_X^- (e.g., $\sigma_{p\text{-}NO_2}^-$). These constants were derived from the reactions of p-substituted phenols and anilines[12,13] and they include not only polar inductive and polar mesomeric effects, but also contributions due to mesomeric interactions between the reactive group and the substituent via the aromatic nucleus. For these shifts in half-wave potential the equation

$$\Delta E_{1/2} = \rho_{\pi,R}\sigma_X^- \tag{30}$$

has been applied. In Chapter III we discuss for which electroactive groups R and for which substituents X the application of σ_X^- values is to be preferred to that of σ_X values. In this way it is possible to demonstrate the participation of mesomeric interaction in certain compounds.

As will be shown in Chapter III, most reductions of benzene derivatives have a nucleophilic mechanism. For such relations the application of the constants σ_X and σ_X^- has given good results. H. C. Brown[14] and his co-workers have shown recently that for electrophilic reactions—such as substitution in an aromatic ring and also reactions in a side chain—mesomeric interactions take place between electron-donating substituents and electron-deficient reaction centers. Brown suggested that for such electrophilic reactions another set of substituent constants σ_X^+ be defined. The standard values of σ_X^+ were determined[12] from 8-chloro-p-cymene solvolysis and verified for about 100 rate constants of various electrophilic reactions.

In polarography the application of the constant σ_X^+ in the equation

$$\Delta E_{1/2} = \rho_{\pi,R}\sigma_X^+ \tag{31}$$

so far, has been limited only to a few cases (*cf.* Chapter III). In all these cases the substituents chosen in the original papers do not enable us to decide which of the equations (31) and (27) gives the better fit.

When the value of the substituent constant σ_X^- is not available, the shift in the half-wave potential due to mesomeric interaction (M_π) can be determined by the use of equation (32).

$$\Delta E_{1/2} = \rho_{\pi,R}\sigma_X + M_\pi \qquad (32)$$

The mesomeric contribution M_π, measured in volts, is the deviation from the linear relation (27), which was obtained for saturated substituents and for substituents which are unlikely to participate in mesomeric effects. The condition of identical reaction mechanism throughout the reaction series must be fulfilled; thus, the half-wave potentials of p-nitroanilines and p-nitrophenols in aqueous solutions cannot be used in this way to determine the value of M_π with the aid of the half-wave potentials of other nitrobenzenes, for the reduction mechanisms are different.

A reaction series in which the reactive group is directly bound to a substituent for which polar substituent constants σ_X^* are used can be treated in a similar way. For substances in which mesomeric, as well as polar, interactions affect the shift in half-wave potential equation (33) can be applied to determine the mesomeric contribution M_π to this shift.

$$\Delta E_{1/2} = \rho_{\pi,R}^*\sigma_X^* + M_\pi \qquad (33)$$

Using this equation for saturated compounds and then comparing the linear relation with the half-wave potentials of unsaturated compounds, we may, from the deviation, determine the approximate value of M_π. The most important example of this type is the comparison of the half-wave potentials of a series of alkyl derivatives with the half-wave potential of the corresponding phenyl derivative. Even though the value $\sigma_{C_6H_5}^* = +0.60$ given by Taft is doubtful to a certain degree (the value $\sigma_{C_6H_5}^* \approx 0.8$ obtained from $\sigma_{C_6H_5}^I$ would be more plausible), this value of $\sigma_{C_6H_5}^*$, derived from ester hydrolysis fits satisfactorily for the half-wave potentials in several series of reducible compounds (cf. Chapter V). On the other hand, for several reaction series a shift toward more positive values of half-wave potential was observed, which corresponds to a contribution of the mesomeric interaction M_π in the substituent effect of the phenyl group.

The shift of half-wave potentials toward positive values due to replacement of an alkyl by a phenyl group, until recently, has been explained in polarographic literature by the effect of conjugation, i.e., by the mesomeric effect. Nevertheless it can be demonstrated for several polarographically active groups that the replacement of alkyl by phenyl affects polarographic reducibility in much the same way as it affects reactivity in ester hydrolysis. It can be deduced that polarographic electrolysis is often influenced substantially by the polar effect of the phenyl group, but is only in a few cases[8,9] is affected substantially by the mesomeric effect of phenyl.

Proof that the deviation from linearity in the direction of more positive potentials in the relation of half-wave potential to σ_X^* is really an approximate measure of the mesomeric effect of the phenyl ring was obtained in the case of 3-phenylsydnones[15] (cf. Chapter IV), for which the contribution of the mesomeric effect was also determined by a consideration of the steric hindrance to coplanarity, i.e., by a completely independent method.

4. Influence of Polar and Steric Effects in Combination

Examples in which both polar and steric effects play a role are encountered in series of aliphatic compounds containing substances with a bulky substituent in the vicinity of the polarographically active group. The shift in the half-wave potential can be then expressed by equation (22). Whereas, even here P can be expressed by equation (28) using the polar substituent constant σ_X^*, the evaluation of the steric contribution S is more complicated. In general, steric effects are not additive, and it seems probable that no generally valid, simple correlation exists between the steric effects of substituents in one reaction series and those of the same substituents in another reaction series (i.e., a linear steric energy relation). It is probable that the linear steric strain energy relation,

$$(\Delta\Delta E_R^{\pm})_1 = \alpha(\Delta\Delta E_R^{\pm})_2 \tag{34}$$

is more generally valid, but the isolation of values of $\Delta\Delta E_R^{\pm}$ from experimental data is difficult.

For certain reaction series, in which the polar effects are small (as indicated by small values of the reaction constants $\rho_{\pi,R}$) or which show a similarity to the standard reaction series used for the derivation of the total steric substituent constants $(E_S)_X$ on the assumption that $\rho_{\pi,R} = 0$, Taft[6] has demonstrated the validity of

$$S = \delta_R(E_S)_X \tag{35}$$

The total steric substituent constant $(E_S)_X$ was determined from rate constants of acid-catalyzed ester hydrolysis (where polar effects are little important) as $(E_S)_X = \log k_A^X / k_A^0$ (where k_A^0 is the rate constant for the derivative in which $X = CH_3$).

The term $\delta_{\pi,R}$ is the steric reaction constant (in volts), expressing the susceptibility of the electrode process to the steric demands of substituents. The value of the steric reaction constant depends on the nature of the electroactive group R and on the composition and temperature of the supporting electrolyte; it is independent of the kind of substituent.

As in the case of mesomeric effects, examples in which steric effects are predominant (and in which $\Delta E_{1/2} = \delta_{\pi,R}(E_S)_X$ would be valid) have not been found in studies of the effect of substituents on shifts in the half-wave potential.

In a few reaction series equation (36) describes the shifts in the half-wave potential better than the simple equation (29).

$$\Delta E_{1/2} = \rho_{\pi,R}\sigma_X + \delta_{\pi,R}(E_S)_X \qquad (36)$$

Examples of the application of equation (36) are given in Chapter V. In evaluating equation (36) as a measure of the steric effect we should not forget that correlations made with a four-parameter equation [like (36)] are always better than correlations made with a two-parameter equation [like (29)].

In general, in the application of expressions involving steric effects, deductions should be made with even more care than in the case of simple polar effects in the aliphatic series.

5. The General Case

An example of the most complicated type of reaction series, in which the shifts in the half-wave potential can be influenced substantially and simultaneously by polar, steric, and mesomeric effects, is afforded by the electrode reactions of *ortho*-substituted aromatic compounds. In these reaction series, equation (20) must be used in its full form.

In the application of equation (20), in the first place substituents are chosen for which only a small mesomeric effect is to be expected. Under such conditions equation (37) is operative:

$$\Delta E_{1/2} = \rho_{\pi,R}\sigma_{o\text{-}X} + \delta_{\pi,R}(E_S)_{o\text{-}X} \qquad (37)$$

Values of $\sigma_{o\text{-}X}$ and $(E_S)_{o\text{-}X}$ were again determined by Taft using rate constants of benzoic ester hydrolysis in acid and alkaline media.

When the applicability of equation (37) to a given reaction series has been proved, the half-wave potentials of derivatives containing substituents showing also mesomeric interaction can be treated by means of equation (20), and the values of M_π can be determined. These correlations have even more limited validity and less experimental support than correlations in aliphatic reaction series. The number of reaction series for which sufficient half-wave potentials are available to allow equations (37) and (20) to be tested is small, so that no conclusions have been reached.[16]

6. The Role and Importance of the General Equation

The role and importance of the general equation (20) in its special forms can be summarized as follows:

1. It enables us to predict the half-wave potentials of substances that have not been studied so far;

2. It enables us to show whether all compounds in a given reaction series are reduced or oxidized by the same mechanism;

3. The sign of reaction constant ρ enables us to distinguish the type of the mechanism;

4. It enables us to check the experimental data;

5. It enables us to distinguish whether the reactive species is protonated or not;

6. It enables us to detect anomalous substituent effects;

7. In some systems, it enables us to distinguish polar, steric, and mesomeric substituent effects and to separate them quantitatively; and

8. It enables us to determine accurate values of the substituent constants σ in a relatively simple manner.

7. The Conditions for the Application of the General Equation

The condition for the application of equation (20) and its special forms can be summarized as follows:

1. The half-wave potentials compared must be obtained under exactly identical conditions. In the interpretation of data from the literature only values measured by one particular author can usually be compared in a reaction series.

2. The mechanism must be similar for all waves compared, and the value of the transfer coefficient α must be the similar or must be

a linear function of the σ constants. Since the electrolysis products and mechanism may not have been identified for all members of the reaction series, it is necessary that (cf. Chapter I): (i) the heights of all waves compared should correspond to the same number of electrons transferred; (ii) the limiting currents should be diffusion-controlled; (iii) the slopes of the waves should not differ substantially; (iv) the number of protons transferred should be the same for all compounds compared (regarding the choice of pH range cf. Chapter I).

3. To establish the validity and limits of the equations such compounds should be chosen in which the polar, steric, and mesomeric effects of substituents vary over as wide a range as possible.

When a deviation of a half-wave potential from the linear relation is to be discussed, or when it is evaluated as a steric or mesomeric contribution to the substituent effect, it is necessary to keep in mind that the general equation (20) and its special forms are only approximate relations. When the errors of measurement of half-wave potentials have a range of about $\pm 5\,\text{mV}$, only deviations greater than about $20\,\text{mV}$ to $40\,\text{mV}$ (depending on the value of $\rho_{\pi,\text{R}}$ or $\rho^*_{\pi,\text{R}}$) are really significant in determining the possible participation of a further effect.

A difficulty very commonly met in the application of the general equation and its special forms, which are summarized in remaining chapters, lies in the smallness of the number of half-wave potentials known in the reaction series studied. In only a few of these reaction series are there ten or more substituents, mostly 5–10. To show the scope of application even reaction series containing only three or four substances with known half-wave potentials have been included when all the data fit the particular equation applied. It must be understood, however, that inferences from such results have limited reliability.

References

[1] N. Mejman, *Zh. Fiz. Khim.* **22**: 1454 (1948).

[2] J. Koutecký, *Chem. Listy* **47**: 323 (1953); *Collection Czech. Chem. Commun.* **18**: 597 (1953).

[3] P. Delahay, *J. Am. Chem. Soc.* **75**: 1480 (1953); **76**: 5417 (1954).

[4] P. Delahay, "New Instrumental Methods in Electrochemistry," p. 82, Interscience Publ., John Wiley & Sons, New York, 1954.

[5] R. Brdička, *Collection Czech. Chem. Commun.* **19**: S 41 (1954).

[6] R. W. Taft, Jr., *Separation of Polar, Steric and Resonance Effects in Reactivity* in "Steric Effects in Organic Chemistry," (ed. by M. S. Newman) John Wiley & Sons, New York, 1956.

[7] J. E. Leffler, *J. Org. Chem.* **20**: 1202 (1955).

[8] P. Zuman, *Collection Czech. Chem. Commun.* **25**: 3225 (1960).

[9] P. Zuman, *Ricerca Sci.* **30**: Contributi teor. sper. polarografia 5, S 229 (1960).

[10] J. D. Roberts, and W. T. Moreland, Jr., *J. Am. Chem. Soc.* **75**: 2167 (1953).

[11] L. P. Hammett, *Chem. Rev.* **17**: 125 (1935).

[12] L. P. Hammett, "Physical Organic Chemistry," p. 184, McGraw-Hill, New York, 1940.

[13] H. H. Jaffé, *Chem. Rev.* **53**: 191 (1953).

[14] H. C. Brown, and Y. Okamoto, *J. Am. Chem. Soc.* **80**, 4979 (1958).

[15] P. Zuman, and D. J. Voaden, *Tetrahedron* **16**: 130 (1961).

[16] P. Zuman, *Collection Czech. Chem. Commun.* **27**: 648 (1962).

III: Benzene Derivatives

1. General

For a benzene derivative whose reduction at the dropping mercury electrode occurs in the side chain the shift in half-wave potential due to the introduction of a *meta* or *para* substituent X is usually given by equation (20) in the form

$$\Delta E_{1/2} = \rho_{\pi,R}\sigma_X \tag{27}$$

When mesomeric interaction between the electroactive group R and a *para* substituent X via the benzene ring is possible, either in the ground state or in the transition state, the following is to be preferred:

$$\Delta E_{1/2} = \rho_{\pi,R}\sigma_X^- \tag{30}$$

Finally, in those so far rare examples in which the reaction constant $\rho_{\pi,R}$ is negative (this being explained by an electrophilic mechanism of the electrode process) equation (31) is used:

$$\Delta E_{1/2} = \rho_{\pi,R}\sigma_X^+ \tag{31}$$

In the case of those *ortho* substituents X which may be expected to have no considerable mesomeric interaction with the electroactive group equation (37) has proved satisfactory:

$$\Delta E_{1/2} = \rho_{\pi,R}\sigma_{o\text{-}X} + \delta_{\pi,R}(E_S)_{o\text{-}X} \tag{37}$$

43

For compounds containing several substituents X in the same benzene ring additivity of the substituent effects is to be expected. Thus, the linear free energy relation can be expressed by

$$\Delta E_{1/2} = \rho_{\pi,R}\Sigma\sigma_X \tag{38}$$

For substances in which two phenyl rings are attached to a bivalent electroactive group, such as a keto or azo group, the shifts in half-wave potential due to substituents in the 3-, 4-, 3'- and 4'-positions can be expressed by

$$\Delta E_{1/2} = \rho_{\pi,R}^{1}\Sigma\sigma_X + \rho_{\pi,R}^{2}\Sigma\sigma_X \tag{39}$$

or

$$\Delta E_{1/2} = \Sigma\rho_{\pi,R}^{i}\sigma_X \tag{39a}$$

Due to the symmetrical shapes of such molecules, it can be expected that $\rho_{\pi,R}^{1} = \rho_{\pi,R}^{2}$, and hence equation (38) is used. For unsymmetrical molecules such as nitrones or α,β-unsaturated ketones equation (39) has to be used.

The applications of these equations in the treatment of shifts in half-wave potential are described in this chapter. Before a survey of the reaction series in which such applications are possible is given, the choice of substituent coefficients σ_X and the principles and methods of selection of half-wave potentials will be discussed.

The possibility of correlating polarographic data with substituent constants by the use of equation (27) has been shown independently for various reaction series by Schultz[1] (1952), Brockman and Pearson[2] (1952), Motoyama and co-workers[3,4] (1952), Jaffé[5] (1953), and the present author[6] (1953), who also has shown the general validity of such relations[7] and discussed the theoretical principles involved, the scope of use of such correlations, and the structural dependence of the reaction constants.[8] Later, attention was paid to applications among benzene derivatives, mainly ketones and aldehydes (see Ref. 9–13), nitro-benzenes,[14] and iodobenzenes.[15,16] Recently, applications of equation (27) have become more frequent in organic polarography (see Ref. 17–24). The increased interest has been reflected also in discussions in review articles and monographs.[25–38]

The early assumption[14,39] that the reversibility of the electrode process involved is the necessary condition of such treatment of half-wave potentials is invalid, as can be understood from the treatment given in Chapter II. To correlate[15] $\log E_{1/2}$ (instead of $E_{1/2}$) with the substituent constant σ_X is theoretically unjustified.

2. Selection of the Values of the Substituent Constants

Values of the substituent constants σ were selected with the aid of recently published collective and critical publications. The values are given in Table III-1.

In cases in which the literature gives several different values, all these alternatives are cited in Table III-1, if none has been proved superior by the use of a sufficient number of half-wave potentials.

3. Selection and Treatment of Half-Wave Potential Values

The selection of the values of half-wave potentials for the verification of the applicability of equation (27) and other linear energy relations was in accordance with the principles and conditions summarized in Chapter II, Section 7.

In most cases only experimental values which had been measured by one single author under identical conditions (according to the worker who measured them) were compared.

When the half-wave potentials compared were pH-dependent, we first compared $E_{1/2}$–pH plots for all the compounds studied. A pH range was selected in which either the half-wave potentials of all compounds compared were no longer pH-dependent, or the slopes, $dE_{1/2}/d$pH, were practically the same for all these compounds, or, in certain cases, the value of $dE_{1/2}/d$pH was a linear function of the substituent constants (see p. 50).

When values of half-wave potential were given for only a few pH values in the original paper, differences in half-wave potential were compared for all these pH values. Before further treatment, cases in which the shift in half-wave potential varied substantially with pH were rejected.

For substances which showed no considerable change in half-wave potential with change in pH over the acidity range studied, values of half-wave potential obtained in supporting electrolytes of the same composition were compared.

Most data used in correlations were obtained in this way. Only in some few cases was it necessary to restrict ourselves for the time being to data given by the author for pH 0 (usually extrapolated values) or for some other deliberately chosen pH value. Hence, even the pioneer work by Brockman and Pearson,[2] in which half-wave potentials were treated at a single pH value [pH 4.7 in the presence of a high concentration (0.02 %) of gelatin], cannot be regarded as satisfactory.

Table III-1
Selected Values of Substituent Constants*

The uncited values are referenced as follows: σ_{m-x}, σ_{p-x} and σ^-_{p-x} from the review by Jaffe[5]. σ_{o-x}, σ^I_X and $(E_s)_{o-x}$ according to Taft[40]. σ^+_{m-x} and σ^+_{p-x} according to Brown and Okamoto[41]. $\sigma^R_X = \sigma_{p-x} - \sigma^I_X$.

Group	σ_{m-x}	σ_{p-x}	σ_{o-x}	$(E_s)_{o-x}$	σ^-_{p-x}	σ^+_{m-x}	σ^+_{p-x}	σ^I_X	σ^R_X
CH_3	-0.07	-0.17	-0.17	0.00	—	-0.07	-0.31	-0.05	-0.12
C_2H_5	-0.05	-0.15			—	-0.06	-0.29₅	-0.05ᵃ	-0.10
$C(CH_3)_3$	-0.12	$\underline{-0.20}$			—	-0.06	-0.26	-0.07	-0.13
CF_3	+0.41₅	$\underline{+0.55}$			+0.73ᵇ	+0.52	+0.61	+0.41	+0.14
CH_2Cl	—	+0.18				+0.14	-0.01	+0.17	+0.01
CH_2CN	—	+0.01							
CH_2CH_2COOH	-0.03	-0.07							
$CH_2C_6H_5$									
OH	+0.12ᶜ	$\underline{-0.37}$ᶜ			—		-0.92	+0.04	-0.62
OCH_3	+0.11₅	$\underline{-0.27}$	-0.39	+0.99	—	+0.05	-0.78	+0.25	-0.52
OC_2H_5	+0.15	$\underline{-0.25}$	-0.35	+0.90				+0.25ᵃ	
$OCH_2C_6H_5$	—	-0.41₅							
OC_6H_5	+0.25ᶜ	-0.03					-0.5	+0.38	-0.41
		$\underline{-0.32}$ᶜ					-0.39ᵇ		
O^-	(-0.71)	(-0.52)							
OCH_2COOH		-0.33ᵈ							
NH_2	-0.16	-0.66			(-0.13)	-0.16	-1.3	+0.10ᵃ	+0.76
$NHCH_3$	-0.30	-0.59						+0.10	
$N(CH_3)_2$	-0.21	-1.05 to -0.21			—		-1.7	+0.10	-0.93
$N(C_2H_5)_2$		-0.70ᵈ							
$NHCOCH_3$	+0.21ᶜ	-0.01₅					-0.6	+0.28ᵃ	-0.29
$NHCOC_6H_5$	+0.22	+0.08					-0.6		
$NHNH_2$	-0.02	-0.55							

Substituent	1	2	3	4	5	6	7
NHOH	−0.04	−0.34					
NH$_3^+$	+0.63	—					
NH$_2$CH$_3^+$	+0.96	—					
N(CH$_3$)$_3^+$	+0.90	+0.66 to 1.11			+0.36	+0.41	+0.86
COOH	+1.01[c]	+0.88[c]		+0.73	+0.32	+0.42	
COOCH$_3$	$+0.31_5$	+0.21 to 0.43		+0.55 to 0.76_5	+0.37	+0.49	+0.30[a]
COOC$_2$H$_5$	+0.37[c]	+0.45[c]		+0.68	+0.37	+0.48	+0.13
CONH$_2$	+0.28	—		+0.63			
CHO	+0.38	+0.22		+1.13			
COCl	+0.38[b]	+0.66[d]					
COCH$_3$	+0.52	+0.52		+0.7 to 0.87		+0.27	+0.25
COC$_2$H$_5$	−0.12[d]						
COC$_6$H$_5$	+0.46						
COC$_6$H$_4$Cl	+0.40[d]						
CN	+0.68[c]	+0.66[c]		+1.00	+0.56	+0.59	+0.07
COO$^-$	+0.10[c] / 0.0[c] / −0.1[c]	+0.13	+0.80	+0.35	−0.03	−0.02	
NO$_2$	+0.78	+0.71	−0.75	+1.04 to 1.27	+0.67	+0.63	+0.15

$\sigma_{o\text{-}NO_2} = +1.22$

* The values underscored are thermodynamic values accurate within ±0.02 σ units.
[a] According to Taft and Lewis.[42]
[b] According to Exner.[43]
[c] According to Brown and Okamoto.[41]
[d] According to Bray and Barnes.[44]
[e] According to Willi and Stocker.[45]

Table III-1 (continued)

The uncited values are referenced as follows: σ_{m-x}, σ_{p-x} and σ^-_{p-x} from the review by Jaffe[5]. σ_{o-x}, σ^I_x and $(E_S)_{o-x}$ according to Taft[40]. σ^+_{m-x} and σ^+_{p-x} according to Brown and Okamoto[41]. $\sigma^R_x = \sigma_{p-x} - \sigma^I_x$.

Group	σ_{m-x}	σ_{p-x}	σ_{o-x}	$(E_S)_{o-x}$	σ^-_{p-x}	σ^+_{m-x}	σ^+_{p-x}	σ^I_x	σ^R_x
NO	—	+0.12							
F	+0.34	+0.06	+0.24	+0.49		+0.35	−0.07	+0.52[a]	−0.46
Cl	+0.37	+0.23	+0.20	+0.18		+0.40	+0.11	+0.47	−0.24
Br	+0.39	+0.23	+0.21	0.00		$+0.40_5$	+0.15	+0.45	−0.22
I	+0.35	+0.27	+0.21	−0.20		+0.36	$+0.13_5$	+0.38	−0.11
SCN	—	+0.70							
Si(CH₃)₃	−0.12 / −0.04[c]	−0.07				+0.01	+0.02	−0.12	+0.05
SCH₃	+0.14	−0.05						+0.25[a]	−0.30
SOCH₃	+0.55	+0.57			+0.73[b]			+0.52[a]	+0.05
SO₂CH₃	+0.65	+0.73			+1.05			+0.59	+0.14
SO₂Cl	+1.20[d]	+1.11[d]							
SO₂NH₂	—	+0.62							
SO₃⁻	+0.15[d]	+0.38							
PO₃H⁻	+0.23	+0.24							
AsO₃H⁻	—	−0.02							
SO₃H	+0.55	—							
C₆H₅	+0.22 / +0.06[c]	+0.01 / +0.64	+1.17	−0.90		+0.11	−0.18	+0.10	−0.09
N=N—C₆H₅	—	—							
C₆H₄—N=N—C₆H₅	+0.14				+1.09				
CH=CH—C₆H₅	−0.26				+0.62				
3,4-(CH₂)₃	−0.48								
3,4-(CH₂)₄	+0.17								
2-naphthyl	+0.04[c]						-0.13_5		
3,4-OCH₂O	−0.16								

The pH-dependent half-wave potentials cannot be incorporated into a set of pH-independent potentials (e.g., at a pH that is not sufficiently high to ensure the pH-independence of the pH-dependent ones); thus, the value for p-hydroxybenzaldehyde cannot be compared with values for other substituted benzaldehydes at pH 13. A simple correction for equilibrium concentration[10] cannot be recommended because of the effect of proton-transfer reactions, recombination, and other complications of the electrode process.

For systems in which the half-wave potential was pH-dependent, only values obtained in well-buffered solutions were compared. When there was a choice, half-wave potentials obtained in the absence of surface-active substances were preferred. When such data were not available, at least care was taken that the concentration of surface-active substance was the same in all solutions compared. Values of "half-wave potential" in cases in which streaming maxima were observed on the curves (e.g., for tolualdehydes[1]) could not be included in a set of correlated half-wave potentials.

In all instances in which sufficient data were available it was proved that values of the transfer coefficient either did not show considerable differences in the reaction series studied (Tables III-2 and III-3) or

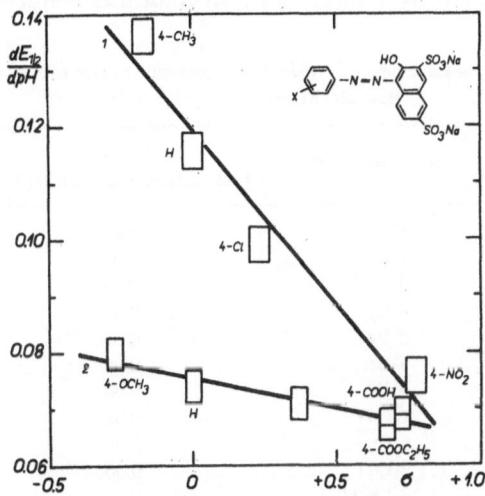

Fig. III-1. Plot of the slope of the pH-dependence of half-wave potentials ($dE_{1/2}/d$pH) of azo dyes against Hammett substituent constants σ_X: 1) Britton–Robinson buffers pH < 4 (Table III-4, No. 52, 53); 2) Clark–Lubs buffers, 0.005% gelatin, pH > 4 (Nos. 54–57).

Table III-2
Values of the Transfer Coefficient α, Calculated from the Relation of $\log[i/(i_d - i)]$
to E in a Reaction Series

Substance	Substituent X	$RT/\alpha nF$ (values from Brockman et al.[2])
$X-C_6H_4CO-C_6H_5$	H	0.096
	4-CH$_3$	0.093
	4-t-C$_4$H$_9$	0.095
	4-OCH$_3$	0.090
	4-Cl	0.100
	3-Br	0.099
$X-C_6H_4\underset{\underset{NNHCONH_2}{\|}}{C}-C_6H_5$	H	0.085
	4-OCH$_3$, α	0.084
	β	0.081
	4-Br, α	0.085
	β	0.085

showed a linear correlation with substituent constants (Fig. III-1, bases on data for azo dyes[19,46]): In the former case, values of transfer coefficients may also be subject to a linear free energy relation with a small proportionality factor, so that the inaccuracy in the determination of the transfer coefficient overshadows its structural dependence. Hence, the

Table III-3
Values of the Slope $dE_{1/2}/d$pH in a Reaction Series of Substituted
Benzaldehydes [$X-C_6H_4-CHO$]

Substituent X	$dE_{1/2}/d$pH (values from Coulson et al.[47])
H	0.064
2-Cl	0.064
3-Cl	0.065
4-Cl	0.064
2-Br	0.063
3-Br	0.066
4-Br	0.066
2-OH	0.064
3-OH	0.063
4-OH	0.063
2-OCH$_3$	0.061
3-OCH$_3$	0.066
4-OCH$_3$	0.065
4-CH$_3$	0.064

half-wave potentials of substances for which the transfer coefficient deviates substantially from the "constant" value or from the linear α versus σ plot should be excluded from the particular reaction series.

Hence, the situation is analogous to that found for homogeneous reactions. Hammett[78] first assumed that a necessary condition for the validity of the linear free energy relation he derived (which formed the basis for all his quantitative empirical correlations between structure and kinetic and similar quantities) is the constancy of the frequency factor (or activation entropy) throughout the reaction series. Leffler[79] later demonstrated that this condition usually is not fulfilled and, moreover, need not be fulfilled. He showed[79] that in several reaction series the temperature-independent component of the reaction constant is also a linear function of structural parameters. In such cases, therefore, the Hammett equation and its modified forms, such as (27), are valid too.

The analogy between the frequency factor in homogeneous reactions and the transfer coefficient in electrode processes, nevertheless, is not complete, for the value of the transfer coefficient affects also the activation enthalpy terms. As in the case of the frequency factor, the accuracy attained in the determination of the transfer coefficient is low. But, whereas, values of the frequency factor are computed from rate constants and their correlation with rate constants or free activation enthalpy involves difficulties,[80] values of transfer coefficients are determined independently, either from the curve slopes or from the pH-dependence.

The values of transfer coefficients obtained from the shape of the wave and from the $E_{1/2}$–pH relation should in general be equal. Unfortunately, there are not enough data available to enable us to make such a comparison. In the only available reaction series (Fig. III-2), according to logarithmic analysis the relation of $dE_{1/2}/dpH$ to α is linear. In the simplest case the slope of this line would be 1.0, but the observed value is 0.44, and for this no explanation at present is known. For phenyl-substituted cinnamic acid[17] values of $RT/\alpha nF$ showed small variations between 0.048 and 0.054, and the values of $dE_{1/2}/dpH$ varied even less (0.063 to 0.065).

Half-wave potentials selected according to these rules were then plotted against values of the substituent constants (σ_X, σ_{o-X}, σ_X^-, and σ_X^+) so as to obtain a straight line with an inclination of about 45°. Experimental data were marked by circles, squares, ovals and rectangles of a size corresponding approximately to the mean deviations

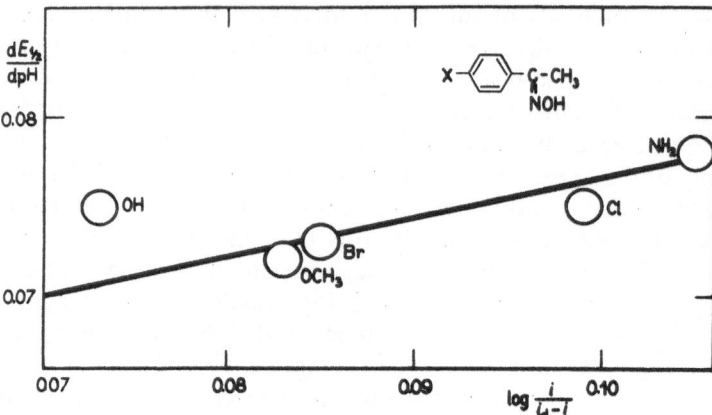

Fig. III-2. Plot of the slope of the pH-dependence of half-wave potentials ($dE_{1/2}/dpH$) of substituted acetophenone oximes against the slope of the logarithmic analysis [$\log i/(i_d - i)$]. Reaction Series No. 20, Table III-4.

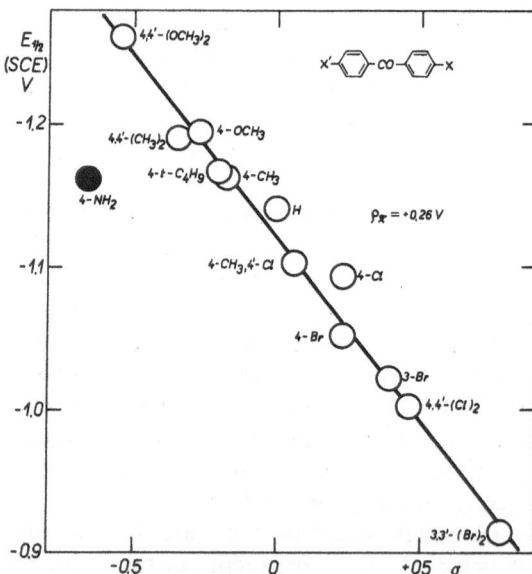

Fig. III-3. Relation of half-wave potentials for the reduction of substituted benzophenones to the sum of the Hammett substituent constants $\Sigma\sigma_X$. Reaction Series No. 22, Table III-4. Full point deviates.

of the two quantities. A regression line, derived by the method of least squares, was plotted next. The line was sometimes calculated, sometimes plotted directly after visual estimation of the trend of the points. Several times, it was proved that the human eye gave results only a little different from those obtained statistically. Because of the possibility to judge deviating values, the visual treatment was sometimes even superior to the rigorous one.

To judge which values of half-wave potential fit the regression line and which should be regarded as deviating a shift of 0.1 to 0.15 σ-units was considered as a "deviation" (Figs. III-3–5). The approximate

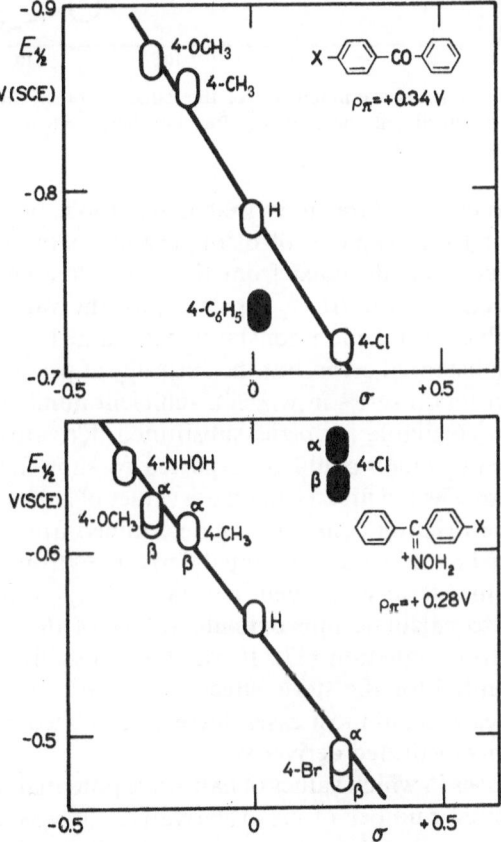

Fig. III-4. Relation of half-wave potentials for the reduction of *a*) substituted benzophenones (Reaction Series No. 24, Table III-4); and *b*) substituted benzophenoneoximes (Reaction Series No. 25), to Hammett substituent constants σ_X. Full points deviate.

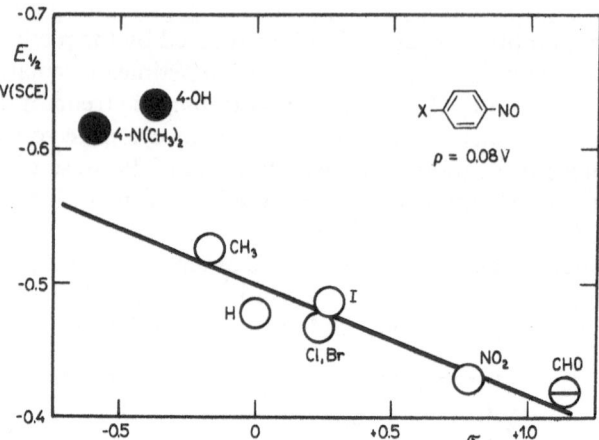

Fig. III-5. Relation of half-wave potentials for the reduction of substituted nitrosobenzenes to Hammett substituent constants σ_X. Reaction Series No. 46, Table III-4. Full points deviate. Halved point: σ_{CHO}^-.

character of such a treatment is well understood, but nevertheless, it was thought to be capable of distinguishing between values that fitted and values that deviated from the linear free energy relation. From the directrix of the $\Delta E_{1/2}$ versus σ plot the value for $\rho_{\pi,R}$ was computed. Values of reaction constants determined in this way may be subject to substantial error, but the accuracy at present is sufficient.

In those reaction series in which a sufficient number of half-wave potentials were available for *ortho*-substituted derivatives, values for those substituents which could be expected to show low mesomeric interaction were selected first. Using these values of half-wave potential, the value of the reaction constant $\rho_{\pi,R}$ determined from the half-wave potentials of *meta*- and *para*-substituted derivatives from equation (27), and values of the steric substituent constants $(E_S)_{o\text{-}X}$ given in Table I, we were able to calculate approximate values of the steric reaction constant $\delta_{\pi,R}$ from equation (37). It was then possible to correct the half-wave potential for the steric effect $[E_{1/2} - \delta_{\pi,R}(E_S)_{o\text{-}X}]$, and such corrected values were, in most cases, further correlated with values for *meta*- and *para*-substituted derivatives.

In those cases in which values of half-wave potential were available for only a limited number of *ortho* derivatives, it was assumed as a first approximation that the value for $\delta_{\pi,R}$ is small and that in equation (37) we can, therefore, restrict ourselves to the first term on the right-hand side.

The question of whether σ_X, σ^-, or σ^+ is to be preferred is discussed in Section 5 of this chapter.

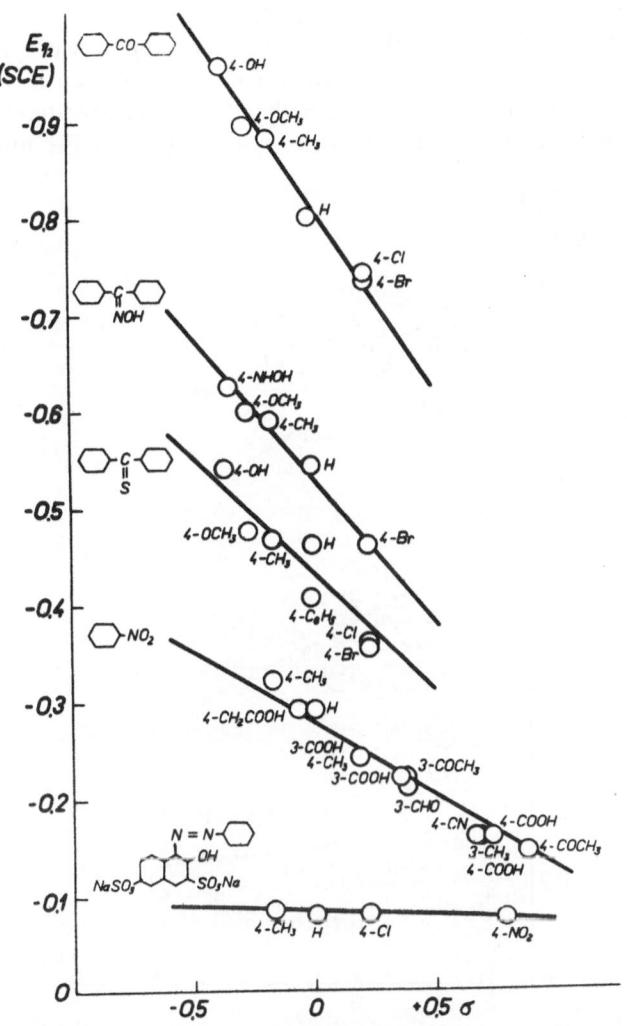

Fig. III-6. Relation of half-wave potentials for the reduction of substituted benzene derivatives in acid media to Hammett substituent constants σ_X. Benzophenones at pH 0, Reaction Series No. 23, Table III-4; benzophenoneoximes at pH 0, Reaction Series No. 25; thiobenzophenones at pH 0, Reaction Series No. 29; nitrobenzenes at pH 2.0, Reaction Series No. 40; azo dyes at pH 2.6, Reaction Series No. 52.

4. Reaction Series for Which Equations (27), (30), (31), and (37) Hold

A survey of reaction series for which the validity of special forms of equation (20) has been demonstrated is given in Table III-4. Some typical examples are depicted in Figs. III-6–8.

In this table and in the following paragraphs the symbol Ar stands for phenyl substituted in the *meta*, *para*, or even in the *ortho* position. E_H is the half-wave potential of the unsubstituted parent compound in a given reaction series, and n denotes the number of

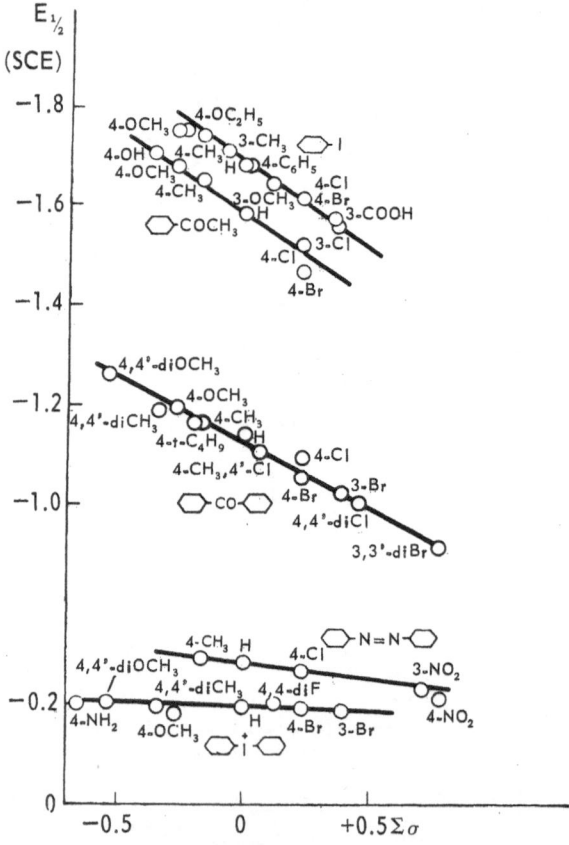

Fig. III-7. Relation of half-wave potentials for the reduction of substituted benzene derivatives to Hammett substituent constants σ_X. Iodobenzenes (pH-independent), Reaction Series No. 62, Table III-4; acetophenones at pH 7.0, Reaction Series No. 16; benzophenones at pH 4.7, Reaction Series No. 22; azo compounds at pH 5.0, Reaction Series No. 55; diphenyliodonium salts at pH 8.6, Reaction Series No. 6–8.

substances for which the validity of the particular equation has been proved. For di- and poly-substituted substances the symbols for the substituents are placed in parentheses. For those reaction series in which equation (30) ought to be used for some *para*-substituted compounds indications are given in parentheses showing whether σ_X or σ_X^- was used. In those cases in which the data available did not allow us to determine the value of the steric reaction constant $\delta_{\pi,R}$ from equation (37) and the second term on the right-hand side of this equation was neglected, the symbols for *ortho* substituents are followed by $\sigma_{o\text{-}X}$ in parentheses to show that only the substituent constant $\sigma_{o\text{-}X}$ was used in the given correction. The symbol $f \neq$ pH is used in cases where the half-wave potential is pH-independent in the pH range studied. When the author who published the original data did not give the experimental values of the half-wave potentials, but only a value extrapolated to a deliberately chosen pH value, for example, to pH 0, this is indicated by the abbreviation "extrapol." The number of electrons transferred in the electrode process is given by 1F, 2F etc., and when gelatin is used this is indicated by the abbreviation "gel." The validity of these relations has been proved so far for 691 values of half-wave potential.

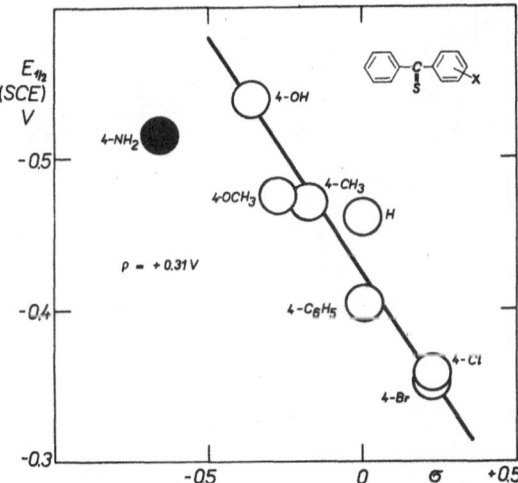

Fig. III-8. Relation of half-wave potentials for the reduction of substituted thiobenzophenones to Hammett substituent constants σ_X. Reaction Series No. 29, Table III-4. Full point deviates.

Table III-4
Reaction Series of Benzene Derivatives for which the Validity of Some Special Forms of Equation (20) has been Proved

No.	Reaction series	Solution	Note	Equation	$\rho_{\pi,R}$ V	E_H(SCE)	n	Substituents — Fitting linear relation	Substituents — Deviating	Ref. to source of $E_{1/2}$
				A Reduction by Nucleophilic Mechanism						
1	Ar—CHO	Buffers pH 0	extrapol. 1st wave, 1F	27, 30	+0.32	−0.78	15	H;3-CH₃;4-CH₃;3-CH₂OH; 4-CH₂OH;3-OH;4-OH; 3-OCH₃;4-OCH₃; (3-OCH₃,4-OH);3,4-OCH₂O; 3-NHCOCH₃;4-NHCOCH₃; 4-COOH(σ);4-CHO(σ);4-CN(σ)	3-NH₃⁺;3-COOH; 3-Cl;4-Cl;3-Br; 4-Br;3-I;4-I	48
2		Buffers pH 13	$f \neq$ pH $\delta_{\pi,R} = +0.17$	27, 30, 37	+0.33	−1.40	24	H;2-OCH₃;3-CH₃;4-CH₃; 2-OCH₃;3-OCH₃;4-OCH₃; 3,4-OCH₂O;4-NH₂;4-N(CH₃)₂; 4-N(CH₃)₃⁺;4-NHCOCH₃; 3-CHO⁻;4-CHO(σ^-);4-CN(σ^-); 4-COO⁻(σ^-);2-Cl;3-Cl;4-Cl; 3-Br;4-Br;3-I;4-I;3-SO₃⁻	3-O⁻ᵃ;4-O⁻ᵃ; (3-OCH₃,4-O⁻)ᵃ; 3-COO⁻;2-Br	10, 48
3		Buffers medium pH	2nd wave $f \neq$ pH	27	+0.34	−1.23	12	H;3-CH₃;4-CH₃;3-OH; 3-OCH₃;4-OCH₃;4-NHCOCH₃; 3-COOH;4-COOH;4-Cl;3-Br; 4-Br;	2-CH₃(σ_{-x});4-OH; 2-OCH₃(σ_{-OCH_3}); 2-Cl(σ_{-Cl});3-Cl; 2-Br(σ_{-Br});3-I;4-I	48
4		Buffers pH 0 50% dioxane	extrapol. 1st wave, 1F	27	+0.29	−0.87	9	H;3-OH;4-OH;3-OCH₃; 4-OCH₃;3-Cl;4-Cl;3-Br;4-Br	4-CH₃; 2-OCH₃(σ_{-OCH_3}); 2-Cl(σ_{-Cl});2-Br(σ_{-Br})	47
5		Buffers pH 2.25 60% EtOH	$\delta_{\pi,R} = +0.15$	27, 37	+0.31	−1.02₂	14	H;3-CH₃;4-CH₃;3-OH;4-OH; 2-OCH₃;2-Cl;3-Cl;4-Cl;3-Br; 4-Br;2-I;3-I;4-I;	2-CH₃;3-OCH₃ᵇ; 3-NH₃⁺	49
6		Buffers pH 3.15 60% EtOH	$\delta_{\pi,R} = +0.15$	27, 37	+0.30	−1.08₂	12	H;3-CH₃;4-CH₃;4-OH; 4-NH₃⁺(σ^{-b});2-Cl;3-Cl;4-Cl; 3-Br;4-Br;3-I;4-I	2-CH₃;3-OH; 3-NH₃⁺;2-I	49

7	Ar—CHO	Buffers pH 2.2 25% EtOH	1F	27	+0.16	−1.01	6	H;4-OCH₃; (3-OCH₃,4-OH);3,4-OCH₂O; 4-NH(CH₃)₂⁺	2-OCH₃($\sigma_{o\text{-}OCH_3}$)	50
8		Buffers pH 12	1F,$f \neq$ pH	27	+0.55	−1.48	6	H;4-OCH₃;4-O⁻; (3-OCH₃,4-O⁻);3,4-OCH₂O; 4-N(CH₃)₂	2-OCH₃($\sigma_{o\text{-}OCH_3}$)	50
9		Buffers; medium pH	2nd wave $f \neq$ pH	27	+0.56	−1.34	6	H;4-OH;4-OCH₃; (3-OCH₃,4-OH);3,4-OCH₂O; 4-N(CH₃)₂	2-OCH₃($\sigma_{o\text{-}OCH_3}$)	50
10		Buffers pH 0.0	extrapol. 1st wave, 1F	27	+0.27	−0.87₃	6	H;3-OH;4-OH;4-OCH₃; (3-OCH₃,4-OH);3,4-OCH₂O	2-CH₃($\sigma_{o\text{-}CH_3}$)	11
11		Buffers pH 7.0		27	+0.26	−1.46	7	H;4-OH;4-OCH₃;4-NH₂; − 3-NHOH;3-Cl;4-Cl	3-CH₃ᶜ;4-CH₃ᶜ	1
12	Ar—COCH₃	0.05 M·N(Et)₄OH, 50% i-PrOH		27, 30	+0.35	−1.69	3	H;4-NH₂;4-CN(σ^-)	4-Br	51
13		Buffers pH 0	extrapol.	27	+0.19	−1.02	3	H;(3,4-diOCH₃);4-Cl	2-naphthyl; 3,4-(CH₂)₄	12
14		Alkaline	$f \neq$ pH	27	+0.22	−1.58	4	H;3,4-(CH₂)₄;4-Cl; 2-naphthyl($\sigma_X = 0.31$)	—	12
15		Buffers pH 0 30% i-PrOH	extrapol.	27	+0.22	−0.99	4	H;(3-CH₃,4-OH);4-OH;4-OCH₃	4-C₆H₅	11
16		Buffers pH 7.0 33% EtOH	1st wave 1F	27	+0.35	−1.58	6	H;4-CH₃;4-Cl;4-OCH₃; 4-Cl;4-Br	—	1
17		Buffers pH 7.0	2nd wave	27	+0.43	−1.72	4	H;4-CH₃;4-Cl;4-Br	—	1

a Half-wave potential is pH-dependent, even at high pH.
b Deviations in the slope of the wave.
c Streaming maxima observed on curves.
d Versus mercury pool electrode.
e Steric hindrance to coplanarity.
f Versus mercurous sulphate electrode.
g Consumption of 4F, another reduction mechanism.

h R¹ =   i R² =   j R² =   k R³ =   m R⁴ =

a Complications due to adsorption phenomena.
p R⁵ = 5-barbituric acid residue.
r R⁶ = 2-thio-5-barbituric acid residue.
s Substituent X directly attached to the ferrocenyl system; σ_p constants used.
t When value of σ_{CHO} is neglected.

Table III-4 (*continued*)

No.	Reaction series	Solution	Note	Equation	$\rho_{z,R}$, V	E_H(SCE)	n	Substituents: Fitting linear relation	Substituents: Deviating	Ref. to source of $E_{1/2}$
18	Ar—COCH₃	Buffers pH 0 50% dioxane	extrapol. 1st wave	27	+0.27	−0.99	12	H; 3-CH₃; 4-CH₃; 4-OH; 3-OCH₃; 4-OCH₃; 2-Cl($\sigma_{o\text{-}Cl}$); 3-Cl; 4-Cl; 2-Br($\sigma_{o\text{-}Br}$); 3-Br; 4-Br;	3-OH; 2-OCH₃($\sigma_{o\text{-}OCH_3}$)	47
19		Buffers pH 0 47.5% EtOH	extrapol.	27	+0.27₅	−1.05₇	5	H; 4-CH₃; 4-C₂H₅; 4-Cl; 4-Br	—	13
20	Ar—C(NOH)CH₃	Buffers pH 0 30% i-PrOH	extrapol.	27	+0.18	−0.78	7	H; 4-CH₃; 4-OH; 4-OCH₃; 4-Cl; 4-Br; 4-NHOH	4-NH₂	52
21		Alkaline 30% i-PrOH	$f \neq$ pH	27	+0.26	−1.82	5	H; 4-CH₃; 4-OCH₃; 4-Cl; 4-Br	4-NH₂; 4-NHOH	52
22	Ar—CO—Ar	Acetate pH 4.7 40% MeOH 0.02% gel.	—	27	+0.26	−1.14	12	H; 4-CH₃; (4,4'-diCH₃); (4-CH₃,4'-Cl); 4-t-C₄H₉; 4-OCH₃; (4,4'-diOCH₃); 4-Cl; (4,4'-diCl); 3-Br; 4-Br; (4,4'-diBr)	4-NH₂	2
23		Buffers pH 0	extrapol.	27	+0.37	−0.87	6	H; 4-CH₃; 4-OH; 4-OCH₃; 4-Cl; 4-Br	—	53
24		Buffers pH 0	extrapol.	27	+0.34	−0.78	4	H; 4-CH₃; 4-OCH₃; 4-Cl	4-C₆H₅	11
25	Ar—C(NOH)—Ar	Buffers pH 0 30% i-PrOH	extrapol.	27	+0.28	−0.54	8	H; 4-CH₃(α); 4-CH₃(β); 4-OCH₃(α); 4-OCH₃(β); 4-Br(α); 4-Br(β); 4-NHOH(α)	4-Cl(α); 4-Cl(β); 4-NH₂(α)	52
26		Alkaline 30% i-PrOH	$f \neq$ pH	27	+0.29	−1.60	7	H; 4-CH₃(α); 4-CH₃(β); 4-Cl(α); 4-Cl(β); 4-Br(α); 4-Br(β)	4-OCH₃(α); 4-OCH₃(β); 4-NH₂(α)	52
27		Acetate pH 4.7 40% MeOH 0.02% gel.	—	27	+0.08	−0.98	5	H; 4-OCH₃(α); 4-OCH₃(β); 4-Br(α); 4-Br(β)	—	
28	Ar—C—Ar ‖ NNHCONH₂	Acetate pH 4.7 40% MeOH 0.02% gel.	—	27	+0.13	−1.10	5	H; 4-OCH₃(α); 4-OCH₃(β); 4-Br(α); 4-Br(β)	—	2

No.	Structure	Buffer	Note	Ref.	$+$	$-$	n	Substituents		Ref.
29	Ar—CS—Ar	Buffer pH 0	extrapol.	27	$+0.31$	-0.46	7	H;4-CH₃;4-OH;4-OCH₃; 4-Cl;4-Br;4-C₆H₅	4-NH₂	53
30	Ar—CH=CHCOCH₃	Buffers pH 8.5	—	27	$+0.20$	-1.37	4	H;3,4-OCH₂O;(4-OH,3-OCH₃); (3-OC₂H₅,4-OH)	(3-OH,4-OCH₃)	54
31		Buffers pH 7.8 NH₄⁺-salt	—	27	$+0.40$	-1.38	4	H;3,4-OCH₂O;(3-OCH₃,4-OH); (3-OC₂H₅,4-OH)	(3-OH,4-OCH₃)	54
32		Buffers pH 10.5 Na⁺-salt	—	27	$+0.28$	-1.38	4	H;(3-OH,4-OCH₃); (4-OH,3-OCH₃);(3-OC₂H₅,4-OH)	3,4-OCH₂O	54
33	Ar—CH=CHCOOH	Buffers pH 0 50% EtOH	extrapol.	27	$+0.32$	-1.23	5	H;4-OCH₃;4-Cl;4-Br:	—	17
34	Ar—CH=CHCOOC₂H₅	Buffers 50% EtOH	$f \neq$ pH	27	$+0.28$	-1.33	5	H;4-OH;4-OCH₃;4-Cl;4-Br	—	17
35	Ar—COOCH₃	Unbuffered pH 7.2	$f \neq$ pH	27,30	$+0.85$	-2.11	3	H;4-Br;4-COOCH₃(σ^-)	—	55
36	Ar—NO₂	Glacial acetic acid 1 M·A:ONH₄	$\delta_{r,R}$ small	27,30, 37	$+0.15$	-0.65_d	23	H;3-CH₃;4-CH₃;(3,4-diCH₃); 3-OH;4-OH;3-NH₂;4-NH₂; 4-N(CH₃)₂;2-NHCOCH₃; 4-NHCOCH₃;3-COOH; 4-COOH(σ^-);3-NO₂;4-NO₂(σ^-); 3,5-diNO₂;2-Cl;3-Cl;4-Cl; 2-Br;3-Br;4-Br;(3,5-diCl)	2-CH₃ᵉ;3-N(CH₃)₂; 3-F:4-F:3-I:4-I	56
37		0.01 M NEt₄Br 66% EtOH		27,37	$+0.50$	-0.93_5	7	H;2-Cl;3-Cl;4-Cl;2-I;3-I; 4-I	—	57
38		Alkaline abs. EtOH	$f \neq$ pH	27	$+0.24$	relat.	6	3-OH;4-OH;3-NH₂;4-NH₂; 3-NHCOCH₃;4-NHCOCH₃	—	58
39		Buffers pH 3.1 10% EtOH	—	27,30	$+0.16$	-0.35	11	H;4-CH₃;3-COOH; 4-COOH(σ^-);3-CHO; 4-CHO(σ^-);3-COCH₃; 4-COCH₃(σ^-);3-CN;4-CN(σ or σ^-);4-I	—	59

Table III-4 (continued)

No.	Reaction series	Solution	Note	Equation	$\rho_{\pi,R}$, V	E_H(SCE)	n	Substituents		Ref. to source of $E_{1/2}$
								Fitting linear relation	Deviating	
40	Ar—NO$_2$	Prideau–Ward Buffer pH 2.0	1st wave	27, 30	$+0.16_5$	-0.29	14	H; 4-CH$_3$; 4-CH$_2$COOH; (3-CH$_3$,4-COOH); (4-CH$_3$,3-COOH); 3-COOH; 4-COOH(σ^-); 3-CHO; 4-CHO(σ^-); 3-COCH$_3$; 4-COCH$_3$(σ^-); 3-CN; 4-CN(σ^-); (3,4-diCOOH)	2-CH$_3$; 2-COOH[e]	14
41		Prideau–Ward Buffer pH 6.8	1st wave	27, 30	$+0.14$	-0.50	6	H; 3-CHO; 4-CHO(σ^-); 3-COCH$_3$; 4-COCH$_3$(σ^-); 4-CN(σ)	3-COO$^-$	14
42		Acetate pH 2.2$_5$ 60% EtOH	$\delta_{\pi,R} \doteq 0$	27, 30, 37	$+0.22$	-0.36_6	23	H; 2-CH$_3$(σ-CH$_3$); 3-CH$_3$; 4-CH$_3$; 3-OH; 4-OH; 3-NH$_2$; 4-NH$_2$; 3-COOH; 4-COOH(σ^-); 3-COOC$_2$H$_5$; 4-COOC$_2$H$_5$(σ^-); 3-CHO; 2-NO$_2$(σ-NO$_2$); 3-NO$_2$; 4-NO$_2$(σ^-); 2-Cl(σ-Cl); 3-Cl; 4-Cl; 2-Br(σ-Br); 3-Br; 4-Br; 3-C$_6$H$_5$	4-CHO (neither σ nor σ^-); 2-I; 3-I; 4-I; 4-C$_6$H$_5$	49
43		Acetate pH 3.2 60% EtOH	$\delta_{\pi,R} \doteq 0$	27, 37	$+0.30$	-0.43	25	H; 2-CH$_3$(σ-CH$_3$); 3-CH$_3$; 4-CH$_3$; 3-OH; 4-OH; 3-NH$_2$; 3-COOH; 3-COOC$_2$H$_5$; 4-COOC$_2$H$_5$; 4-CHO; 2-NO$_2$(σ-NO$_2$); 3-NO$_2$; 4-NO$_2$; 2-Cl(σ-Cl); 3-Cl; 4-Cl; 2-Br(σ-Br); 3-Br; 4-Br; 2-I(σ-I); 3-I; 4-I; 3-C$_6$H$_5$	3-CHO; 4-NH$_2$; 4-C$_6$H$_5$	49
44		HCl-KCl pH 1.0 10% EtOH	—	27, 30	$+0.14$	-0.22	9	H; 3-OH; 4-OH; 3-OCH$_3$; 4-OCH$_3$; 3-COOH; 4-COOH(σ^-); 3-COOCH$_3$; 4-COOCH$_3$(σ^-)	2-OCH$_3$(σ-OCH$_3$)	60
45		McIlvaine, pH 4.0 10% EtOH	—	27, 30	$+0.17$	-0.41_5	9	H; 3-OH; 4-OH; 3-OCH$_3$; 4-OCH$_3$; 3-COOH; 4-COOH(σ^-); 3-COOCH$_3$; 4-COOCH$_3$(σ^-)	2-OCH$_3$(σ-OCH$_3$)	60

No.	Substrate	Conditions		Ref.			n	Substituents (σ)	Substituents	Ref.
46	Ar—NO	Buffer pH 7.0	1st wave 2F	27, 30	+0.08	−0.48f	7	H; 4-CH$_3$; 4-CHO(σ^-); 4-NO$_2$(σ); 4-Cl; 4-Br; 4-I	4-OHe; 4-N(CH$_3$)$_2$	61
47	Ar—N=N—C$_6$H$_5$	Buffer pH 5.5	—	27	+0.22	−0.33	5	H; 3-CH$_3$; 4-OCH$_3$; 4-COOC$_2$H$_5$; 4-SO$_3^-$	4-NH$_2$; 4-N(CH$_3$)$_2$	62
48		Buffer pH 7.0	—	27	+0.26	−0.51	5	H; 3-CH$_3$; 4-OCH$_3$; 4-COOC$_2$H$_5$; 4-SO$_3^-$	4-NH$_2$; 4-N(CH$_3$)$_2$	62
49		0.1 M AcOH, 0.1 M AcONa 75% EtOH		27	+0.17	−0.32	3	H; 4-OH; 4-OCH$_3$	—	63
50	Ar—N=N—R^{1h}	0.1 M AcOH, 0.1 M AcONa, 75% EtOH		27	+0.08$_4$	−0.46$_5$	5	H; 3-OCH$_3$; 4-OCH$_3$; 3-SO$_3^-$; 4-SO$_3^-$	—	63
51	Ar—N=N—R^{2j}	0.1 M AcOH, 0.1 M AcONa, 75% EtOH		27	+0.10$_5$	−0.36$_5$	3	H; 4-OCH$_3$; 4-SO$_3^-$	—	63
52	Ar—N=N—R^{3k}	Buffer pH 2.6		27	+0.008	−0.08	4	H; 4-CH$_3$; 4-NO$_2$; 4-Cl	—	19
53		Buffer pH 3.3		27	+0.03	−0.16	4	H; 4-CH$_3$; 4-NO$_2$; 4-Cl	—	19
54		Buffer pH 4.1		27	+0.07	−0.25	4	H; 4-CH$_3$; 4-NO$_2$; 4-Cl	—	19
55		Buffer pH 5.0		27	+0.07$_6$	−0.33	5	H; 4-CH$_3$; 3-NO$_2$; 4-NO$_2$; 4-Cl	—	19
56		Buffer pH 6.1		27	+0.04	−0.42	3	H; 4-CH$_3$; 4-Cl	—	19
57		Buffer pH 7.1		27	+0.16	−0.50	3	H; 4-CH$_3$; 4-Cl	—	19
58		Clark–Lubs, pH 3.5 0.005% gel		27, 30	+0.06$_5$	−0.25	5	H; 4-OCH$_3$; 3-Cl; 4-COOH(σ^-); 4-COOC$_2$H$_5$(σ^-)	—	46
59		Clark–Lubs, pH 7.0 0.005% gel	Eq. (37) unsuitable	27, 30	+0.12	−0.53	17	H; 3-CH$_3$; 4-CH$_3$; 3-OCH$_3$; 4-OCH$_3$; 3-OC$_2$H$_5$; 4-OC$_2$H$_5$; 4-NH$_2$; 4-NHCOCH$_3$; 4-COOH; 4-COOC$_2$H$_5$; 3-Cl; 4-Cl; 4-Br; 4-SO$_3^-$; 4-SO$_2$NH$_2$; 2-naphthyl(σ_x = +0.17)	2-CH$_3$($\sigma_{o\text{-CH}_3}$); 2-OCH$_3$($\sigma_{o\text{-OCH}_3}$); 2-OC$_2$H$_5$($\sigma_{o\text{-OC}_2\text{H}_5}$); 2-Cl($\sigma_{o\text{-Cl}}$); 2,3-diCH$_3$; 2,4-diCH$_3$($\sigma_{o\text{-CH}_3}$)	46

Table III-4 (*continued*)

No.	Reaction series	Solution	Note	Equation	$\rho_{\pi,R}$, V	E_H(SCE)	n	Substituents (Fitting linear relation)	Substituents (Deviating)	Ref. to source of $E_{1/2}$
60	Ar—N=N—R³	Clark–Lubs, pH 9.05, 0.005% gel	—	27, 30	+0.16	−0.66	5	H; 4-OCH₃; 3-Cl; 4-COO⁻(σ^-); 4-COOC₂H₅(σ^-)	—	46
61	Ar—N=N—R⁴ᵐ	Clark–Lubs, pH 7.0, 0.005% gel	—	27, 30	+0.10	−0.73	10	H; 4-CH₃; 4-OCH₃; 4-OC₂H₅; 4-NH₂; 4-NHCOCH₃; 4-COOC₂H₅(σ^-); 3-Cl; 4-Cl; 4-SO₃⁻	—	46
62	Ar—CH=CH—Ar	0.1 M N(C₂H₅)₄I 75% dioxane	$\delta \neq 0$	27, 37	+0.25	−2.17	14	H; (2,4,6,2',4',6'-hexaCH₃); (2-CH₃,4'-N(CH₃)₂); (3-CH₃,4'-N(CH₃)₂); (4-CH₃,4'-N(CH₃)₂); (2-CH₃,4'-N(CH₃)₂); (2,5-diCH₃,4'-N(CH₃)₂); 4-OH; (4-OH,4'-NH₂); 4-OCH₃; (4,4'-diOCH₃); 4-NH₂; 3-N(CH₃)₂; 4-N(CH₃)₂	(4-N(CH₃)₂,1-naphthyl); (4-N(CH₃)₂,2-naphthyl)	64
63	Ar—I	Acetate pH 7 90% EtOH	$\delta_{\pi,R} = 0.2$	27, 37	+0.31	−1.68	19	H; 2-CH₃; 3-CH₃; 4-CH₃; (2,4-diCH₃); (2,5-diCH₃); (2,6-diCH₃); 2-OCH₃; 3-OCH₃; 4-OCH₃; 2-OC₂H₅; 4-OC₂H₅; 4-NHCOCH₃; 3-COOH; 4-COOH; 3-Cl; 4-Cl; 4-Br; 4-C₆H₅	2-Cl; 2-Br	15
64		0.01 M NEt₄Br 66% EtOH	$\delta_{\pi,R} = 0.2$	27, 37	+0.31	−1.62	9	H; 2-CH₃; 3-CH₃; 4-CH₃; 3,6-diOCH₃; 3-Cl; 4-Cl; 1-naphthyl; 2-naphthyl	2-Cl	65
65		0.01 M NEt₄Br 66% EtOH		27	+0.57	−1.65	13	H; 2-CH₃; 3-CH₃; 4-CH₃; 3-NH₂; 3-Cl; 4-Cl; 3-Br; 4-Br; 3-I; 4-I; 3-C₆H₅; 4-C₆H₅	3-OH; 4-OH; 4-NH₂(f = pH); 2-Cl; 2-Br; 2-I	49

No.	Compound	Conditions	Method	Ref.			n	Substituents		Ref.
66	Ar—I	0.1 M NEt₄ phosphate pH 8.6 50% EtOH	from Ar—$\overset{+}{\text{I}}$—Ar	27, 37	+0.12	-1.69_5	12	H; 2-CH₃; (2,4,6-triCH₃); (4,4'-diCH₃); (4,4'-di-i-C₃H₇); 4-OCH₃; (4,4'-diOCH₃); 2-NH₂ (from NO₂); 4-NH₂; 4-NH₂ (from 4-NO₂); 4-F; (4,4'-diF)	(4,4'-diC₆H₁₁); 3-NH₂ (from NO₂); 2-COOH; (3,3'-diCOOH); 3-Br; 4-Br; (4,4'-diBr)	16
67	Ar—Cl	0.05 M NEt₄OH 85% EtOH	—	27, 37	+0.20	—	3	2-Cl($\sigma_{o\text{-Cl}}$); 3-Cl; 4-Cl	—	66
68	Ar—$\overset{+}{\text{I}}$—Ar	0.1 M NEt₄ phosphate, pH 8.6, 50% EtOH	—	27, 37	+0.02	-0.20	10	H; 2-CH₃($\sigma_{o\text{-CH}_3}$); (4,4'-diCH₃); (2,4,6,2',4',6'-hexaCH₃); 4-OCH₃; (4,4'-diOCH₃); 4-NH₂; (4,4'-diF); 3-Br; 4-Br	(4,4'-di-i-C₃H₇)ᵃ; (4,4'-diC₆H₁₁)ᵃ; 2-COOH; (3,3'-diCOOH); (4,4'-diCl); (4,4'-diBr); (4,4'-diI)	16
69	Ar—NCS	Clark–Lubs, pH 2.2, 50% MeOH	1st wave	27	+0.29	-1.03	9	H; 4-CH₂COOH($\sigma_x = 0.01$); 4-COOH; 4-COOC₂H₅; 4-Cl; 3-Br; 4-Br; 4-I; 4-N = NC₆H₅	3-COOH; 3-N(CH₃)₂; 4-N(CH₃)₂	24
70		Clark–Lubs, pH 2.2, 50% MeOH	2nd wave	27	+0.29	-1.15	8	H; 3-COOH; 4-COOH; 4-COOC₂H₅; 4-Cl; 3-Br; 4-Br; 4-I	3-N(CH₃)₂; 4-N(CH₃)₂	24
71		Sörensen–Walbum, pH 9.1, 50% MeOH	—	27	+0.33	-1.18	8	H; 4-CH₂COO⁻($\sigma_x = 0.01$); 3-COO⁻; 4-COOC₂H₅; 4-Cl; 3-Br; 4-Br; 4-I	4-COO⁻; 3-N(CH₃)₂; 4-N(CH₃)₂; 4-N=NC₆H₅	24
72	Ar—SO₂C₆H₄NO₂(p)	Buffers pH 0 50% EtOH	extrapol.	27	$+0.03_6$	-0.02	4	H; 3'-CH₃; 4'-OCH₃; 3'-Cl	4'-Brᵇ; 2'-Cl($\sigma_{o\text{-Cl}}$)	67
73	Ar—SO₂C₆H₄NO₂(o)	Buffers pH 0 50% EtOH	extrapol. $\delta_\pi \dot{=} 0$	27, 37	+0.08	-0.09	5	H; 2'-CH₃($\sigma_{o\text{-CH}_3}$); 4'-CH₃; 2'-Cl($\sigma_{o\text{-Cl}}$); 3'-Cl	4'-Brᵇ	67
74	Ar—SC₆H₄NO₂(o)	Buffers pH 0 50% EtOH	extrapol.	27, 37	+0.13	-0.12	4	H; 2'-CH₃($\sigma_{o\text{-CH}_3}$); 3'-CH₃; 4'-CH₃	4'-OCH₃; 4'-NH₂	67
75	Ar—AsO₃H₂	Buffer pH 3.0	—	27, 37	+0.28	-1.27	6	H; 4-CH₃; 4-OH; 4-OCH₃; 4-NH₂; 4-NHCOCH₃	(2,4-diCl)	68

Table III-4 (continued)

No.	Reaction series	Solution	Note	Equation	$\rho_{\pi,R}$, V	E_H(SCE)	n	Substituents — Fitting linear relation	Substituents — Deviating	Ref. to source of $E_{1/2}$
76	Ar—AsO$_3$H$_2$	Buffer pH 1.0	—	27	+0.04	−0.95	3	H;3-NH$_2$;4-NH$_2$	—	69
77		Buffer pH 3.0	—	27	+0.01$_5$	−1.17$_5$	3	H;3-NH$_2$;4-NH$_2$	—	69
				B Reduction by Electrophilic Mechanism						
78	Ar—NO$_2$	conc. H$_2$SO$_4$ (98, 65%)	σ_X^+ σ_X fit equally well	27, 31, 37	−0.18	−0.25^d	8	H;4-CH$_3$;3-COOH;3-COCH$_3$; 3-NH$_3^+$;3-NO$_2$;4-F; 2-SO$_2$NH$_2$($\sigma_{p\text{-}SO_2NH_2}$)	3-Cl;4-Cl	70
79	Ar—I—Ar	0.1 M NEt$_4$ phosphate pH 8.6; 50% EtOH	σ for one substituent 2nd wave from No. 68; $f \neq$ pH	27, 31	−0.03$_5$	−1.14	9	H;4-CH$_3$;4-OCH$_3$;4-NH$_2$;4-F; 4-Cl;3-Br;4-Br;4-I	(2,4,6-triCH$_3$)^e	16
80	Ar—N=N—N—R^{5p}	1 M NH$_3$, 1 M NH$_4$Cl, 0.01% gel.	—	27, 30	−0.29	−0.56	6	H;3-COO$^-$;4-COO$^-$(σ^-); (3-OH;4-COO$^-$); 4-COOCH$_2$N(C$_2$H$_5$)$_2$($\sigma_X = 0.45$); 2-naphthyl($\sigma_X = 0.31$)	4-SO$_3^-$	71
81	Ar—N=N—N—R^{6r}	1 M NH$_3$, 1 M NH$_4$Cl, 0.01% gel.		27, 30	−0.08$_6$	−0.56	5	H;3-COO$^-$;4-COO$^-$(σ^-); (3-OH;4-COO$^-$); 4-COOCH$_2$N(C$_2$H$_5$)$_2$($\sigma_X = 0.45$); 4-SO$_3^-$	2-naphthyl	71
				C Oxidations and Other Anodic Waves						
82	Ar—CH$_2$OH	0.5 M NaClO$_4$ acetonitrile	Pt electrode $\delta_{\pi,R}$ small	27, 37	+0.91	—	7	4-CH$_3$;2-OCH$_3$;4-OCH$_3$;2-Cl; 3-Cl;4-Cl;4-Br	3-OCH$_3$;4-I; 2-naphthyl	72

No.	Ar-X	Electrolyte/solvent	Method	Ref.	Value 1	Value 2	n	Substituents		Ref.
83	Ar—CH(OH)CH$_3$	0.5 M NaClO$_4$ acetonitrile	Pt electrode	27	+1.22	—	3	4-OCH$_3$;4-Cl;4-I	2-naphthyl	72
84	Ar—CH(OH)—Ar	0.5 M NaClO$_4$ acetonitrile	Pt electrode	27	+0.56	—	3	4-OCH$_3$;(4,4'-diOCH$_3$); (4,4'-diCl)	—	72
85	Ar—OH	Buffer pH 7.0 aqueous and EtOH	Pt electrode	27, 30	+0.52	+0.53	9	H;4-CH$_2$CH(NH$_2$)COO$^-$;4-OH; 4-OCH$_3$;4-NH$_2$;4-NHCOCH$_3$; 4-COOCH$_3$(σ^-);4-COC$_2$H$_5$(σ); 4-NO$_2$(σ)	—	73
86	Ar-ferrocene	0.2 M LiClO$_4$ acetonitrile	Pt electrode $E_{1/4}$, chrono-potentiometry	27	+0.13	+0.37	6	H;4-OCH$_3$;4-CH$_3$CO;4-Cl; 4-Br;4-NO$_2$	—	22
87	X-ferrocene[a]	0.2 M LiClO$_4$ acetonitrile	Pt electrode $E_{1/4}$, chrono-potentiometry	27, 30	+0.26 (+0.42)^y	+0.34	14	H;CH$_3$;C$_2$H$_5$;C$_3$H$_7$;i-C$_3$H$_7$; s-C$_4$H$_9$;1,1'-diCH$_3$;1,1'-diC$_2$H$_5$; C$_5$H$_5$;diC$_6$H$_5$;CH$_3$CO(σ^-); CHO(σ^-);(CH$_3$)$_3$Si;di(CH$_3$)$_3$Si	CHO(σ_{p-CHO})	22
88		0.2 M LiClO$_4$ acetonitrile	Pt electrode $E_{1/4}$, chrono-potentiometry	27, 30	+0.32	+0.19	7	H;C$_2$H$_5$;1,1'-diC$_2$H$_5$; COOH(σ^-);COCH$_3$(σ^-); 1,1'-diCOCH$_3$(σ^-); COC$_6$H$_5$(σ^- = 0.8)	—	74
89		0.5 M HClO$_4$ 50% EtOH	Pt electrode	27, 30	+0.30	+0.20	6	H;NH$_2$;NHCOCH$_3$; COCH$_3$(σ^-);CHO(σ^-);CN(σ^-)	—	23
90	Ar—NHNH$_2$	Buffer pH 0	extrapol.	27	+0.02	+0.23$_5$	4	H;4-CH$_3$;4-Cl;4-NO$_2$	4-Br^b	20
91		Buffer pH 13	extrapol.	27	+0.08	−0.37$_5$	4	H;4-CH$_3$;4-Cl;4-NO$_2$	4-Br^b	20

5. Range of Validity of Equation (27); Significance of Deviations for the Discussion of the Electrode Process Mechanism

The vast majority of the reaction series for which a sufficient number of half-wave potentials have been measured for appropriately chosen derivatives show good correlation when equation (27) is used. Only three reaction series show almost no correlation at all: *para*-substituted benzyl bromides,[75] *para*-substituted benzenediazonium hydrogen sulfates,[76] and diaryl disulfides.[77] The first two of these series show analogous behavior. It seems that competitive reaction mechanisms are involved, one for electron-attracting and one for electron-donating substituents. This problem will be considered in more detail in the discussion of the sign and magnitude of the reaction constant (Section 11 of this chapter). For diaryl disulfides the waves of unsubstituted diphenyl disulfide and its methyl derivatives have $\alpha = 0.110$ to 0.090, whereas, those of its methoxy derivatives have $\alpha = 0.030$. Values for other such compounds were not available. Hence, the fundamental conditions for the application of equation (20), and, therefore, also of (27), were not fulfilled. It is not surprising that these values were not correlated by equation (27).

Deviations in the half-wave potentials from the straight line were observed in other reaction series for substances for which substantial changes in the value of the transfer coefficient α occur. For example, for chlorobenzophenone oxime (Reaction Series No. 25) (Fig. III-4), which shows a substantial deviation, a markedly smaller value of α was observed than for other derivatives studied. For benzaldehydes[48] (Reaction Series No. 1) at pH 0 those compounds showing shifts in half-wave potential which correspond to equation (27) have values of $dE_{1/2}/d\mathrm{pH}$ between 0.060 and 0.070 (with the single exception of *p*-hydroxybenzaldehyde, for which the value 0.076 was found). Among those which do not fit equation (27), the halogen derivatives have a slope of 0.058 (*p*-iodobenzene, on the contrary, has the higher value of 0.071), and the groups 3-COOH, 3-NH$_3^+$, and 4-CN have the values 0.084, 0.077, and 0.050, respectively.

A comparison of the reaction series given in Table III-4 indicates that the most frequent deviations are observed for *para*-amino derivatives (discussed in Section 7 of this chapter), halogen derivatives, and charged substituents.

The deviations found for halogen derivatives are often accompanied by changes in the value of the transfer coefficient. Halogen atoms are deformable and in some electrode reactions they are assumed to act as electron bridges. Hence, the possibility cannot be excluded that for halogen derivatives the orientation of the molecule in the transition state differs from the orientation in other members of the reaction series. It should be possible to approach the problem as in the case of amino compounds, i.e., by the computation of special polarographic substituent constants $\sigma_{\pi, X}$ and their verification using sufficient experimental material.

Substituents carrying unit charge, such as O^-, COO^-, and NH_3^+, also sometimes deviate from the linear relation, as in homogeneous reaction kinetics. These deviations can be explained by the influence of the unit charge on the orientation of the molecule at the moment of impact, or by the effect of the strong coulombic field of the ionized grouping on the transition state.

However, such deviations usually are not large enough to prevent us from distinguishing the charge of the electroactive species. That is to say, the difference in the values of the substituent constant σ_X for the charged and the uncharged substituent is of such a magnitude (Table III-5) that it is possible to distinguish whether the half-wave potential can be better correlated using the substituent constant for the charged or for the uncharged substituent.

Thus, in 90% ethanol containing acetate buffer at "pH" 7.0 the shifts in the half-wave potentials of m- and p-iodobenzoic acids fit the linear relation obtained for the reduction of the $C-I$ bond in other substituted iodobenzenes (Reaction Series No. 63) when values of

Table III-5
Substituent Constants for Charged and Uncharged Substituents

Substituent	σ_m	σ_p
Uncharged		
NH_2	−0.16	0.66
OH	+0.12	−0.37
COOH	+0.35	+0.32
Charged		
NH_3^+	+063	—
O^-	−0.7	−0.5
COO^-	−0.1	0.0

$\sigma_{m\text{-COOH}}$ and $\sigma_{p\text{-COOH}}$ are used. On the other hand when values for $\sigma_{m\text{-COO}}^{-}$ and $\sigma_{p\text{-COO}}^{-}$ were used, no correlation was found. It can be concluded, therefore, that free m- and p-iodobenzoic acids are reduced as neutral molecules, even when the pH is so high that in aqueous solution almost only anions would be expected to be present. The presence of ethanol results in increase in the pK values of both iodo-benzoic and acetic acids and is probably responsible for the observed effect. Similarly, it can be demonstrated by the use of $\sigma_{p\text{-NH}_3^+}$ (Table III-7 p. 75) that under the given conditions p-iodoaniline is most probably reduced in the protonated form.

Another example is provided by the azo dyes in Reaction Series No. 58 and 60. At pH 3.5 and 7.0 values of $\sigma_{p\text{-COOH}}$ fit the linear correlation, whereas, at pH 9.05 $\sigma_{p\text{-COO}}^{-}$ and not $\sigma_{p\text{-COOH}}$ ought to be used. It can be inferred that the reactive form changes with change in pH.

It should be kept in mind that recombination and other reactions antecedent to the electrode process proper can influence the course (slope) of the wave and the values of the half-wave potential not only in the pH range in which the wave height changes with pH, but even in a pH range (or in a concentration range of another reactant taking part in the formation of the depolarizer) in which the limiting current is still governed by diffusion. The half-wave potentials of such waves often cannot be compared with the half-wave potentials of substances that do not undergo the chemical reaction antecedent to the electrode process proper. Such effects of antecedent reactions enable us to explain why the half-wave potentials of nitrobenzoic and o-nitrotoluic acids[14] fit equation (27) if measured in acid media, whereas, values measured at higher pH-values deviate.* Only in sufficiently acidic solutions is the free acid (which is the electroactive form in these cases) present in the bulk of the solution. At higher pH the free acid is formed in the neighborhood of the electrode by the recombination reaction. Therefore, it is possible to compare the half-wave potentials of nitro-benzoic acids with those of other derivatives either at pH < 3 or at pH > 11, at which the acid is present as the anion both in the bulk of the solution and when reduced at the electrode.

Finally, deviations from the linear relation expressed in equation (27) are observed when a change occurs in the number of electrons

---

* In this pH range the half-wave potentials differ from those of other derivatives in their pH-dependence.

transferred in the reduction. For example, the half-wave potentials of 3'- and 4'-aminoacetophenones[1] do not fit the linear relation between $E_{1/2}$ and σ_X that applies to other members of Series No. 16, because under the conditions used the amino derivatives are reduced in one two-electron step, whereas the other derivatives are reduced in two one-electron steps. Similarly, equation (27) cannot be expected to be valid for three benzaldehyde derivatives[81] for which, under the given conditions, $n = 0.6$, $n = 2.0$, and $n = 0.8$ have been found for the numbers of electrons transferred. Correspondingly, the half-wave potentials of N,N-dimethyl-p-nitrosoaniline and p-nitrosophenol,[61] which are reduced to the amines via the quinonoid form with the uptake of four electrons, do not follow the linear relation which holds for other substituted nitrosobenzenes (Reaction Series No. 46) (Fig. III-5), which are reduced in the first step to phenylhydroxylamine with the uptake of two electrons.

When there is a deviation from the linear relation (27), we can conclude that for the deviating substance the mechanism of the electrode process is different from that applying to the rest of the reaction series. On the other hand, it can also be deduced that all compounds in the given reaction series that follow the linear relation (27) are reduced by the same mechanism. This conclusion is a probable one, but it is not necessarily true. Accidental effects and the compensation of some effects by others acting in the opposite direction can result in a linear relation, apparently in accordance with equation (27), even when the reduction of a particular substance differs in mechanism from the reduction processes in the rest of the reaction series. Thus, it seems improbable that p-aminophenol is oxidized by the same mechanism as unsubstituted phenol or p-OH-C_6H_4-$COOCH_3$. In spite of this, the half-wave potentials of all three compounds[73] follow equation (27) (Reaction Series No. 85). Similarly, the half-wave potentials of p-nitrophenols fit equation (27) as applied to substituted nitrobenzenes at pH 1.0 and 4.0 (Reaction Series No. 44 and 45), even though p-nitrophenol takes up six electrons in the electroreduction, whereas, the other nitrobenzenes take up only four.

It can be concluded, therefore, that even though equation (27) helps us to detect deviations in reaction mechanism in a reaction series, the results of the application of this equation should not be the only criterion. The linear free energy treatment should be supplemented by other types of measurement.

6. Additivity of Substituent Effects

For substances containing several substituents in the *meta* and *para* positions of the same benzene ring the additivity of the substituent effects is demonstrated in several examples in Table III-4.

The data available for symmetrical compounds containing two benzene rings, such as stilbene, benzophenone, or azobenzene, with substituents in both rings are usually limited to mono- and di-substituted compounds. So far, it has been impossible to decide whether equation (39) or (38) should be applied, even when the second equation describes the available data with sufficient accuracy.

An interesting example of the application of the additivity of substituent effects was found for the half-wave potentials of diaryliodonium salts.[16] The half-wave potentials of the first waves, correspond-ing to the reduction of $Ar\overset{(+)}{-}I-Ar$ cations, are influenced by substituents in both benzene rings, and their effects are additive and in accordance with equation (38) (Reaction Series No. 68). The second, more negative waves of these compounds were assumed[16] to correspond to the reduc-tion of the Ar-I-Ar radical. However, the half-wave potentials of these waves are practically identical for diaryliodonium derivatives substi-tuted by a substituent X in one ring and for derivatives substituted symmetrically by the same substituent X in both rings. The half-wave potentials of monosubstituted compounds fit equations (27) and (31). The half-wave potentials of disubstituted compounds fit the same straight line when they are correlated with σ_X [instead of $2\sigma_X$, as required by equation (38)]. This can be interpreted on the view that it is not the radical Ar-I-Ar, for which substitution in both rings would be of importance, that is involved in the electrode process, but a radical containing only one phenyl group. The small absolute value of the reaction constant diminishes the reliability of this explanation.

7. Application of the Constants σ_X^- and the Determination of New Values of the Substituent Constants σ_X

When a reducible group is in mesomeric interaction with the benzene nucleus, which contains a *para* substituent that is also in substantial mesomeric interaction with the nucleus, mesomeric interac-tion of the substituent with the reducible group via the benzene ring is possible. In these cases substituent constants σ_{p-X}^- derived from the reactions of *para*-substituted phenols and anilines[5,18] are used for electron-accepting substituents, and equation (30) is applied.

Having determined the value of $\rho_{\pi,R}$ from equation (27) using the half-wave potentials of *meta*-substituted derivatives (and of those

Table III-6

Reaction Series in which Mesomeric Interaction Occurs between the Electroactive Group and the Substituent and Equation (30) [or (27)] is Applied

Reaction series	No.[a]	σ_{p-X}^- fits	σ_{p-X} fits
ArCHO	1	CHO	COOH, CN
	2	CHO, CN, COO$^-$	—
	3	—	COOH
ArCOCH$_3$	12	CN	—
ArCOOCH$_3$	35	COOCH$_3$	—
ArNO$_2$	36	NO$_2$, COOH	
	39	COOH, CHO, COCH$_3$	(CN)[b]
	40	COOH, CHO, COCH$_3$	CN
	41	CHO, COCH$_3$	CN, COO$^-$
	42	COOH, COOC$_2$H$_5$, NO$_2$	CHO
	43	(NO$_2$)[b]	COOH, CHO, (NO)[b]
	44	COOH, COOCH$_3$	—
	45	COOH, COOCH$_3$	—
ArNO	46	CHO	NO$_2$[c]
ArN=NAr	47–51	—	COOC$_2$H$_5$
	52–55	—	(NO$_2$)[b]
	59	COOH, COOC$_2$H$_5$	—
	61	COOC$_2$H$_5$	—
ArI	63	—	COOH
	64	—	COO$^-$
ArNCS	69	—	COOH, COOC$_2$H$_5$
	70	—	COOH, COOC$_2$H$_5$
Het[d]—N=N—Ar	80	COO$^-$	—
	81	COO$^-$	—
Ar-ferrocene	86	—	CH$_3$CO, NO$_2$
X-ferrocene[e]	87	CH$_3$CO, CHO	—
	88	COOH, CH$_3$CO, (CH$_3$CO)$_2$, COC$_6$H$_5$	—
	89	COCH$_3$, CHO, CN	—
ArNHNH$_2$	90	NO$_2$	—
	91	NO$_2$	—

[a] See, Table III-4.
[b] Doubtful.
[c] Change in mechanism of the electrode process.
[d] Heterocyclic ring.
[e] Substituent X directly attached to the ferrocene ring.

para-substituted derivatives in which the substituents have no appreciable mesomeric interaction with the electroactive group), we must decide whether $\sigma_{p\text{-}X}$ or $\sigma_{p\text{-}X}^-$ is better in accord with the $E_{1/2}$ versus σ plot. Values of σ_X^- were used for correlations of half-wave potentials by Tirouflet[14] for nitrobenzenes and by Grabowski[9,10] for benzaldehydes without stating any theoretical reasons. Table III-6 lists some reaction series with an indication of whether $\sigma_{p\text{-}X}^-$ or $\sigma_{p\text{-}X}$ gives the better correlation in particular cases. Table III-6 shows that the substituent constants σ_X^- and equation (30) may be used for the reduction of ArCHO (at higher pH, at which protonation does not play a role), ArCOCH$_3$, ArCOOCH$_3$, and possibly also for ArN = NAr, ArNO, and ArNHNH$_2$, in which the conditions for mesomeric interaction between the substituent and the reducible group via the ring are fulfilled (in the ground or in the transition state).* The observed deviations, mainly for CN, should be studied further.†

On the other hand, $\sigma_{p\text{-}X}^-$ values should not be used for ArI and ArNCS, in which mesomeric interaction cannot play any substantial role.

An interesting example is provided by ferrocene derivatives. When the substitution occurs directly in the ferrocene ring (Reaction Series No. 87, 88, and 89), mesomeric interaction between the substituent and the reactive center in the ferrocene system occurs, and $\sigma_{p\text{-}X}^-$ should be used. On the other hand, when the substituent is separated by a benzene ring, no mesomeric interaction occurs between the substituent and the ferrocene nucleus via the benzene ring, which is shown by the fact that σ_X, but not σ_X^-, values fit the linear relation (Fig. III-9). This can be interpreted as resulting from the relative positions of the benzene and ferrocene rings, which exclude mutual mesomeric interaction.

Since deviations have been observed in many reaction series in the case of *p*-amino groups when the constants $\sigma_{p\text{-}NH_2}$ and $\sigma_{p\text{-}NH_3^+}$ have been used, and since these shifts have all been in the same direction (see, e.g., Figs. III-3 and -10), approximate values of the special polarographic constant $\sigma_{p\text{-}X}^\pi$ were computed (Table III-7). It will be seen that the mean

---

* In the definition of the substituent constant $\sigma_{p\text{-}X}^-$, it is supposed that the shift of electrons in the molecule is directed from the side chain to the substituent. This condition is not fulfilled in some examples listed in Table III-6 as fitting equation (30). The explanation of the applicability of $\sigma_{p\text{-}X}^-$ in these examples is not known.

† A change in the reduction mechanism of some *p*-cyano derivatives (resulting in the reduction of the cyano group to CH$_2$NH$_2$) has been detected recently. This explains at least some of the observed deviations.

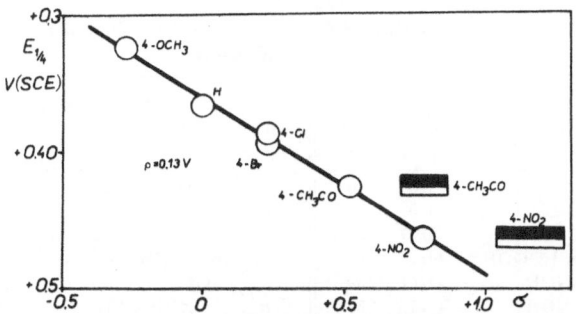

Fig. III-9. Relation of quarter-wave potentials ($E_{1/4}$) for the oxidation of aryl ferrocenes with the substituent in the phenyl ring to Hammett substituent constants σ_X. Reaction Series No. 86, Table III-4. Halved points: σ_X^- values used.

deviations of the values of $\sigma_{p\text{-NH}_3^+}^\pi$ obtained are of the same order as those of other substituent constants. Hence, at least for polarographic reductions, $\sigma_{p\text{-NH}_2}^\pi$ and $\sigma_{p\text{-NH}_3^+}^\pi$ are better suited than the tabulated [5,40] data for $\sigma_{p\text{-NH}_2}$ and $\sigma_{p\text{-NH}_3^+}$.

In addition to these constants, a few values of other substituent constants σ_X and σ_X^- were obtained from polarographic measurements. Since most of the values obtained from Reaction Series No. 1–91 and summarized in Table III-8 were obtained from only a single reaction series, they should be regarded only as approximate and informative.

8. Ortho Derivatives—Application of Equation (37)

The number of reaction series for which the value of the reaction constant $\rho_{\pi,R}$ has been determined and for which, furthermore, a

Table III-7

Approximate Values of the Substituent Constants $\sigma_{p\text{-X}}^\pi$ for Amino Groups Determined from Polarographic Measurements

Reaction series	No.[a]	$\sigma_{p-\text{NH}_2}^\pi$	Reaction series	No.[a]	$\sigma_{p-\text{NH}_3^+}^\pi$
ArC(=NOH)CH$_3$	21	−0.10	ArC(=NOH)Ar	25	−0.30
ArCOAr	22	−0.16	ArCSAr	29	−0.27
ArC(=NOH)Ar	26	−0.10	ArN=NAr	47	−0.23
ArN=NAr	48	−0.16	Ar—I	63	−0.32

[a] See Table III-4.

Table III-8
Approximate Values of the Substituent Constants σ_X and σ_X^- Determined from Polarographic Measurements

Constant	Group	Value	Reaction series No.[a]	Constant	Group	Value	Reaction series No.[a]
σ_m	CH_2OH	0.0	40	σ_m	$C(=NOH)CH_3$	+0.49	39
σ_p	CH_2OH	0.0	1	$\sigma_p^{(-)}$	$C(=NOH)CH_3$	+0.61	39
σ_p	CH_2COOH	−0.07	40	σ_m	$N=C(CH_3)NHC_6H_5$	+0.29	38
σ_p	$CH=CHCOOH$	+0.9	40	σ_p	$N=C(CH_3)NHC_6H_5$	+0.08	38
σ_m	$CH=NOH$	+0.61	39	σ_m	SO_3^-	+0.41	59
σ_p	$CH=NOH$	+0.73	39	σ_p	$SO_2NHCOCH_3$	+1.00	59

[a] See Table III-4.

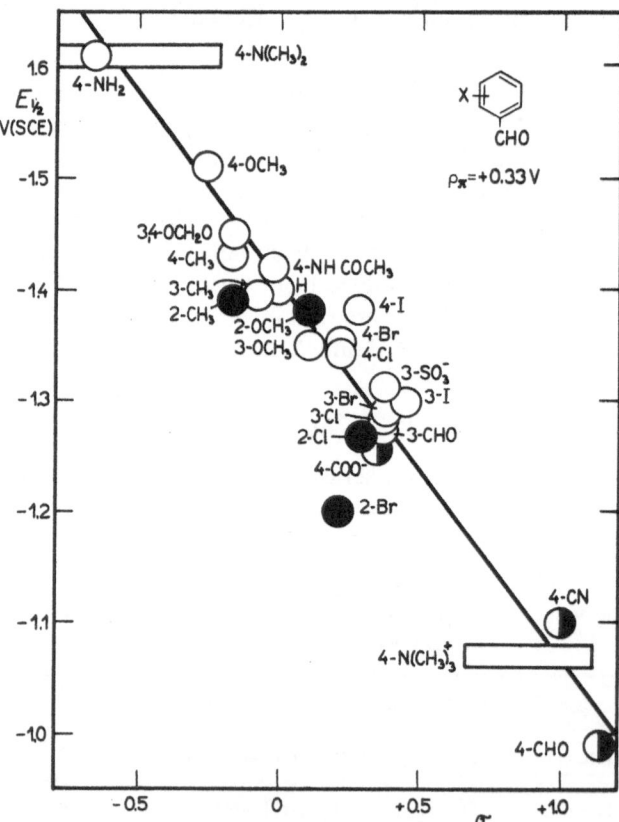

Fig. III-10. Relation of half-wave potentials for the reduction of substituted benzaldehydes at pH 13 to Hammett substituent constants σ_X. Reaction Series No. 2, Table III-4. Full points, *ortho* derivatives; halved points, σ_X^- values used.

sufficient number of half-wave potentials of *ortho* derivatives (at least four to make it possible to calculate the steric reaction constants $\delta_{\pi,R}$) have been recorded is rather limited.

The value of $\delta_{\pi,R}$ has been determined from equation (37) for the following reaction series: The half-wave potentials of substituted benzaldehydes (Reaction Series No. 2) at pH 13; here $\delta_{\pi,CHO} = +0.17$ V fits for 2-CH_3, 2-Cl, and 2-OCH_3, but deviation is observed for 2-Br (Fig. III-10). The half-wave potentials of substituted nitrobenzenes in acetate buffers of pH 2.2_5 containing 60% ethanol (Reaction Series No. 42); here for $\delta_{\pi,NO_2} \approx 0.0$ V the shifts in half-wave potential for 2-CH_3, 2-Cl, 2-Br, and 2-NO_2 follow equation (37), but the value for 2-I deviates. The half-wave potentials of substituted iodobenzenes in an acetate buffer of pH 7.0 containing 90% of ethanol (Reaction Series No. 63) (Fig. III-11); here for $\delta_{\pi,I} = 0.2$ V a linear correlation was found for 2-CH_3, 2-OCH_3, 2-OC_2H_5, 2,4-di-CH_3, 2,5-di-CH_3, 2,6-di-CH_3, 2,4,6-tri-CH_3, and 3,6-di-OCH_3, but deviations for 2-Cl and 2-Br were observed.

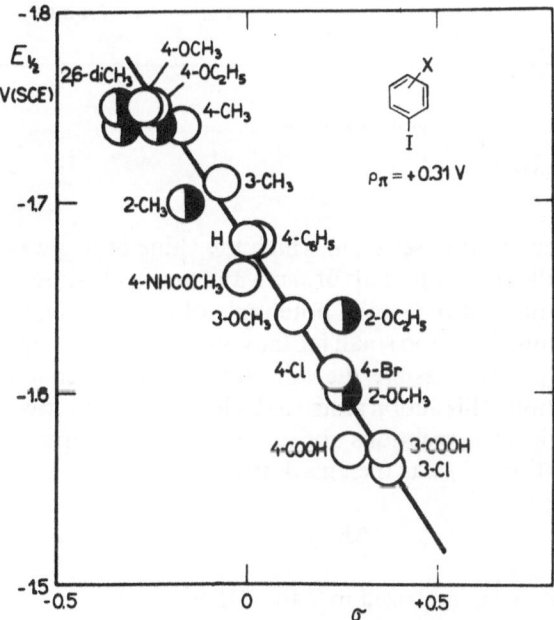

Fig. III-11. Relation of half-wave potentials for the reduction of substituted iodobenzenes to Hammett substituent constants σ_X. Reaction Series No. 63, Table III-4. Halved points: *ortho* derivatives.

Table III-9

Application of Values of the Substituent Constant $\sigma_{o\text{-}X}$ in Equation (40) [$\delta_{\pi,R}$ in Equation (37) was set equal to zero]

Reaction series	No.[a]	Linear correlation fitted by	Derivations observed for
ArCOCH$_3$	18	2-Cl, 2-Br	2-OCH$_3$[b]
ArNO$_2$	36	2-Cl, 2-Br	2-CH$_3$[b]
	37	2-Cl, 2-Br	—
	40	—	2-CH$_3$[b]
	42	2-CH$_3$, 2-NO$_2$, 2-Cl, 2-Br	2-I
	43	2-CH$_3$, 2-NO$_2$, 2-Cl, 2-Br, 2-I	—
ArN=NAr	59		2-CH$_3$,[b] 2-OCH$_3$,[b] 2-OC$_2$H$_5$,[b] 2-Cl[b] (2,3-di-CH$_3$)[b] (2,4-di-CH$_3$)[b]
ArCH=CHAr	62	2-CH$_3$, (2,5-di-CH$_3$), (2,4,6-tri-CH$_3$)	—
ArI	63	2-CH$_3$, (2,4,6-tri-CH$_3$)	—
	64	2-CH$_3$	2-OCH$_3$,[c] 2-Cl[c]
ArCl	67	2-Cl	
ArAsO$_3$H$_2$	75		2,4-diCl[c]
ArCH$_2$OH[d]	82	2-Cl, 2-OCH$_3$	—

[a] See Table III-4.
[b] Steric hindrance to coplanarity is possible.
[c] $\delta_{\pi,R}$ is not very small.
[d] Oxidation waves.

In other reaction series, in which the value of $\rho_{\pi,R}$ was determined using the half-wave potentials of *meta*- and *para*-substituted compounds but the number of half-wave potentials of *ortho* derivatives that had been determined was too small for the computation of the steric reaction constant $\delta_{\pi,R}$ to be possible, the value for $\delta_{\pi,R}$ was taken to be zero as an approximation. This action seems to be justified by the low values found for $\delta_{\pi,R}$ in the four reactions above. Only the first term on right-hand side of equation (37) was then used, that is

$$\Delta E_{1/2} = \rho_{\pi,R}\sigma_{o\text{-}X} \tag{40}$$

The results are summarized in Table III-9.

As expected, equation (37) cannot be applied to those *ortho* substituents which give substantial mesomeric interaction, and most *o*-hydroxy and *o*-amino derivatives are excluded. Moreover, for such substances a change in the course of the reduction is sometimes observed (e.g., for $ArNO_2$).

The number of reaction series tested is too small for us to decide whether the application of the substituent constants $\sigma_{o\text{-}x}$ and $(E_S)_{o\text{-}x}$, which express polar and steric effects in ester hydrolysis, is justified for the treatment of structural effects in electroreduction. It is clear that the introduction of another parameter δ would result in a better correlation than the original Hammett-type equation; however, it is questionable whether the correction δE_S should be introduced at all; it may be preferable to use special $\sigma_{o\text{-}x}$ constants instead of $\sigma_{p\text{-}x}$ (used for *ortho* substituents as a first approximation) so as to follow the deviations. This idea is extended in the definition of the "polarographic *ortho*-shift" (p. 84).

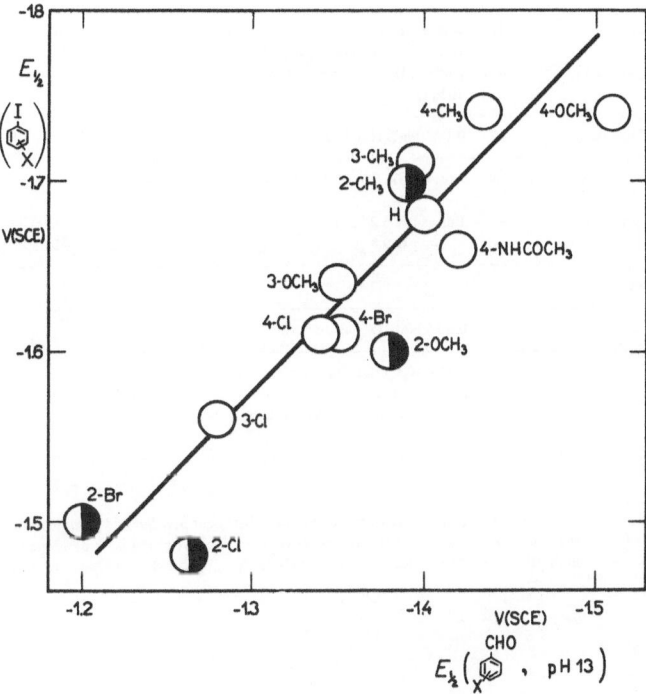

Fig. III-12. Relation of half-wave potentials for the reduction of substituted iodobenzenes (Reaction Series No. 62, Table III-4) to half-wave potentials for the reduction of substituted benzaldehydes (Reaction Series No. 2). Halved points: *ortho* derivatives.

Table III-10

Reaction series	No.	Solution	CH_3	OH	OCH_3
Ar—CHO	1	Buffer pH 0; first wave; 1F	+0.06	+0.04	+0.09
	2	Buffer pH 13; $f \neq$ pH	+0.04	+0.17[a]	+0.13
	3	Buffer, medium pH; second wave; $f \neq$ pH	+0.14	+0.30	+0.10
	4	Buffer pH 0, extrapolated; 50 % dioxane		+0.07	+0.11
	7	Buffer pH 2.2; 25 % ethanol		+0.08	+0.10
	8	Buffer pH 12; 25 % ethanol		+0.05	+0.10
	9	Buffer, medium pH; second wave; $f \neq$ pH		+0.25	+0.10
	10	Buffer pH 0, extrapolated; 30 % i-propanol	+0.06	+0.05	
	—	HCl, Na acetate; 60 % ethanol; pH 1.73	$+0.05_8$	$+0.06_6$	
		pH 2.25	+0.06	$+0.07_6$	
		pH 2.59	$+0.06_2$	$+0.06_8$	
		pH 3.16	$+0.04_4$	$+0.06_6$	
Ar—COCH$_3$	18	Buffer pH 0, extrapolated; 50 % dioxane		+0.03	+0.12
	—	Buffer pH 6.6; first wave; 10 % ethanol			
	—	Buffer pH 6.6; second wave; 10 % ethanol			
Ar—CH=CH—Ar	62	0.1 M N(C$_2$H$_5$)$_4$I; 75 % dioxane	+0.06		
Ar—SS—Ar	—	Acetate pH 7.0; 95 % ethanol	+0.04		+0.02
Ar-tetrazolium	—	Buffer pH 3.2		0.00	
	—	pH 9.8		+0.08	
Ar-pyridinium	—				
Ar—CH$_2$OH[e]	82	0.5 M NaClO$_4$; acetonitrile			+0.03
Ar—NO$_2$	36	1 M CH$_3$COONH$_4$; glacial acetic acid	−0.05	+0.07	
	37	0.1 M N(C$_2$H$_5$)$_4$Br; 66 % ethanol			
	—	Buffer pH 0			
	38	Alkaline, $f \neq$ pH; absol. ethanol		+0.17	

Ar—stands for the benzene ring bearing the substituent in *ortho* and *para* position; $f \neq$ pH— the half-wave potential is pH-independent; [a] the half-wave potential is pH-dependent even for high pH-values; [b] the value for COO$^-$; [c] C-phenyl substituted; [d] possibility of steric hindrance

Difference of the Half-Wave Potentials of *ortho*- and *para*-Substituted Benzene Compounds (Δo)

NH$_2$	NHCOCH$_3$	COOH	CHO	OCOCH$_3$	Cl	Br	I	SCN	Reference to source of $E_{1/2}$
0.00	+0.02	−0.01	−0.03	+0.06	+0.06	+0.06	+0.06		48
+0.04	+0.10	−0.08[b]	−0.09		+0.08	+0.15	+0.13		48
	+0.16			+0.15	+0.15	+0.15	+0.11		48
					+0.08	+0.08			47
									50
									50
									50
									11
					+0.08$_4$	+0.04	+0.05$_1$		49
					+0.08$_1$	+0.04	+0.06$_2$		49
			—		+0.08$_8$	+0.06	+0.05$_8$		49
+0.06$_8$					+0.08$_0$	+0.05	+0.06$_6$		49
					−0.04	−0.05			47
								−0.26	82
								−0.03	82
									64
									77
		−0.15[c,d]							83
		−0.20[c,d]							83
+0.09									84
					+0.05				72
+0.00$_5$[f]					−0.03	−0.04$_5$			56
					+0.06$_4$		−0.04		57
+0.00$_5$[f]									85
+0.04	+0.13		+0.09[g]						58, 86, 87

of coplanarity; [e]anodic oxidation wave; [f]value for NH$_3^+$; [g]value for N=C(CH$_3$)NHC$_6$H$_5$; [h]value for OC$_2$H$_5$; [j]value for CH=NNHCONH$_2$; [k]value for CH=NNHC(=NH)NH$_2$; [m]value for C$_6$H$_5$; [n]value for O$^-$.

Table III-10 (*continued*)

Reaction series	No.	Solution	CH$_3$	OH	OCH$_3$
Ar—NO$_2$	40	Prideau–Ward buffer pH 2.0			
	—	Britton–Robinson buffer pH 7.0; 80 % ethanol			+0.03[h]
	—	Buffer pH 3.1			
		pH 4.8			
		pH 8.7			
	—	Buffer pH 4.0			
		pH 6.0			
		pH 8.0			
	—	McIlvaine buffer; 50 % ethanol;			
		pH 1.3	−0.05		
		pH 4.9	−0.06		
		pH 8.4	−0.04		
	—	Buffer pH 10.1			
		pH 12.0			
	43	HCl—CH$_3$COONa; 60 % ethanol;			
		pH 1.7	−0.02	+0.16[p]	
		pH 2.2$_5$	−0.02	+0.19[p]	
		pH 3.2	−0.08	+0.20[p]	
	44	HCl—NaCl; pH 1.0; 10 % ethanol		+0.07$_5$	+0.05
	45	McIlvaine; pH 4.0; 10 % ethanol		+0.11$_5$	0.00
Ar—NO$_2$,3—COOH	40	Prideau–Ward buffer pH 2.0	−0.12		
Ar—I	63	Acetate pH 7.0; 90 % ethanol	+0.04	+0.11[h]	+0.15
	64	0.01 M N(C$_2$H$_5$)$_4$Br; $f \neq$ pH; 66 % ethanol	+0.00$_4$	+0.16	−0.16[n]
	—	0.01 M N(C$_2$H$_5$)$_4$Br; 66 % ethanol	+0.00$_2$	+0.19$_5$	
Ar—SCN	—	Buffer, $f \neq$ pH			
Ar—AsO$_3$H$_2$	76	Buffer pH 1.0			
	77	pH 3.0			
Ar—AsO(OH)Ar	—	Buffer pH 3.0			
Ar—N=N—Ar	47	Buffer pH 5.5			
	48	pH 7.0			
Ar—N=N—Ar	59	Clark–Lubs; pH 7.0	+0.04		+0.04
C$_6$H$_5$—N=N—Ar	—	0.1 M CH$_3$COOH; 0.1 M CH$_3$COONa; 75 % ethanol		−0.10[d]	
p-OCH$_3$C$_6$H$_4$—N=N—Ar	—	0.1 M CH$_3$COOH; 0.1 M CH$_3$COONa; 75 % ethanol		−0.11[d]	
p-SO$_3$HC$_6$H$_4$—N=N—Ar	—	0.1 M CH$_3$COOH; 0.1 M CH$_3$COONa; 75 % ethanol		−0.10[d]	

NH$_2$	NHCOCH$_3$	COOH	CHO	OCOCH$_3$	Cl	Br	I	SCN	Reference to source of $E_{1/2}$
									Δo
		-0.10^d	-0.03						14
									88
								-0.02	89
								$+0.05$	89
								$+0.02$	89
			-0.04^j	$-0.11^{d,k}$					90
			-0.04^j	$+0.06^k$					90
			$+0.08^j$	$+0.03^k$					90
									91
									91
									91
			-0.01						92
			$+0.02$						92
$+0.07^p$		-0.12	-0.05^p	-0.12	$+0.04$	-0.03	-0.01	-0.05^m	49
$+0.06^p$		-0.11	-0.07^p	-0.11	$+0.03$	-0.02	-0.01	$-0.07_5{}^m$	49
$+0.07^p$		-0.12	-0.02^p	-0.12	-0.01	-0.05	-0.01	-0.07^m	49
		-0.06		-0.05_5					60
		-0.08		-0.05_5					60
									14
	$+0.07^m$				$+0.13$	$+0.11$	$+0.10$		15
-0.14_5		-0.06^b			$+0.20$				57, 65
$+0.16$	$+0.10_5{}^m$				$+0.20$	$+0.23$	$+0.22$		49
$+0.13$									89
$+0.02^f$									69
$+0.01^f$									69
$-0.17^{d,f}$									69
		$+0.04$							62
		$+0.04$							62
				$+0.03^h$		$+0.04$			46
									63
									63
									63

An alternative treatment of the shifts in half-wave potential can be carried out with the aid of specific polarographic substituent constants introduced for the polar and steric effects of *ortho* substituents in nucleophilic polarographic reductions. Since in this type of reaction one partner (the electrode or the electron) remains constant, it is reasonable to expect that both mesomeric and steric conditions will remain sensibly constant for a particular reaction series; it was expected that substituent constants so defined would show less deviation in the shifts in half-wave potential.

The possibilities of this kind of treatment are shown in Fig. III-12, where the half-wave potentials of substituted benzaldehydes[48] (Reaction Series No. 2) are compared with the half-wave potentials of the correspondingly substituted iodobenzenes (Reaction Series No. 63) on analogy with the treatment used in the derivation of equation (49) (Section 13 of this Chapter). This comparison is equivalent to choosing one of the series as a standard for the computation of the values $\sigma_{\pi,o\text{-}X}$ and $(E_S)_{\pi,o\text{-}X}$. No substantial deviations were observed even for those halogen derivatives which did not fit equation (37). This observation suggests a parallelism between the two series in the influence of steric and other *ortho* effects.

According to the precepts of theoretical organic chemistry the polar effect of a particular substituent operates with approximately equal strength from the *ortho* and the *para* positions. In agreement with this view the values of $\sigma_{o\text{-}X}$ generally are not very different from those of $\sigma_{p\text{-}X}$ (see Table III-1). If we assume this to be true, another approximate value characterizing the magnitude of the *ortho* effect can be obtained by comparing the half-wave potentials of *ortho*- and *para*-substituted compounds. This treatment is possible even in cases where a mesomeric interaction between the polarographically active group R and the substituent X takes place. Moreover, it enables us to characterize the *ortho* effect even in reaction series in which the lack of a sufficient number of proper values of half-wave potentials prevents the computation of the steric reaction constant $\delta_{\pi,R}$, or even of the polar reaction constant $\rho_{\pi,R}$. Approximate values for the nonpolar *ortho* effect, described as the "polarographic ortho-shift" Δ_o, are defined by the difference

$$(E_{1/2})_{o\text{-}X} - (E_{1/2})_{p\text{-}X} = \Delta_o \tag{41}$$

This means that

$$\Delta_o = \rho_{\pi,R}(\sigma_{o\text{-}X} - \sigma_{p\text{-}X}) + \delta_{\pi,R}(E_S)_{o\text{-}X} + [M_{\pi,o\text{-}X} - M_{\pi,p\text{-}X}] \tag{42}$$

If the group under study shows no substantial mesomeric effect, the bracketed expression in equation (42) can be neglected. If $\sigma_{o\text{-}X} \approx \sigma_{p\text{-}X}$, which is fulfilled for CH_3, OC_2H_5, CN, NO_2, Cl, Br, and I (but apparently not for OCH_3 and F, see Table III-1), $\Delta_o \approx \delta_{\pi,R}(E_S)_{o\text{-}X}$. Thus, the difference in the half-wave potentials of *ortho*- and *para*-substituted compounds, values of which are given in Table III-10, is a measure of the nonpolar contribution to the *ortho* effect.

It is evident from a comparison of Tables III-9 and III-10 that equation (40) is obeyed when Δ_o is small. This implies that in these cases $\delta_{\pi,R}$, and also $\delta_{\pi,R}(E_S)_{o\text{-}X}$, is small.

The material gathered in Table III-10 is too limited to allow a detailed discussion. It is only possible to state that, whereas in some reaction series (No. 43, 63, and 64) the value of the polarographic *ortho* shift Δ_o is significantly dependent on the nature of the substituent, in others (e.g., in No. 3, corresponding to the reduction of a free radical, or in No. 59) Δ_o is almost independent of the nature of the substituent. In certain types of reduction process Δ_o seems to be only slightly dependent on the buffer solution used, while in others (mainly those compounds in which the degree of ionization of the substituent or the polarographically active group can change) Δ_o depends on pH.

Finally, it is noteworthy that in several cases the values of Δ_o in Table III-10 are positive, i.e., the *ortho* derivatives are reduced at more positive potentials than their *para* isomers. Such *ortho* shifts we call positive *ortho* shifts.

The largest positive *ortho* shifts are observed in halogen compounds. It could be inferred that a direct field effect, i.e., the influence of the external electric field of the halogen, plays a part here. However, two opposing effects seem to contribute to this *ortho* effect—the intensity of the electric field and the atomic volume of the halogen (i.e., a direct polar effect and a steric effect)—for the *ortho* shift Δ_o shows no typical trend in the sequence Cl, Br, I.

The largest negative *ortho* shifts are observed when the functional group can undergo marked mesomeric interaction with the aromatic ring and when both the substituent and the polarographically active group are bulky. Such behavior is explained by steric hindrance to coplanarity, which is discussed in the next section.

9. Hindrance to Coplanarity Due to *ortho* Substituents

The π- or p-electron interaction, which leads to conjugation effects, is only fully possible when the interacting bonds are coplanar. A bulky substituent attached in the vicinity of the conjugated bonds may prevent them from achieving coplanarity and so limit their interaction by its steric effect. This effect, which more or less prevents the mutual interaction of conjugated systems, is called steric hindrance to coplanarity.

Steric hindrance to coplanarity in substituted nitrobenzenes and azo compounds, where its influence has been clearly and unambiguously demonstrated, will be discussed first, and some other examples will follow.* Finally, it will be shown that in derivatives of acetophenone, benzophenone, and benzaldehyde the effect of steric hindrance to coplanarity or polarographic behavior cannot be taken as proved.

A typical symptom (characteristic) of steric hindrance to coplanarity is the nonadditivity of the effects of additional *ortho* substituents, which is frequently accompanied by a change in the value of the transfer coefficient α.

Among nitro compounds, in which steric hindrance to coplanarity was originally detected,[93] nitrotoluenes will be discussed first. The polar effect of the methyl group is small, and the identical values given (Table III-1) for $\sigma_{o\text{-CH}_3}$ and $\sigma_{p\text{-CH}_3}$ allow us to conclude that $(E_S)_{o\text{-CH}_3} = 0$. Moreover, the methyl group can be assumed to have no substantial mesomeric interaction with the benzene ring.

Since the polarographic effect does not change appreciably in the pH range 1.3 to 8.4, we shall restrict ourselves to values measured[91] at pH 4.9:

$X—C_6H_4—NO_2$	$E_{1/2}$ (V, S.C.E.)
4-CH$_3$	-0.58
2-CH$_3$	-0.64
2,4,6-tri-CH$_3$	-0.76

The *ortho* shift is negative ($\Delta_o = -0.06$ V), and since the polar effects of *o*- and *p*-methyl groups are almost equal, and on the basis of other considerations discussed earlier, this whole difference can be ascribed to steric hindrance to coplanarity.

In order to compute the contribution of the second *o*-methyl group in nitromesitylene the effect of the 4-methyl group must be first estimated. The value of the reaction constant for this reaction series is

---

* This effect also has been demonstrated for 3-substituted phenylsydnones (Chapter IV) and considered for azulene ketones (Chapter VI) and quinonoid compounds (Chapter VIII).

$\rho_{\pi,NO_2} = 0.16$ V and, hence, $(\Delta E_{1/2})_{4\text{-}CH_3} = \rho_{\pi,NO_2}\sigma_{p\text{-}CH_3} = 0.16 \times$ $-0.17 \approx -0.03$ V. From the total difference between the 2-CH$_3$ and 2,4,6-tri-CH$_3$ derivative $(-0.12$ V$)$ the effect of the *p*-methyl group can be substracted, and the difference -0.09 V corresponds to the effect of the second methyl group, of which about -0.03 V is due to its polar effect. Thus, about -0.06 V again remains for the increase in steric hindrance to coplanarity due to the presence of the second *o*-methyl group. It seems that the presence of the first *o*-methyl group $(\Delta_{o_1} = 0.06$ V$)$ does not remove all the mesomeric interaction between the benzene ring and the nitro group, since the presence of another *o*-methyl group $(\Delta_{o_2} = -0.06$ V$)$ produces approximately the same effect. However, the available polarographic data do not allow us to conclude that the presence of two methyl groups completely prevents all interaction at the moment that reduction occurs.

The shift in the potential is accompanied by a change in the value of the transfer coefficient (see Fig. 4 in Ref. 91).

Similarly, steric hindrance to coplanarity influences the half-wave potentials of *p*-dinitrobenzenes.[94] From the value of the reaction constant $\rho_{\pi,NO_2} = 0.35$ V (at pH above 10, at which the half-wave potentials are pH-independent) the value for the shift due to the polar effect of an *o*- or *p*-methyl group $(\Delta E_{1/2})_{o\text{-}CH_3} = -0.06$ V was computed, whilst monomethyl-*p*-dinitrobenzene actually gave a shift of a comparable magnitude $(-0.07$ V$)$ when related to the potential of unsubstituted *p*-dinitrobenzene. For tetramethyl-*p*-dinitrobenzene, however, instead of the expected shift of -0.24 V a substantially greater shift $(-0.35$ V$)$ was observed. The increase in the shift by -0.11 V can be attributed to steric hindrance to the mesomeric interaction of the two nitro-groups with the benzene ring and their mutual interaction through the ring. The contribution for one nitro group $(\Delta_o = -0.05_5$ V$)$ is comparable with that found for mononitrobenzenes.

		R^1	R^2	$E_{1/2}$(pH 2)	ΔH	Δ_o
	I	H	H	-0.22 V		
	II	CH$_3$	H	-0.36 V	-0.14	-0.12
	III	H	CH$_3$	-0.24 V	-0.02	

Another example of negative *ortho* shifts due to methyl groups can be found in the half-wave potentials of nitrotoluic acids.[14] Compounds II and III both have the nitro group in the *meta* position relative to the carboxy group. The substitution of a methyl group in the *para* position with respect to the nitro group in III causes a shift of approximately the value expected for its polar effect

$$[(\Delta E_{1/2})_{p\text{-}CH_3} = \rho_{\pi,NO_2}\sigma_{p\text{-}CH_3} = 0.16 \times -0.17 = -0.03 \text{ V}].$$

The effects of the methyl and carboxy groups are therefore additive, and the resulting shift in half-wave potential shows no substantial mutual interaction.

Thus, the additional shift of -0.12 V observed with the substance II must be caused by a nonpolar *ortho* effect, which can be identified with the steric hindrance to the coplanarity of the nitro group with the benzene ring. The substantially greater negative *ortho* shift observed here, as compared with simple *o*-nitrotoluenes ($\Delta_o \approx -0.06$ V), cannot be explained by hindrance to the mesomeric interaction between the nitro and the carboxy groups, which are in *meta* orientation. It can be assumed that, due to the demonstrated interaction between the nitro and methyl groups, the free rotation of the methyl group is hindered. Such a "frozen" methyl group also may influence the carboxy group, e.g., by hindering its coplanarity with the benzene ring (buttressing effect). This would cause a decrease in the polar resonance effect of the carboxy group on the nitro group. Since *m*-COOH shifts the half-wave potential of the nitro group by $+0.05$ V (at certain pH values[14]), the direction as well as the magnitude of the shift observed for II is in agreement with the above explanation. Other examples of the modification of the effect of the carboxy group by an adjacent methyl group are cited below.

Further evidence for the occurrence of steric hindrance to coplanarity in the above compounds can be adduced from a comparison of the half-wave potentials[14] of compounds IV and V:

		R^1	R^2		pH:1	5.4	6.8
IV	NO$_2$	H	$E_{1/2}$	-0.09	-0.32	-0.39
V	H	NO$_2$	(V, SCE.)	-0.09	-0.34	-0.40

In these compounds, in which the nitro group is always in the *meta* position relative to the carbonyl group, the difference in the potentials of the compound IV, with the nitro group in the *ortho* position with respect to $-CH_2-$, and of compound V, with the nitro group in the *para* position, does not exceed 0.02 V over the whole pH range studied. It is concluded that the steric effect of the methyl group is lost, once its free rotation is prevented by ring formation.

Nitrobenzoic and nitrotoluic acids present a more complicated situation, as shown by the following half-wave potentials[14]:

					$E_{1/2}(\text{pH 2})$	ΔH	$\Delta_o(\text{COOH})$
	VI	Nitrobenzene			-0.27		
COOH		R^1	R^2				
R¹ structure	VII	NO^2	H		-0.26	$+0.01$	-0.10
	VIII	H	NO_2		-0.16	$+0.11$	

		R^1	R^2	$E_{1/2}(\text{pH 2})$			$\Delta_o(\text{COOH})$
COOH, CH₃ structure	IX	H	NO_2	-0.15			
	X	NO_2	H	-0.28			-0.13

As a first approximation it is again supposed that the polar effect of the carboxy group is roughly the same in both the *para* and the *ortho* positions. The shift $\Delta_o(\text{COOH})$ observed with the acid VII can then be attributed to a negative *ortho* effect. This effect could be caused also by hindrance to coplanarity, e.g., by the field effect. With the *o*-nitro-benzoic acid (VII) two possible mechanisms for such an effect exist: either (*i*) the nitro group is shifted from the position coplanar with the benzene ring with consequent reduction in the mesomeric interaction between these two fragments of the molecule, or (*ii*) the carboxy group is shifted and its polar resonance effect, therefore, is lessened. If this second mechanism operates alone, then of the total polar effect of the carboxy group on the reduction of the nitro group ($\Delta H = 0.11$ V) only $+0.01$ V would be contributed by the polar inductive effect. This seems unlikely in the light of the usual behavior of the COOH group in chemical reactions.

Mechanism (*i*) is also supported by the effect of the methyl group in the *ortho* position relative to the carboxy group. For compounds containing such an *o*-methyl group the comparison (IX and X) reveals a greater negative *ortho* shift [$\Delta_o(\text{COOH}) = -0.13$ V]. It seems that either simultaneous effects of nitro and methyl groups on the coplan-arity of the carboxy group occur or the effect of the carboxy group on the nitro group is enhanced as a result of the effect of the methyl group on the carboxy group. This is a relatively rare case of the steric effect of a substituent in a benzene ring acting from a distance on the reactive group.

Another example is presented by the nitrophthalic acids XI and XII, the half-wave potentials of which were measured by Tirouflet[14]:

		R^1	R^2	$E_{1/2}$(pH 9)	Δ_o
	XI	NO_2	H	-0.84	-0.14
	XII	H	NO_2	-0.70	

(Structure: benzene ring with COO^-, COO^-, R^1, R^2 substituents)

In XI the nitro group is *ortho* and *meta* relative to the COO^- groups and in XII it is *para* and *meta* relative to these groups. For these compounds, therefore, approximately the same shifts due to the polar effect only can be expected. The observed negative *ortho* shift for the 3-nitro derivative XI can be explained by steric hindrance to the coplanarity of the NO_2 or COO^- group, or of both these groups. In addition to these effects the field effect of the COO^- group may be involved, and the observed dependence of the value of the negative *ortho* shift on pH seems to support this view.*

Another group of substances for which the influence of steric hindrance to coplanarity can be regarded as established is the azo compounds. Since the effects are practically pH-independent between pH 3 and 8.3, we shall restrict ourselves to a discussion of half-wave potentials[91] at pH 6 (in McIlvaine buffers containing 50% ethanol):

$R^1-N=N-R^2$	R^1	R^2	$E_{1/2}$(pH 6)	Δ(o-CH$_3$)	Δ(2,4,6-tri-CH$_3$)
XIII	phenyl	phenyl	-0.42		
XIV	phenyl	o-tolyl	-0.45	-0.03	
XV	phenyl	mesityl	-0.56		-0.14
XVI	o-tolyl	o-tolyl	-0.51	-0.06	
XVII	o-tolyl	mesityl	-0.65		-0.23

The nonadditivity of the shifts prove that the observed effects are indeed caused by steric hindrance to coplanarity. The substitution

---

* Half-wave potentials at lower pH values unfortunately are unsuitable for quantitative discussion, since in this pH region antecedent chemical reactions play a part.

by the first methyl group in the *ortho* position in XIV causes a shift of -0.03 V compared with the parent substance XIII. For this reaction series $\rho_{\pi,R} = +0.17$ V and $\sigma_{o\text{-}CH_3} = 0.17$, so that $(\Delta E_{1/2})_{o\text{-}CH_3} = -0.02_9$ V; this whole shift of -0.03 V can be attributed to the polar contribution. For the trimethyl derivative XV the polar shift would be -0.09 V; the shift observed is -0.14 V; so that 0.05 V can be ascribed to steric hindrance to coplanarity.

The effects of substitution in both rings are not additive. When both rings are substituted steric hindrance to coplanarity is observed even when only one methyl group is present in each ring. Instead of a shift of -0.03 V due to the polar effect only, -0.06 V was observed for XVI, -0.03 V being attributable to steric hindrance to coplanarity.

A more pronounced disturbing effect on coplanarity was observed for the azo compound XVII, where instead of the shift of -0.12 V expected for four methyl groups a shift of -0.23 V was observed. Two methyl groups in one phenyl ring and one in the other results in a shift of -0.11 V, attributable to hindrance to coplanarity.

In none of the observed examples can it be deduced that the mesomeric interaction of the phenyl ring with and via the azo group is completely eliminated. This is shown by the following data, obtained by subtracting -0.03 V for the polar effect of each methyl group.

	X^1	X^2	Shift for the nonpolar *ortho* effect	Shift for the second o-CH$_3$
XIV	o-CH$_3$	—	0	
XV	Di-o-CH$_3$	—	-0.05	-0.05
XVI	o-CH$_3$	o-CH$_3$	-0.03	
XVII	Di-o-CH$_3$	o-CH$_3$	-0.11	-0.08

The nonpolar shifts in half-wave potential also can be expressed graphically:

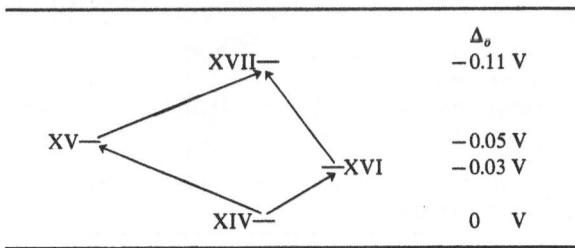

It may be concluded that (*i*) only the second *o*-methyl group, whether in the same ring as the other *o*-methyl (B) or in a different ring (A), causes significant hindrance to coplanarity; (*ii*) substitution by two *o*-methyl groups in one ring (B) results in greater steric hindrance to coplanarity than their substitution in different rings (A); (*iii*) the effect of the introduction of a second *o*-methyl group into one ring is greater when the other ring is *o*-tolyl (D) than when it is phenyl (B); (*iv*) finally, the effect of the substitution of *o*-methyl is greater when the second ring is mesityl (C) than when it is phenyl (A).

Generally speaking, the effect of steric hindrance to coplanarity steadily increases with an increasing number of methyl groups and, therefore, mesomeric interaction is not removed completely even in compound XVII.

Prévost, Souchay, and Malen[91] consider that steric hindrance to coplanarity occurs even in the case of acetophenones and benzophenones substituted by methyl groups in the phenyl rings. The following values of half-wave potential were found[31] in McIlvaine buffers containing 50% ethanol:

R-COCH$_3$	R	pH	$E_{1/2}$	$\Delta(o\text{-CH}_3)$	$\Delta(2,4,6,\text{-tri-CH}_3)$
XVIII	phenyl	1.3	-1.14		
XIX	*o*-tolyl		-1.27	-0.07	
XX	mesityl		> -1.34		> -0.20
XVIII	phenyl	4.9	-1.33		
XIX	*o*-tolyl		-1.40	-0.07	
XX	mesityl		> -1.56		> -0.23
XVIII	phenyl	8.4	-1.58		
XIX	*o*-tolyl		-1.65	-0.07	
XX	mesityl		-1.80		-0.22

R^1—CO—R^2	R^1	R^2	pH	$E_{1/2}$	$\Delta(\text{tri-CH}_3)$	$\Delta(\text{hexa-CH}_3)$
XXI	phenyl	phenyl	1.3(1F)	-0.92		
XXII	phenyl	mesityl		-1.11	-0.19	-0.29
XXIII	mesityl	mesityl		-1.21		
XXI	phenyl	phenyl	8.4(2F)	-1.31		
XXII	phenyl	mesityl		-1.46	-0.15	
XXIII	mesityl	mesityl		-1.61		-0.30

The data cited seem insufficient to prove that steric hindrance to coplanarity occurs in these cases. In the paper[91] no values are given for 4'-methylacetophenone and 4-methylbenzophenone. If we accept the approximate value of $\rho_{\pi,R} = 0.35$ V for both series (see Table III-4), the expected shift in half-wave potential due to the polar effect of an *o*- or *p*-methyl group will be $\rho_{\pi,R}\sigma_{o,p\text{-}CH_3} = 0.35 \times -0.17 = -0.06$ V. This value is in good agreement with the value of -0.07 V found for XIX, and with the approximate values for one methyl group in XX (about -0.07 V), in XXII (about -0.06 V), and in XXIII (about -0.05 V). Moreover, the data[91] cited above show additivity of the effects of methyl groups. Thus in acetophenone one methyl causes a shift of -0.07 V, while the introduction of three methyl groups results in a shift of about -0.22 V. Similarly, for benzophenone the substitution of three methyl groups in XXII causes about one-half of the shift due to six methyl groups in XXIII. Such additivity would hardly be expected if steric hindrance to coplanarity were operating.

The transfer coefficient is not given in the paper cited,[91] but Fig. 3 in this paper[91] shows that there is no substantial difference between the shapes of the polarographic waves of XXI, XXII, and XXIII. This is contrary to the behavior observed in cases in which steric hindrance to coplanarity undoubtedly takes place.

For substituted benzaldehydes,[91] and also for iodobenzenes (as can be deduced from the positive or small *ortho* shifts in Table III-10), steric hindrance to coplanarity is not exhibited by *o*-methyl groups. For the diiodotyrosine derivatives XXIV, in which the substituent X is in the *ortho* position relative to the two iodine atoms, the half-wave potentials[95] show no correlation either with $\sigma_{p\text{-}X}$ or with $\sigma_{o\text{-}X}$. Here a primary steric effect cannot be excluded.

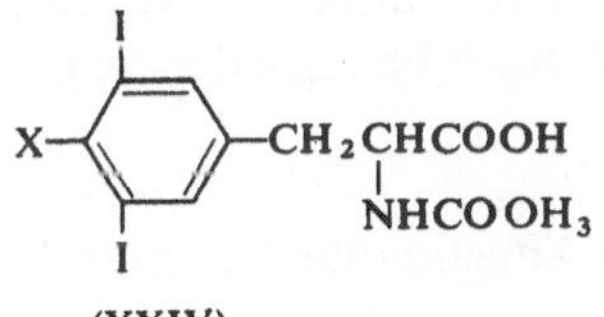

(XXIV)

It can be concluded that the effect of steric hindrance to coplanarity arising from substituents in the *ortho* position has been proved for nitro and azo compounds. However, for carbonyl compounds the steric conditions during polarographic reductions are different, and correlations with spectra and dipole moments[91] should be extended with care. The possibility cannot be excluded that in polarography a leveling effect due to the electrode surface tends to impose coplanarity, i.e., interaction between the electrode surface and the electroactive compound results in the formation of a more planar form than that present in the solution.

10. The Question of Hydrogen Bonds in *ortho*-Substituted Compounds

In *o*-nitrophenols, *o*-nitroanilines, and *o*-nitrophenylhydroxyla-mines,[96] reduction occurs (mainly at low pH) at more positive or only slightly more negative potentials than for the analogous *meta* derivatives and at substantially more positive potentials than for the *para* derivatives. This difference has been explained[96] by hydrogen bonding. In our opinion, however, the difference between *ortho* and *meta* derivatives arises from a change in the reduction mechanism. Whereas *ortho* hydroxy and amino nitro compounds are reduced to amines, the *meta* derivatives are reduced to substituted phenylhydroxylamines. The difference between the *ortho* and the *para* isomers, resulting in a positive *ortho* shift, is due either to a difference in the extent of the mesomeric interaction of the substituents with the nitro group (corresponding to different σ_X^- values), or to another kind of *ortho* effect, i.e., to the field effect. A similar marked positive *ortho* shift was observed for *o*-hydroxy and *o*-amino derivatives of benzaldehyde and iodobenzene, and also for *o*-methoxy, *o*-amino, and *o*-dialkylamino derivatives (Table III-10). It scarcely would be possible to accept an explanation based on hydrogen bonding for these several types of reducible groups and for all these various substituents, especially since water-containing solutions were studied.

In our opinion, a contribution of hydrogen bonding to the shift in half-wave potential cannot be regarded as proved in any case so far reported.

11. Factors Affecting the Reaction Constant $\rho_{\pi,R}$

Whereas, considerable attention has been paid in the literature to the values of substituent constants σ_X for various substituents and to various types of interaction with the reactive center, to the range of validity of the linear free energy relations, and to deviations from these correlations, the factors affecting the values of reaction constants ρ have been studied relatively rarely, see Refs. 5, 7, 40.

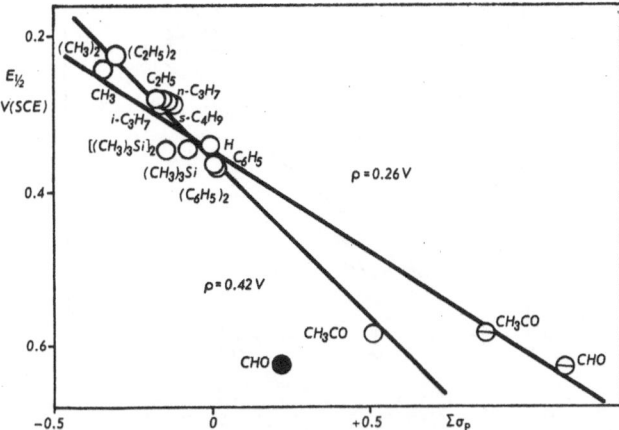

Fig. III-13. Relation of quartef-wave potentials $(E_{1/4})$ for the oxidation of ring-substituted ferrocenes on the sum of the Hammett substituent constants $\Sigma\sigma_p$. Reaction Series No. 87, Table III-4. Full point: deviates; halved points: σ_X^- values used. For steeper line with $\rho_{\pi,R} = 0.42$ V σ_X values were considered. For the less steep line with $\rho_{\pi,R} = 0.26$ V σ_X^- values were used.

An attempt has been made[5] to express the effectiveness with which the electrical effect is transmitted from the substituent to the reactive center, which determines the value of ρ. A relation was sought for the dependence of ρ on the length of the side chain, on the polarizibility of the side chain, on the electron density at the reaction center, and on reaction conditions. In homogeneous kinetics, it is rather difficult to find a sufficient number of reactions of the same type carried out under comparable conditions, and it is, therefore, not an easy task to predict the approximate value, and sometimes even the sign, of the reaction constant ρ. Similarity in the course of the reaction and in reaction conditions makes polarography especially suitable for this purpose.

    The determination of the value of ρ from experimental data will be mentioned first. The principles were already stated in Section 3 of this chapter: the slope of the $E_{1/2}$–σ relation is measured and expressed in volts. It is evident from Figs. III-3–11 that there usually is no difficulty in measuring the slope. In some cases difficulties may arise, especially when the substances studied do not differ much in their substituent constants. An example of this type may be the behavior of ferrocenes substituted directly in the ferrocene ring. In the plotting of the regression line for Reaction Series No. 87 there were two possibilities: By the use of $\sigma_{p\text{-}CH_3CO}$ and $\sigma_{p\text{-}CHO}$ a value of $\rho_{\pi,R} = 0.42$ V was found and the value

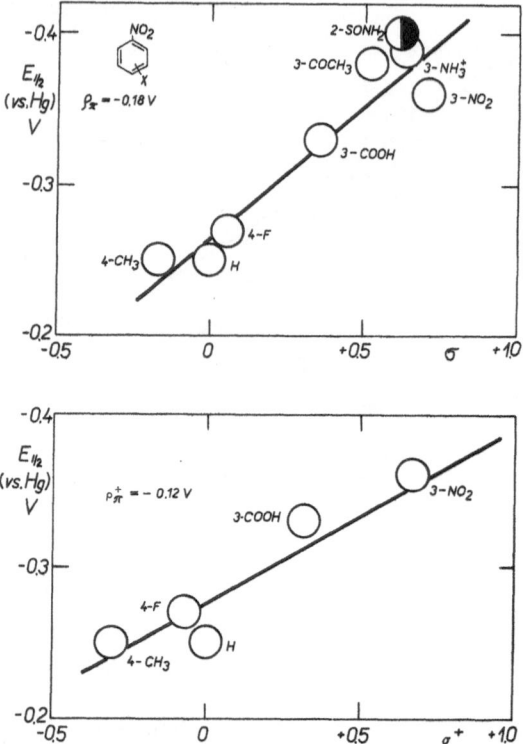

Fig. III-14. Reduction of nitro compounds in concentrated sulphuric acid media. Reaction Series No. 78, Table III-4. *a*) Dependence of half-wave potentials on Hammett substituent constants σ_X; *b*) dependence of half-wave potentials on substituent constant for electrophilic reactions σ_X^+. Halved point: *ortho*-derivative.

for the potential of the aldehyde deviated (Fig. III-13). By the use of $\sigma_{p\text{-CH}_3\text{CO}}^-$ and $\sigma_{p\text{-CHO}}^-$ a linear correlation with $\rho_{\pi,R} = 0.26$ V was obtained. As no earlier information existed about the preference of σ_X or σ_X^- values in this series, it was not possible to decide from these data only. In Reaction Series No. 88 and 89 values of potentials of further substituents were included and values $\rho_{\pi,R} = 0.32$ V and $\rho_{\pi,R} = 0.30$ V, respectively, were found; these gave correlations with σ_X^-, but not with σ_X values (Fig. III-13). Thus, it can be concluded that the application of σ_X^- and $\rho_{\pi,R} = 0.26$ V is to be preferred for Reaction Series No. 87.

We will now turn to the sign of the constant $\rho_{\pi,R}$. It is assumed[5,78] that a positive value of ρ indicates that the reaction is facilitated by a low electron density at the reaction center. On the other hand, reactions

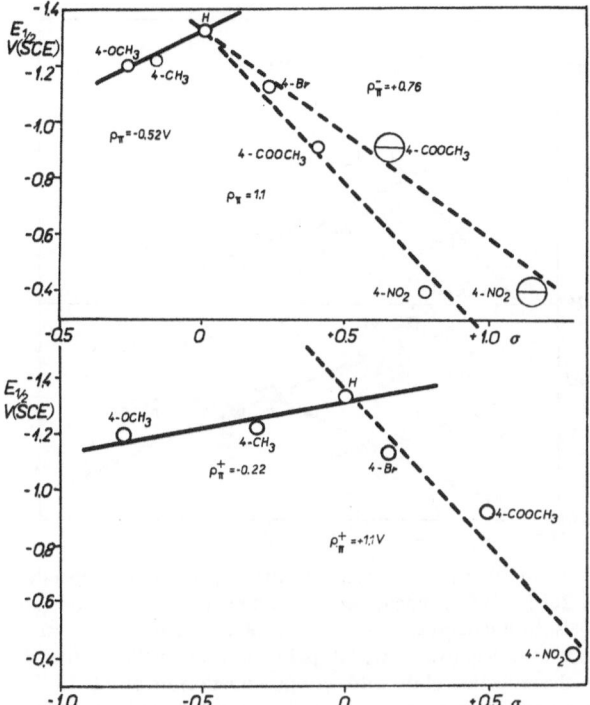

Fig. III-15. Reduction of phenyl-substituted benzyl bromides in methanolic media containing 0.25 M lithium methoxide. Half-wave potentials from Ref. 97. *a*) Dependence on the Hammett substituent constants σ_X: *b*) dependence on the substituent constants for electrophilic reactions σ_X^+. Halved points: σ_X^- values used. For the steeper line in *a*) σ_X, for the less steep with $\rho_{\pi,R}^- = 0.76$ V σ_X^- values were used.

promoted by high electron density on the reactive group have a negative value of ρ.

If with increasing positive value of the substituent constant σ_X (i.e., with increasing electrophilic character in the substituent) the half-wave potential is shifted to more positive values, then in accordance with usage in homogeneous kinetics the sign of the reaction constant ρ is defined as positive.

In the first 77 reaction series given in Table III-4, which correspond to reduction processes, the sign of the reaction constant $\rho_{\pi,R}$ is positive. It appears that in all these cases the mechanism of the potential-determining step is similar, i.e., it is a nucleophilic reaction in which the reduced substance is the electrophilic component. The most probable nucleophilic agent is the electron (or the electrode). The possibility of

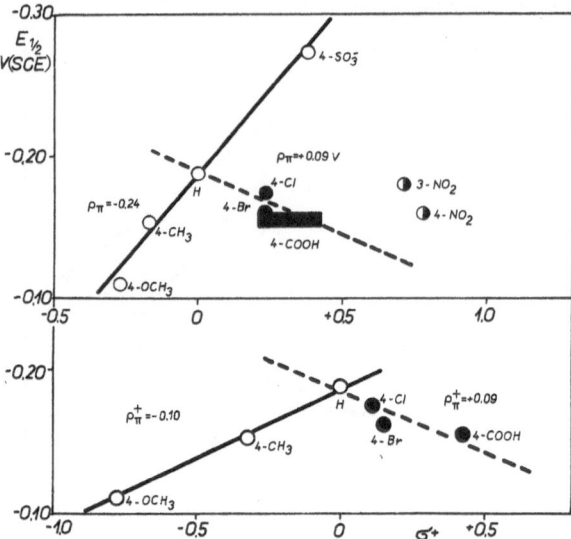

Fig. III-16. Reduction of benzenediazonium salts at pH 4. pH-independent half-wave potentials from Ref. 98. *a*) Dependence on the Hammett substituent constants σ_X; *b*) dependence on the substituent constants for electrophilic reaction σ_X^+. Circles correspond to regression line with negative slope, full points to a line with positive slope. Halved points deviate—the reduction of the nitro group is possibly involved.

the primary importance of an electrophilic attack by, for example, a proton prior to electron transfer thus is excluded.*

There are only four examples of aromatic reaction series for which negative values of the reaction constant $\rho_{\pi,R}$ have been found (Table III-4, No. 78-81). The first of these reactions, the reduction of substituted nitrobenzenes in concentrated sulfuric acid,[70] fits satisfactorily among other electrophilic reactions in concentrated sulfuric acid.[41] The second reaction series corresponds to substituent effects on the second reduction wave of diaryliodonium salts.[16] It has been deduced in Section 6 of this chapter that this wave probably does not correspond to the reduction of the Ar-I-Ar radical, but rather to the reduction of a radical containing a single phenyl group. The formation of this radical could be the electrophilic reaction determining the potential of the second wave. No explanation can be given for the fact that heterocyclic

---

* It is worth mentioning that, whereas, such a nucleophilic mechanism operates for nitrobenzenes and iodobenzenes, nitroalkanes and secondary and tertiary alkyl halides are reduced by another, probably electrophilic mechanism (see Chapter V).

azo compounds (Reaction Series No. 80 and 81) give a negative reaction constant, whereas azobenzene and naphthaleneazobenzene derivatives give a positive constant.

For electrophilic reaction (for electron donating substituents), H. C. Brown[41] suggested the application of the special constants $\sigma_{p\text{-}X}^{+}$, in which the mesomeric interaction between the reaction center and substituent is incorporated, in equation (31).

Figure III-14 shows that for nitrobenzenes equation (31) with the constants σ_X^{+} and equation (27) with the constants σ_X give almost equally satisfactory correlation. In the case of the other three reactions mentioned the small number of data available, and the unsatisfactory choice of derivatives studied, made it impossible to determine which equation [(31) or (27)] should be applied.

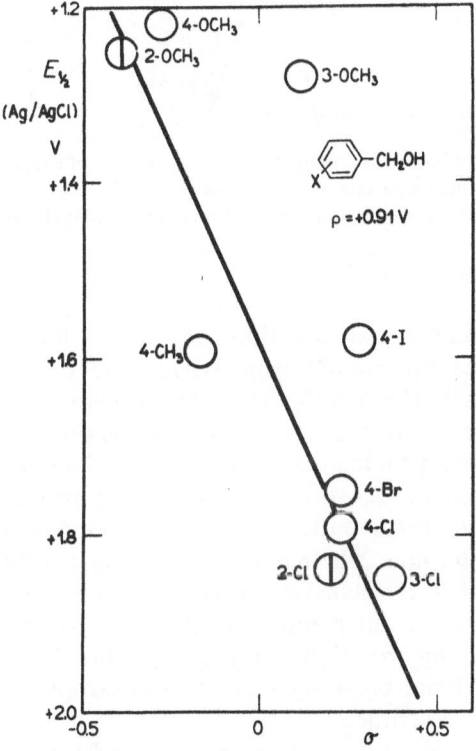

Fig. III-17. Relation of half-wave potentials for the oxidation of substituted benzyl alcohols to Hammett substituent constants σ_X. Reaction Series No. 82, Table III-4. Halved circles: *ortho* derivatives.

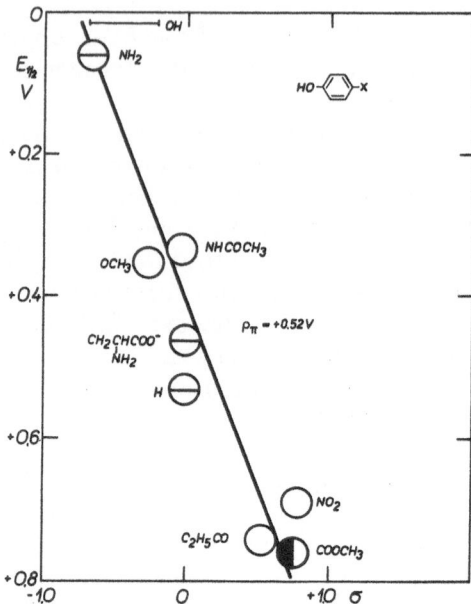

Fig. III-18. Relation of half-wave potentials for the oxidation of substituted phenols to the Hammett substituent constant σ_X. Reaction Series No. 85, Table III-4. Circles: ethanolic solution; halved circles: aqueous solutions. Halved point for methoxycarbonyl group: $\sigma^-_{COOCH_3}$ was used.

A rather complicated situation was found for reductions of substituted benzyl bromides[97] (Fig. III-15) and benzenediazonium hydrogen sulfates[98] (Fig. III-16). The relation between the half-wave potentials and the substituent constants, which is similar in both reaction series, can be approximated by two straight lines: with a positive slope for electrophilic substituents, but with a negative slope for electron donating substituents. Again, the scarcity of data made it impossible to determine whether the use of the constants σ^+_X with equation (31) or of the constants σ_X with equation (27) is to be preferred. This behavior can be interpreted as the result of two competing potential-determining reactions: charge transfer for electrophilic substituents, and bond-breaking[97] or other electrophilic reaction for electron donating substituents.

The positive sign of the values of $\rho_{\pi,R}$ found for all ten reaction series (No. 82–91) corresponding to oxidation processes (Figs. III-17 and III-18), and for mercury compound formation which have been reported up to now, is surprising. The discussion of the appropriate mechanism must be postponed until more experimental data are available.

As mentioned above, values of the reaction constants $\rho_{\pi,R}$ depend not only on the nature of the polarographically active group, but also on reaction conditions. The effects of depolarizer concentration, drop time, and temperature have not yet been determined carefully enough to permit the discussion of these effects on the value of $\rho_{\pi,R}$. The decisive factor is the composition of the supporting electrolyte, which can affect the composition (e.g., the ionization state) of the electroactive particle and also, through change in the composition of the double layer, even the electrode process proper.

The effect of pH will be discussed first. When the slope $dE_{1/2}/dpH$ remains constant in the given reaction series (in a chosen pH range), the reaction constant $\rho_{\pi,R}$ remains constant in this pH range. Several examples of such behavior have been demonstrated.

For substances which give different values of $dE_{1/2}/dpH$ in different pH ranges, in each of which $dE_{1/2}/dpH$ remains the same for all substances involved, the value of $\rho_{\pi,R}$ was found to vary from one pH range to another. These changes reflect changes in the reactive species with pH (e.g., ionization) as well as changes in the reaction mechanisms.

For reaction series for which $dE_{1/2}/dpH$ does not remain constant, but changes linearly with the substituent effects (Fig. III-1), a linear relation can be expected between the measured value of $\rho_{\pi,R}$ and the pH. Such a relation has been established for the reduction of azo compounds.

The effect of cations on polarographic reductions[99] has rarely been studied for reaction series containing a large number of derivatives. Marked effects of cations on the value of the coefficients $\rho_{\pi,R}$ have been observed, especially in two reaction series discussed below.

Table III-11

Effect of Cations on the Reaction Constants for the Reduction of Substituted Benzylideneacetones

0.25 M solution of	$\rho_{\pi,R}(V)$
$N(CH_3)_4Br$	$+0.20$
NaBr	$+0.28$
NH_4Cl	$+0.40$

Table III-12

Effect of Neutral Salts on the Reaction Constant $\rho_{\pi,R}$ and the Half-Wave Potential E_H of Unsubstituted Nitrobenzene

Solvent	0.1 M solution of	$\rho_{\pi,R}(V)$	$E_H(V)$
Dimethylformamide	$N(C_2H_5)_4I$	0.42	-1.12_7
	CsI	0.38_5	-1.09_5
	$NaClO_4$	0.35_5	-1.06
	NaI	0.35_5	-1.06
	LiCl	0.36	-1.04_3
Acetronitrile	NaI	0.22	-1.03

When the half-wave potentials of reduction waves of substituted benzylideneacetones[54] in solutions containing 0.04 M Britton–Robinson buffer, 0.25 M neutral salt, 80% ethanol, and 0.014% gelatin (Reaction Series No. 30–32) were compared in a pH region in which the

Table III-13

Selected Values of the Reaction Constant $\rho_{\pi,R}$ and Half-Wave Potentials of the Unsubstituted Compounds E_H

Compound	pH < 3		pH 5–8 (and unbuffered)		pH > 11	
	$\rho_{\pi,R}$	E_H	$\rho_{\pi,R}$	E_H	$\rho_{\pi,R}$	E_H
ArCHO	+0.32	−0.78	+0.26	−1.46	+0.33	−1.40
ArCOCH$_3$	+0.22	−1.00	+0.35	−1.58	+0.22	−1.58
ArC(NOH)CH$_3$	+0.18	−0.78	—	—	+0.26	−1.82
ArCOAr	+0.37	−0.80	+0.26	−1.14	—	—
ArC(NOH)Ar	+0.28	−0.54	—	—	+0.29	−1.60
ArCSAr	+0.31	−0.46	—	—	—	—
ArCH=CH—COCH$_3$	—	—	+0.28	−1.37	—	—
ArCOOCH$_3$	—	—	+0.85	−2.11	—	—
ArNO$_2$	+0.16	−0.29	+0.14	−0.50	+0.24	—
ArNO	—	—	+0.08	−0.20	—	—
ArN=NAr	+0.10	−0.08	+0.16	−0.51	—	—
ArCH=CH—Ar	—	—	+0.25	−2.17	—	—
ArI	—	—	+0.30	−1.68	+0.30	−1.68
ArCl	—	—	—	—	+0.20	−2.62[a]
ArÏAr	+0.02	−0.20	+0.02	−0.20	+0.02	−0.20
ArAsO$_3$H$_2$	+0.28	−1.27	—	—	—	—

[a] Extrapolated value.

half-wave potentials were practically independent of pH, the values given in Table III-11 were found for the reduction constants.

In the original paper the slopes of waves are not given, but the reproduced curves[54] show a change in the slope in the presence of cations. In the presence of ammonium ions a tendency for the separation of two waves is visible on the curves, the one-electron step being shifted toward more positive potentials. The marked change in the course of the electrode process is here manifested by a change in the value of reaction constant $\rho_{\pi,R}$.

The second reaction series consists of substituted nitrobenzenes, polarographed in solutions of various salts in dimethylformamide and acetonitrile.[100] By the use of the half-wave potentials of the one-electron reduction step (resulting in the formation of nitrobenzene anion radical) the reaction constants given in Table III-12 were computed.

An inspection of Table III-12 reveals that both the reaction constant and the half-wave potential of the parent compound are markedly affected by the cation and the kind of solvent used, but neither of these quantities is influenced by the kind of anion present.

These examples demonstrate that special forms of equation (20) can be used for solutions other than aqueous ones. In most common applications the solvent is water or a mixture of water and an organic solvent, but anhydrous solvents have been used, such as glacial acetic acid (Reaction Series No. 36), concentrated sulfuric acid (No. 78), absolute ethanol (No. 38), and acetonitrile (No. 82, 84, and the above examples).

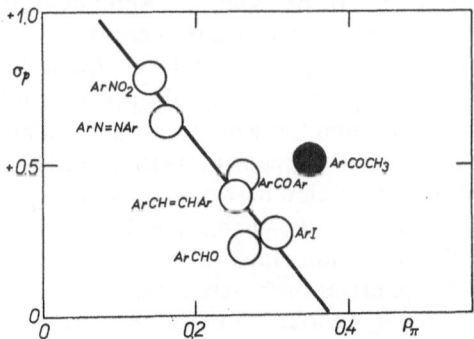

Fig. III-19. Relation of the Hammett substituent constant of the reducible group σ_{p-R} to the value of the reaction constant $\rho_{\pi,R}$ obtained for the same group R at pH 5 to 8. Full point deviates; similarly also values for nitrosobenzenes and methyl benzoates.

Since the polarographic method enables us to determine the reaction constants of numerous nucleophilic reactions for which one partner (i.e., electron or the electrode) is the same in all rate-determining steps, polarographic data are useful in the comparison of the effect of the reducible group on the value of the corresponding reaction constant $\rho_{\pi,R}$.

For this purpose three pH regions were selected: acid solution (pH < 3), the medium pH range (pH 5–8) (values obtained in unbuffered solution for systems whose half-wave potentials are pH-independent are included here), and alkaline solutions (pH > 11) (the rest of the pH-independent systems are included here). Selected values are collected in Table III-13, which gives also the half-wave potentials of unsubstituted members of the reaction series (E_H).

Hine[101] has recently shown that for an equilibrium $AX_1 \rightleftharpoons AX_2$, to which the Hammett equation can be applied, equation (43) can be used. Here the proportionality constant τ_p is dependent on the reaction conditions but not on the type of reaction studied. For unconjugated substituents for *meta*- and *para*-substituted substances its value is 1.14 (*cf.* Ref. 80). This relation between the reaction constant ρ and the substituent constant σ can be applied only to equilibrium constants when both $\sigma_{p\text{-}X_1}$ and $\sigma_{p\text{-}X_2}$ are known.

$$\rho_p^X = \tau_p(\sigma_{p\text{-}X_1} - \sigma_{p\text{-}X_2}) \tag{43}$$

In view of the fact that for reactions rates—and, thus, also for the half-wave potentials of irreversible systems—the value of $\sigma_{p\text{-}X_2}$, which corresponds to the transition state, is unknown, a simple relation between $\rho_{\pi,R}$ and $\sigma_{p\text{-}X}$ cannot be expected. Actually, a trend[8] has been found in the correlation between reaction constants $\rho_{\pi,R}$ for reactions measured at pH 5-8 and $\sigma_{p\text{-}X}$ (Fig. III-19). Apart from $ArCOCH_3$ (Fig. III-19), $ArNO$ and $ArCOOCH_3$ did not fit the linear relation. In acid and alkaline solution the scattering was even greater.

These findings are in agreement with equation (43) insofar as there is no simple linear relation between $\rho_{\pi,R}$ and $\sigma_{p\text{-}X_1}$. The nonlinearity is caused by the fact that the changes in the coefficient $\sigma_{p\text{-}X_2}$, which corresponds to the transition state, do not fully parallel the changes in $\sigma_{p\text{-}X_1}$, which characterizes the reactant state of the molecule.

On the other hand, we found it useful to correlate the value of the reaction constant $\rho_{\pi,R}$ with the value of the half-wave potential of the unsubstituted parent compound, which characterizes the reducibility of the given group R. Examples of the linear relation observed are

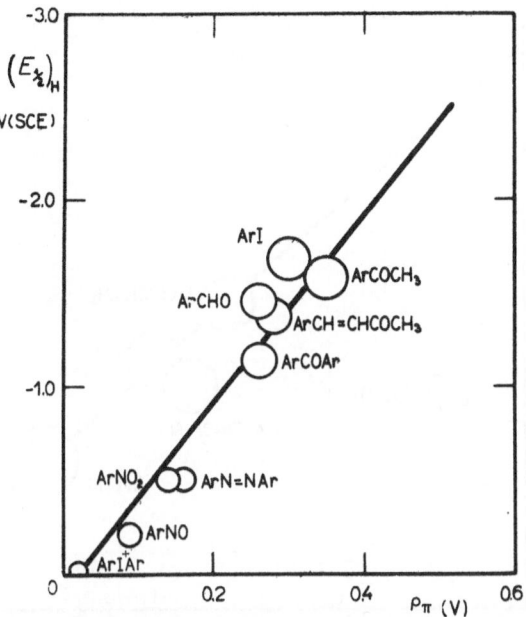

Fig. III-20. Relation of the half-wave potentials $(E_{1/2})_H$ of parent unsubstituted compounds containing the reducible group R to the values of the reaction constant $\rho_{\pi,R}$ obtained for benzene derivatives containing the same reducible group R at pH 5 to 8 and in unbuffered media (for pH-independent systems).

given in Figs. III-20 and III-21. This relation allows us to transform the three-parameter equation (27), written in the form (27a), into a two-parameter form (44).

$$(E_{1/2})_X = \rho_{\pi,R}\sigma_X + E_H \qquad\qquad (27a)$$

$$(E_{1/2})_X = \rho_{\pi,R}\sigma_X + \varkappa\rho_{\pi,R} + K \qquad\qquad (44)$$

where $\varkappa$ and K are constants which are independent of the kind of nucleophilic aromatic reaction series, but may depend on the reaction conditions.

In acid solutions and in the middle pH range the relation of $\rho_{\pi,R}$ to E_H is such that the reaction is the more susceptible to the influence of the substituent, the more negative the potential at which the reduction proceeds. At higher pH, when it is the unprotonated form that is reduced, the reaction constants are nearly all the same and do not show any significant trend. Further experimental evidence is necessary.

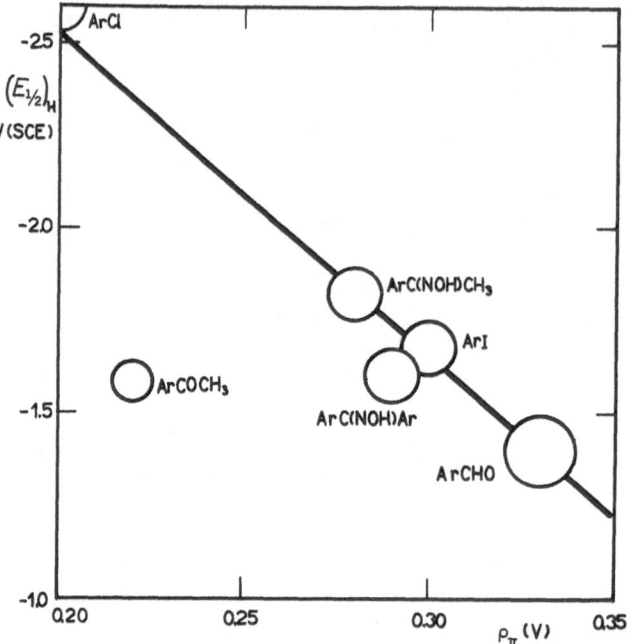

Fig. III-21. Relation of the half-wave potentials $(E_{1/2})_H$ of the parent unsubstituted compounds containing the reducible group R to the values of the reaction constant $\rho_{\pi,R}$ obtained for benzene derivatives containing the same reducible group R at pH > 11.

The application of equation (44) is not limited to the comparison of various electroactive groups, but even can be used for the comparison of reaction constants obtained for reaction series of compounds with the same electroactive group, polarographed under varying conditions.

Thus, for the azo compounds in Reaction Series No. 52–57 and 58–60, in which, as stated earlier (p. 101), the reaction constant is a function of pH, linear correlation was found between these constants and the half-wave potential of the unsubstantiated parent compound at the particular pH (Fig. III-22).

Similarly, for the reduction of nitrobenzenes in dimethylformamide in the presence of various neutral salts,[100] where the reaction constants are a function of the nature of the cation of the neutral salt (Table III-12), a linear relation was found between the reaction constants and the half-wave potentials of the parent unsubstituted compound measured in the solutions containing the particular cations (Fig. III-23).

For the value of the reaction constant Hammett (see Ref. 5) suggested the expression

$$\rho = \frac{B_1/D + B_2}{RT\,d^2} \tag{45}$$

where R is the gas constant, T is the absolute temperature, D is the dielectric constant of the solvent, and d is the distance between the substituent and the reaction center. B_1 is attributed to purely electrostatic interaction between the benzene derivative and the medium, and B_2 is assumed to be a measure of the susceptibility of the reaction to changes in the electric charge on the reaction center caused by substitution.

If, to the first approximation, the half-wave potential of an irreversible system is regarded as a measure of the polarizability of the bond undergoing reduction, then the relation found

$$\rho_{\pi,R} = \frac{E_H - K}{\varkappa} \tag{46}$$

is a relation between a measure of the polarizability of the bond to be reduced and the tendency for the effect of the substituent to be transferred to the bond undergoing reduction.

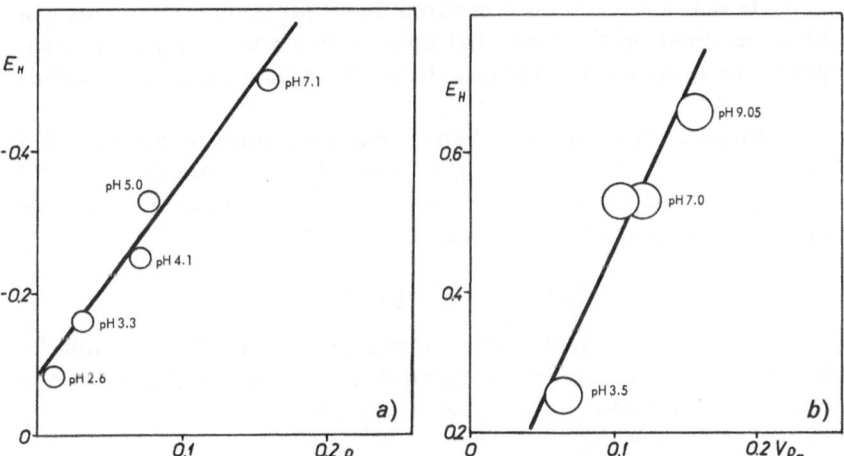

Fig. III-22. Reduction of azo compounds. Relation of the half-wave potentials of the parent unsubstituted compounds E_H at a given pH value to the values of the reaction constants $\rho_{\pi,R}$ obtained at the same pH value. *a)* Reaction Series No. 52–55 and 57; *b)* Reaction Series No. 58–60, Table III-4.

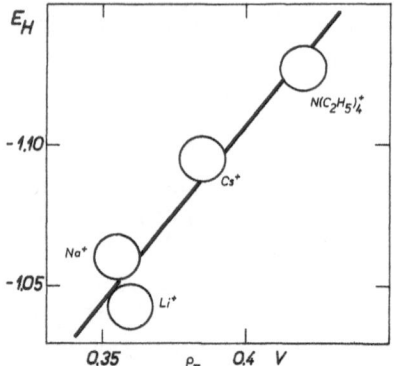

Fig. III-23. Reduction of substituted nitrobenzenes in dimethylformamide. Relation of the half-wave potential of the parent unsubstituted compound E_H in a solution containing a given cation to the value of reaction constant $\rho_{\pi,R}$ obtained in the same supporting electrolyte (Table III-12).

As far as we know, no relation analogous to equation (44) has been proposed on the basis of the data of homogeneous kinetics. This may be explained because in polarography more comparable data—obtained under similar conditions and with one reagent—are available.

The establishment of the validity of equation (44) would seem to indicate that the reactions compared are really comparable. This can be understood on the view that analogous electron-transfer nucleophilic processes occur in the rate-determining steps in all the reactions compared.

When functions of antecedent chemical reactions are incorporated in the values of half-wave potential, the validity of equation (27) and similar relations shows that for such chemical reactions the principle of linear free energy relations must again be valid.

12. Other Empirical Relations

When equation (27) is transformed into (47) or (48), the shifts in half-wave potential can be compared directly with dissociation or rate constants determined in the same reaction series.

$$\Delta E_{1/2} = \rho' \log \frac{K}{K_0} \tag{47}$$

$$\Delta E_{1/2} = \rho'' \log \frac{K}{K_0} \tag{48}$$

An application of this type is the comparison of the half-wave potentials of substituted azo compounds[19] (Reaction Series No. 52–57) with the rates of decolorization of these compounds after irradiation[19] with the aid of equation (48). As a linear relation exists between these two properties it is possible to make deductions on the light stability of such dyes (even on keratin fibers) from measurements of half-wave potential.

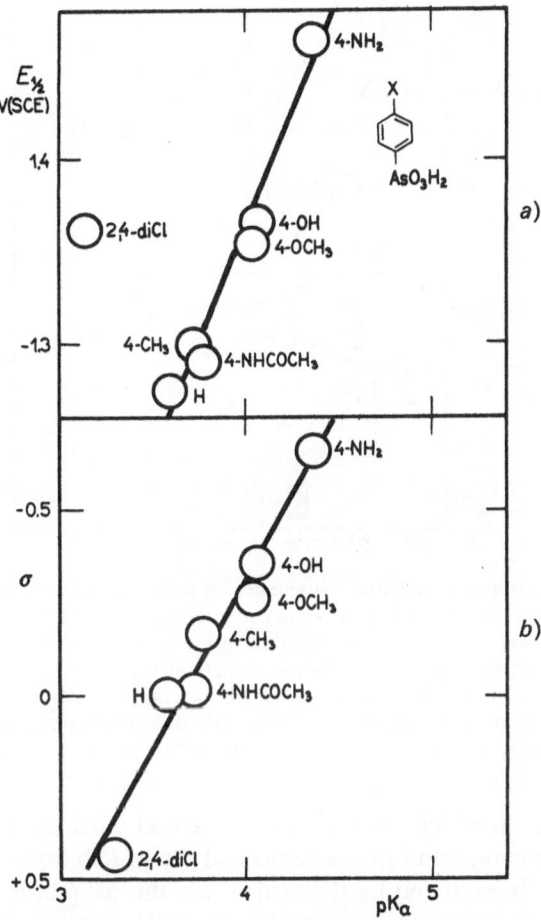

Fig. III-24. Substituted benzenearsonic acids. *a)* Relation of the half-wave potentials of reduction waves (Reaction Series No. 75, Table III-4) to the logarithms of acid dissociation constants pK_a; *b)* relation of the Hammett substituent constants σ_X to the logarithms of acid dissociation constants pK_a.

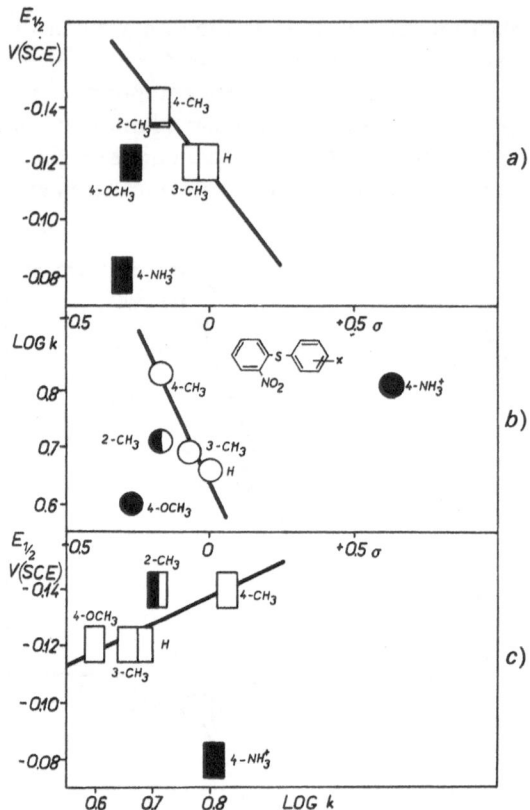

Fig. III-25. Substituted *o*-nitrophenyl sulfides. *a*) Relation of half-wave potentials for the reduction of the nitro group to Hammett substituent constants σ_X; *b*) relation of logarithms of rate constants for the reaction with titanous chloride to Hammett substituent constants σ_X; *c*) relation of half-wave potentials for the reduction of the nitro group to the logarithms of rate constants for the reduction of the nitro group by titanous chloride. Full point: deviates; halved points: *ortho* derivatives. Reaction Series No. 74, Table III-4, half-wave potentials and rate constants from Ref. 67.

In cases in which the half-wave potential deviates from equation (27) for one compound in the series and the possibility of a change of mechanism is excluded as the reason for the deviation, comparison with equilibrium or rate constants by the use of equation (47) or (48) can determine the origin of the deviation. The half-wave potential which deviates from the $E_{1/2}$ versus σ plot can fit the log K versus $E_{1/2}$ or log k versus $E_{1/2}$ plot or can again show deviation.

Thus, for acetophenone oximes[52] (Reaction Series No. 20) the effect of the p-amino group on reduction is similar to its effect on dissociation, and the substituent constant $\sigma_{p\text{-NH}_2}^- = -0.10$ (Table III-7) shows a correlation even with shifts in dissociation constant. Similarly, the value for 2,4-dichlorobenzenearsonic acid[68] fits equation (47), but not (27). The *ortho* substituent, therefore, causes similar relative effects on reduction and on dissociation (Fig. III-24). On the other hand, for benzophenone oximes (Reaction Series No. 25) correlation by equation (47) is definitely worse than by equation (27). Thus, especially for β-isomers, dissociation is influenced by factors other than the dissociation of analogously substituted benzoic acids and the polarographic reduction of benzophenone oximes; these other factors probably derive from the interaction of the oxime group with phenyl rings in the dissociation.

The polarographic reduction of substituted nitrobenzenes was compared with the rate of their reduction by titanous chloride at a very early period.[3,102] From some recent measurements,[67] the following conclusions can be drawn.

For o-nitrophenyl phenyl sulfides (Reaction Series No. 74) deviations were observed for $p'\text{-OCH}_3$ and $p'\text{-NH}_3^+$ derivatives from both $E_{1/2}$ versus σ_X and $\log k$ versus σ_X plots (Fig. III-25). The p'-methoxy derivative fitted the $E_{1/2}$ versus $\log k$ plot, whereas, the p'-amino derivative deviated even from this plot. Such a comparison allows us to conclude that for the p'-methoxy group a special effect of this group (not included in $\sigma_{p\text{-OCH}_3}$) is operating both in polarographic reduction and the reaction with titanous chloride. On the other hand, the p'-amino group exerts different effects on the polarographic reduction and on the reduction with titanous chloride. These effects, both of which also differ from the effects of the amino group in the *para* position in benzoic acid, therefore, change according to the reaction involved. One possible explanation is a transfer of mesomeric effects in the transition state, involving p- and d-orbitals, which could be variable according to the transition complex concerned. Whereas, in polarographic reductions the o-methyl group had the same effect on shifts in half-wave potential as the p-methyl group, differences were observed in the kinetic data. Hence, in the reaction with titanous chloride there is a certain *ortho* effect which is absent in electrochemical reduction.

Similarly, for the analogous sulphones[67] (Reaction Series No. 73) p'-Br deviates from the $E_{1/2}$ versus σ_X plot, and o'-CH$_3$ and o'-Cl from $\log k$ versus σ_X plot. Hence, the more negative half-wave potential for the p'-bromo derivative is caused by a specifically polarographic effect. Similarly, as in the previous example, reaction with titanous chloride is more susceptible to *ortho* effects than polarographic reduction.

It is well understood that the validity of such deductions, especially in the two last-mentioned reaction series, is restricted by the limited number of compounds in the reaction series, but these deductions should demonstrate the reasoning that is possible in such comparisons.

It can be deduced from these examples that comparisons of correlations in accordance with equations (27) and (47), or (48) are very useful. The possibilities can be summarized as follows.

1. Deviation from the $E_{1/2}$ versus log k plot is observed. (*i*) Deviation from $E_{1/2}$ versus σ and good correlation with log k versus σ are observed; a specific effect is involved in the polarographic electrolysis; (*ii*) Deviation from log k versus σ and good correlation with $E_{1/2}$ versus σ are observed; there is a specific effect, which is involved not in the polarography, but in the kinetics or equilibria; (*iii*) No correlation, either with $E_{1/2}$ versus σ or with log k versus σ, is observed; an effect is involved which differs from the effect of the particular substituent on benzoic acid dissociation and differs from one reaction series to another; it is specific, nonadditive substituent effect; it can be a steric, sometimes a mesomeric, but never a polar effect.

2. Good $E_{1/2}$ versus log k correlation is observed. (*i*) Neither $E_{1/2}$ versus σ nor log k versus σ shows good linear relationship; an effect is involved which is characteristic for the given reaction series, but not included in the effect of the particular substituent on benzoic acid dissociation, and which has the same relative effects on the polarography and the rate of the chemical reaction; (*ii*) Both $E_{1/2}$ versus σ and log k versus σ give good correlations; only polar and other effects assumed to be operating in the derivation of equation (27) are involved.

It should be kept in mind that the substituent constants σ have been shown to express the general properties of the substituent. Deviations from equation (27) and other similar forms of equation (20) indicate a deviation from this expression of the general properties of the substituent. The application of equations (47) and (48) allows us to discuss the reason for such deviations in substituent effects.

In addition to rate and equilibrium constants, for which the Hammett equation and similar linear free energy relations were originally devised, several other physicochemical constants have been correlated[5,40] by the use of various modified forms of this equation. In several cases, the arguments advanced in support of the validity of

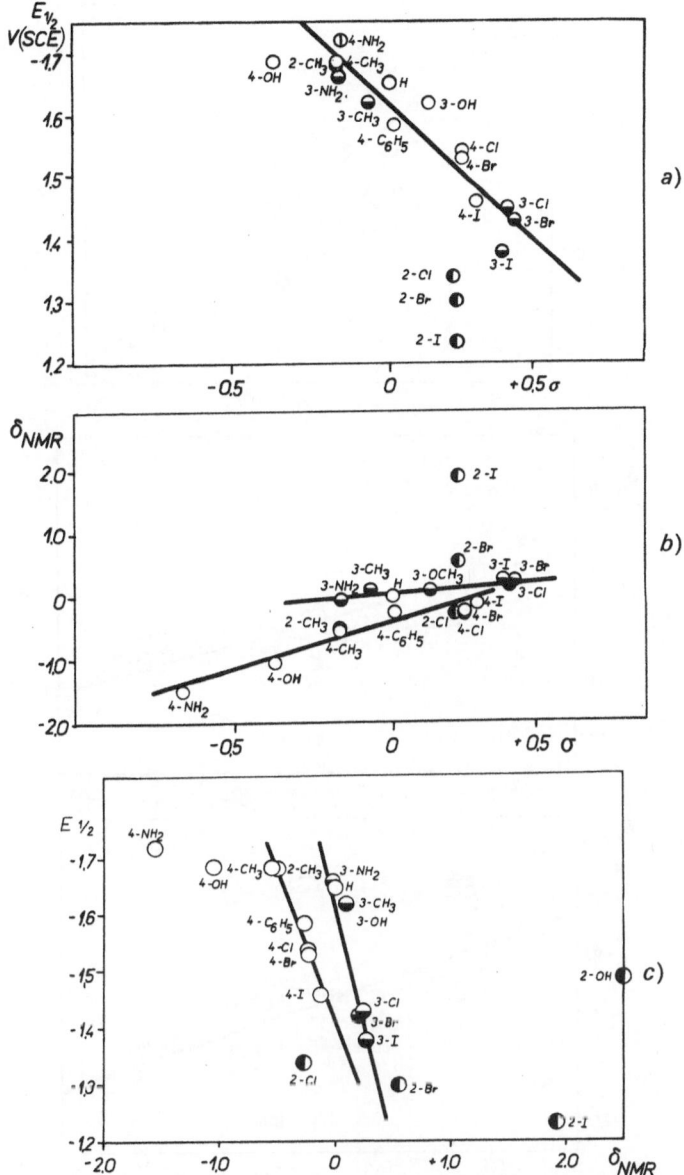

Fig. III-26. Substituted iodobenzenes. *a*) Relation of the half-wave potentials for the reduction of the C—I bond to Hammett substituent constants σ_X; *b*) relation of the chemical shifts δ_{NMR} for substituted fluorobenzenes to Hammett substituent constants σ_X; *c*) relation of half-wave potentials for the reduction of the C—I bond on the chemical shifts δ_{NMR} for substituted fluorobenzenes. ○ *para*; ◐ *meta*; ◑ *ortho* derivatives; ⊕ $\sigma_{p\text{-}NH_2}^{\pi}$ used. Half-wave potentials and chemical shifts from Ref. 103.

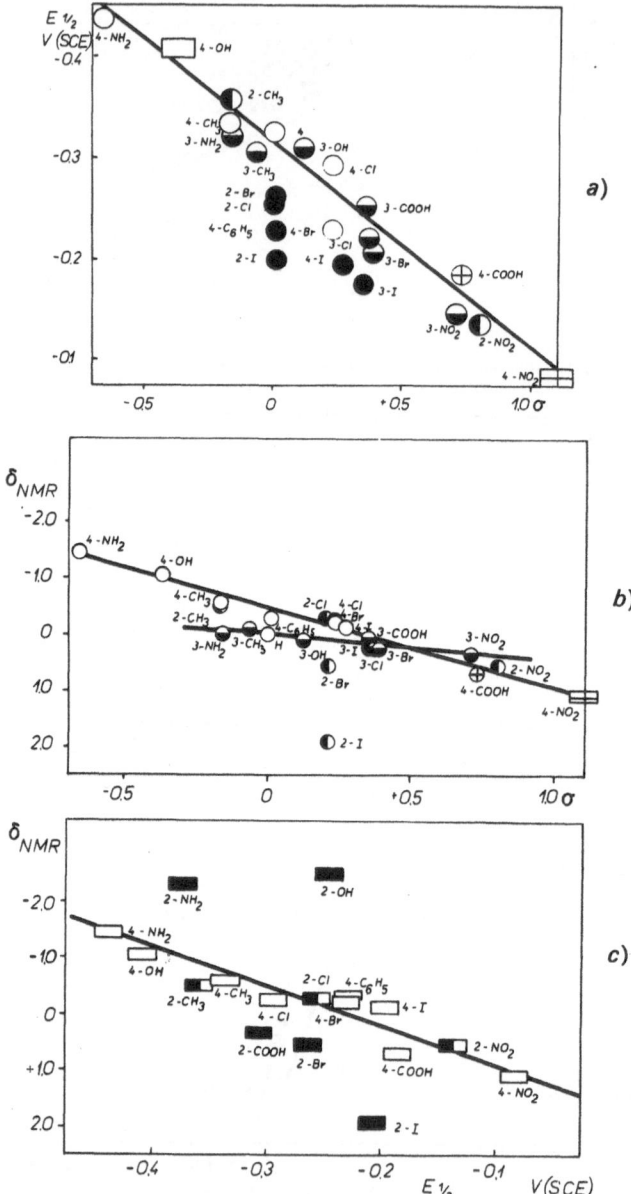

Fig. III-27. Substituted nitrobenzenes. *a*) Relation of the half-wave potentials for the reduction of the nitro group to Hammett substituent constants σ_X; *b*) relation of the chemical shifts δ_{NMR} to Hammett substituent constants σ_X; *c*) relation of the chemical shifts δ_{NMR} to half-wave potentials. ○ *para*, ◔ *meta*; ◑ *ortho*; ⊕ σ_{p-X}^-; ● *deviate*. Half-wave potentials and chemical shifts from Ref. 103.

such equations were more intricate than those concerned with the correlation of polarographic half-wave potentials, which can be considered as most nearly related to kinetic and equilibrium data. It is apparent that a linear relation between the values of a particular physicochemical quantity and the substituent constants σ, on the one hand, and a linear relation between half-wave potentials and σ in the same reaction series, on the other, implies also a linear relation between such physicochemical data and polarographic data. A correlation of polarographic with other physicochemical data possesses all the above-mentioned advantages of the correlations of polarographic with kinetic and equilibrium data.

In accordance with the deductions drawn in the preceding paragraph, a correlation exists[103] between half-wave potentials and chemical shifts in nuclear resonance spectra (δ_{NMR}). For substituted iodobenzenes (Fig. III-26) the $\delta_{NMR}-\sigma$ correlation is split into two linear portions, one for *meta*, and one for *para* derivatives (including 2-methyl and 2-chloro derivatives correlated with σ_{o-X} and $\delta_{\pi,R} \approx 0$). Hence, the $E_{1/2}$ versus δ_{NMR} plot is also split into two linear portions (Fig. III-26). All *ortho* halogen derivatives deviate from the $E_{1/2}$ versus σ plot. Comparison with δ_{NMR} versus σ and $E_{1/2}$ versus δ_{NMR} plots shows that this deviation is not inherent in the polarography, but is an *ortho* effect of the large halogen atom. No correlation was found for 2-OH when σ_{p-OH} was used (σ_{o-OH} is not available). The *p*-Amino derivatives fit the δ_{NMR} versus σ plot. Derivation from $E_{1/2}$ versus δ_{NMR} shows that an effect is operating in the polarography which is not expressed in σ_{p-NH_2}. Nevertheless, this effect is satisfactorily expressed by $\sigma^{\pi}_{p-NH_2}$ (Table III-7), and *p*-iodophenol shows deviations from $E_{1/2}$ versus σ and $E_{1/2}$ versus δ_{NMR}, but not from δ_{NMR} versus σ. Thus, a specific effect is included in the polarographic reduction, probably the dissociation of the phenolic hydroxyl as indicated by the pH-dependence of the half-wave potential.

For substituted nitrobenzenes half-wave potentials[49] at pH 1.7 were chosen* (Fig. III-27). Again, *meta* derivatives give a separate

---

* Both for substituted nitrobenzenes and benzaldehydes, whose half-wave potentials are pH-dependent, the correlation of values obtained in unbuffered solutions[103] cannot be recommended. The uncertainty involved in the electrode process is more important than the change in the ionization of the substituent. Moreover, even though it is stated in the Experimental part of the paper cited[103] that unbuffered solutions were used, the data given in Table III of this paper[103] include values of half-wave potentials measured[49] at pH 1.7 for nitrobenzenes substituted in the *meta* and *para* (but not *ortho*!) positions.

δ_{NMR} versus σ line and, therefore, also $E_{1/2}$ versus δ_{NMR} line (necessarily linear and, therefore, not shown in the $E_{1/2}$ versus δ_{NMR} plot in Fig. III-27). For 4-COOH and 4-NO$_2$ $\sigma_{p\text{-}X}^-$ values are used both for $E_{1/2}$ versus σ and δ_{NMR} versus σ plots. The contribution of mesomeric interactions between the nitro group to be reduced and these two substituents via the benzene ring therefore affects both polarographic and NMR measurements, as shown in the $E_{1/2}$–δ_{NMR} correlation, to which 4-COOH and 4-NO$_2$ give a good fit. The p-Nitrophenols and p-nitroanilines are known to be reduced by six electrons in acid solutions, whereas other nitrobenzenes require four electrons. The fitting of values to $E_{1/2}$ versus σ and $E_{1/2}$ versus δ_{NMR} plots can be interpreted (with more certainty than fitting to only $E_{1/2}$ versus σ plots) on the view that the potential-determining step is the same for all the nitrobenzenes involved. The uptake of further electrons must occur in subsequent reactions. For 2-NO$_2$ and 2-CH$_3$ $\sigma_{o\text{-}X}$ (and $\delta_{\pi,R} \approx 0$) can be used, 2-Br and 2-I again deviate from all three plots due to the specific *ortho* effects of bulky groups. For 2-COOH and 2-OH, for which values of $\sigma_{o\text{-}X}$ are not available, deviation from $E_{1/2}$ versus δ_{NMR} shows that the *ortho* effects of these groups operating in polarographic reduction and in NMR are different. For 2-Cl, 4-C$_6$H$_5$, 3-I, and 4-I deviations were observed for the $E_{1/2}$ versus σ plot. The correlation shown in the $E_{1/2}$ versus δ_{NMR} plot must be regarded as due to compensation of errors, for all these data fit the δ_{NMR} versus σ plot; such a possibility must always be borne in mind. Specific polarographic effects should be looked for in explanation of deviations from $E_{1/2}$ versus σ plots. The behavior of substituted benzaldehydes has been said[103] to be analogous to that of iodobenzenes.

Nuclear quadropole interaction frequencies, which are linear functions of substituent constants, have been shown[104] to be proportional to half-wave potentials in the case of substituted iodobenzenes.

Spectral data which show[40,41] linear correlations with σ or σ^+ constants can be correlated with half-wave potentials. For infrared spectra linear correlation has been found[105] between the frequency of the carbonyl group and half-wave potentials. These frequencies should be correlated with σ_X^+, whereas, polarographic reductions should be correlated with σ_X values, so that the comparison of values for substituents that have substantially different values of σ_{X_1} and $\sigma_{X_1}^+$ is of importance. Such groups as p-NH$_2$ and p-OCH$_3$ show deviations from the linear $E_{1/2}$ versus σ plot; other compounds of this type were not included in the group of substances studied.

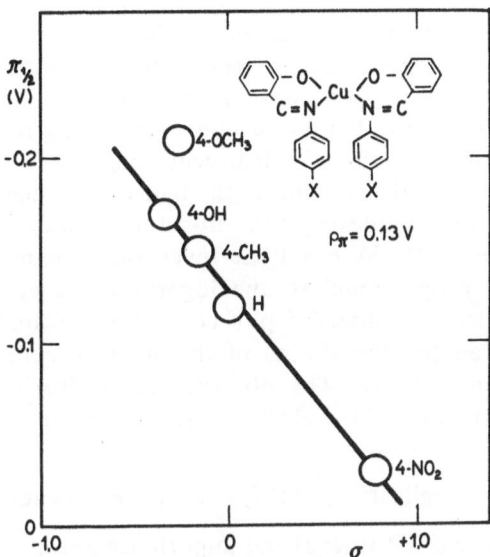

Fig. III-28. Relation of half-wave potentials of complexed cupric copper to Hammett substituent constants σ_X (*cf.* Ref. 8).

In the ultraviolet spectrum, correlation has been found between the half-wave potential and the wavelength of the first maximum in the long-wave region. Such correlations have been shown to apply in the case of oximes[52] (Reaction Series No. 20 and 25), nitrosobenzenes[61] (Reaction Series No. 46), and nitrobenzenes.[106,107]

In addition to polarographic half-wave potentials, other quantities, characterizing electrolysis by related electrochemical methods, can be correlated with structure. In Table III-4, $E_{1/4}$ values obtained chronopotentiometrically have been included. Linear free energy relations are obviously obtained for peak potentials on i versus E curves oscillographically, especially by the use of single-sweep methods. Such applications were demonstrated by Tirouflet[26] for substituted nitrobenzenes and benzaldehydes.

Another oscilloscopic technique, using controlled current electrolysis and measurement of the dE/dt versus E curves, also has been applied in structural studies. It has been shown recently that indentation potentials on the dE/dt versus E curves are linear functions of substituent constants.[108] Such depolarization potentials have been correlated with σ_{p-X} for substituted ferrocenes.[109,110]

Finally, it is found that equation (27) is obeyed even in systems in which the reduction occurs at a central metal ion bound to a chelating group in the side chain of a benzene ring containing various substituents. The shifts in the half-wave potential for the reduction of the metal are linear functions of substituent constants.[8] An example of such a correlation is the relation of the half-wave potential of copper, reduced from its complexes with substituted salicylaldehyde anils, to σ_X values (Fig. III-28). For these reversible systems the half-wave potentials are proportional to the logarithms of complex-stability constants. Hence, the observed plot corresponds actually to a linear relation between the logarithms of the stability constants and the substituent constants σ_X. The observed linear free energy relation indicates the mobility of the electrons in the chelate ring.

13. Applications of (27) and Related Equations

In Section 6 of Chapter II the importance and principal applications of equation (20) and its special forms were briefly summarized. Such relations can be used in the study of mechanisms and the course of electrode reactions and in structural studies.

The application of linear free energy relations in studies of the mechanism of electrode processes has been mentioned several times in the preceding sections. It has been shown how equation (27) can be used to detect deviations from the mechanism in general operation in the reaction series. Changes in $dE_{1/2}/d$pH, transfer coefficient α, number of electrons transferred n, or the electrolysis product can be detected. Comparison with equilibrium or rate constants, or other physical quantities allows us to distinguish polarographic phenomena from extraordinary substituent effects. It also has been shown how it is possible to make deductions (from the sign of the reaction constant) concerning the type of mechanism by which the electrode process proper goes. In some cases, it is possible with the use of the constants σ_X^- to detect mesomeric interaction between the electroactive group and the substituent and to detect and evaluate other mesomeric and steric effects. In some cases, it is possible to determine the degree of ionization of the substituent in the electrode process. Most of the information obtained in this way should not be taken as final, but rather as a guide for more detailed study.

In this section the order in which two electroactive groups are reduced, the prediction of half-wave potentials, the determination of potential ranges for monosubstituted benzene compounds, applications in structural studies, analytical applications, and the planning of future work will be discussed.

Two Electroactive Groups. Sometimes it is possible to determine which of the groups R^1 or R^2 is reduced at the more positive potential in the molecule R^1R^2-Y-X. To decide whether R^1 or R^2 is electroactive in the more positive step, and, therefore, which of these two groups (now written as X^2 or X^1) affects the reduction in the capacity of a substituent, equation (27) can be applied. First, group 1 is treated as the polarographically active group characterized by the reaction constant ρ_{π,R^1} and group 2 as a substituent to which constant σ_{X_2} applies. Second, group 2 is assumed to be electroactive (ρ_{π,R_2}) and group 1 to be the substituent (σ_{X_1}). It is then necessary to decide which of the two relations is fitted most satisfactorily by the observed values of $\Delta E_{1/2}$.

The application of such relations can be demonstrated for the case of the reduction of *m*-nitrobenzaldehyde. This substance is reduced at pH 7 in two steps.[1] The first wave corresponds to the reduction of the nitro group to the corresponding hydroxylamino derivative. It was necessary to decide whether it is the NHOH or the CHO group that is reduced in the second wave. It was established that the shift in the half-wave potential $\Delta E_{1/2}$ corresponds to $\rho_{\pi,CHO}\sigma_{3-NHOH}$ and not to $\rho_{\pi,NHOH}\sigma_{3-CHO}$. Hence, it can be concluded that the intermediate $3\text{-NHOH} \cdot C_6H_4CHO$ is reduced with formation of $3\text{-NHOH} \cdot C_6H_4CH_2OH$. The reduction of the latter occurs at such high negative potentials that its wave is overlapped by the reduction of the supporting electrolyte.

On the other hand, three waves are observed for *p*-nitrobenzaldehyde under identical conditions. The half-wave potential of the third, most negative, wave fits the $\Delta E_{1/2}$ versus $\sigma_{4\text{-}NH_2}$ plot (with $\rho_{\pi,CHO}$) but not $\Delta E_{1/2}$ versus $\sigma_{4\text{-}CH_2OH}$ (with $\rho_{\pi,NHOH}$). The third wave thus corresponds to the reduction of $4\text{-}NH_2 \cdot C_6H_4CHO$. The amino group must be formed in the second step, which corresponds to the reduction of $4\text{-NHOH} \cdot C_6H_4CHO$ to $4\text{-}NH_2 \cdot C_6H_4CHO$. Thus, the intermediate formation of $4\text{-NHOH} \cdot C_6H_4CH_2OH$ could not be established in this case.

The detection of the reduced group is possible even for molecules containing two identical polarographically active groups R and a substituent X in an unsymmetrical arrangement. As in the above examples, we try to decide which value of the substituent constant σ_X

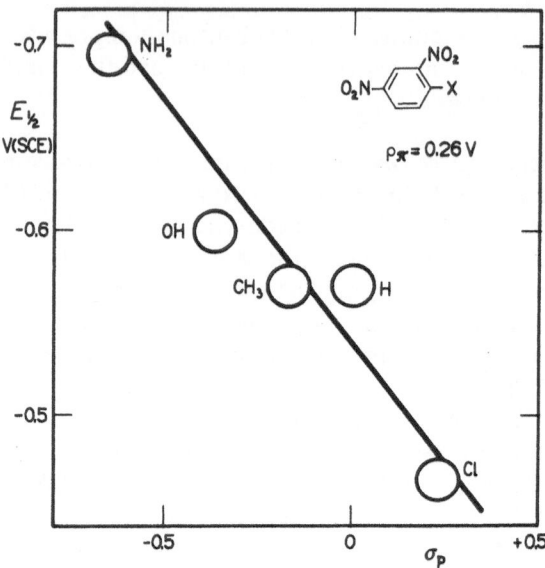

Fig. III-29. Relation of half-wave potentials for the reduction of substituted *m*-dinitro-benzenes to the Hammett substituent constant $\sigma_{p\text{-}X}$. The first wave at pH 11 in 20% methanol was measured.

gives a better fit to a linear plot of $\Delta E_{1/2}$ versus σ_X. Thus, for mono-substituted *p*-dinitrobenzenes a substituent in the benzene ring is in the *meta* position with respect to one nitro group, but in *ortho* position with respect to the other. Hence, for the first wave of 2-methoxy-1,4-dinitrobenzene the shift $\Delta E_{1/2}$ in the half-wave potential relative to that of unsubstituted p-dinitrobenzene fits the equation

$$\Delta E_{1/2} = \rho_{\pi,\mathrm{NO_2}} \sigma_{m\text{-}\mathrm{OCH_3}}$$

but not $\Delta E_{1/2} = \rho_{\pi,\mathrm{NO_2}} \sigma_{o\text{-}\mathrm{OCH_3}}$. It can be deduced that the nitro group in *meta* position relative to the methoxy group is reduced first. The product is *N*-(3-methoxy-4-nitrophenyl) hydroxylamine.

Similarly, the half-wave potentials of 1,3-dinitrobenzenes[111] substituted in position 4, fit the equation $\Delta E_{1/2} = \rho_{\pi,\mathrm{NO_2}} \sigma_{p\text{-}X}$ (Fig. III-29). Since in the reduction of nitro groups considerable *ortho* effects have been observed and $\sigma_{o\text{-}X} \approx \sigma_{p\text{-}X}$ cannot be used here as an approximation, the reduction of the 3-nitro group in the *ortho* position relative to the substituent cannot be assumed to take place. Reduction of the nitro group in position 1 and the formation of a 3-(hydroxyamino) 6-substituted nitrobenzene is assumed.

Prediction of Half-Wave Potentials. There are four possible methods of predicting the value of a half-wave potential of a substance that has not been studied.

If values of the reaction constant and of the half-wave potential of the unsubstituted parent compound have been given in the literature for the medium concerned, knowing the substituent constant we can determine the shift in the half-wave potential using equation (27).

Otherwise, the reaction constant must be determined or estimated. For the determination of a reaction constant the measurement of four or five half-wave potentials for properly chosen substances is usually sufficient. *Meta* substituents and *para* substituents such that no considerable mesomeric interaction with the electroactive group can be expected should be preferred. The substituents chosen for the determination of the reaction constant $\rho_{\pi,R}$ should cover the broadest possible range of σ_X values and half-wave potentials. The substituents also must fulfill the condition, characteristic for polarography, that the substituent must not be reduced at a more positive potential than the electroactive group R.

A second method of predicting half-wave potentials does not depend on the use of tabulated values of substituent constants. It amounts to the determination of substituent constants from polarographic measurements. Equation (27) is transformed into

$$(E_{1/2})_{1,X} - (E_{1/2})_{1,H} = \rho'_\pi[(E_{1/2})_{2,X} - (E_{1/2})_{2,H}] \qquad (49)$$

The validity of such an expression is referred to in Section 8 of this chapter (see Fig. III-11). We can, thus, choose a reference series (e.g., ArCHO) and measure 5–10 half-wave potentials of members of this series (subscript 1). Then, after determining 3–5 values of half-wave potentials for another reaction series (subscript 2) we may predict the approximate values of the half-wave potentials of even these compounds of the second series for which the value of the substituent constant is not available. The choice of standard substance for the determination of ρ'_π is governed by the same rules as for the determination of $\rho_{\pi,R}$.

Finally, it is possible to estimate the value of the reaction constant $\rho_{\pi,R}$ from its relation to the half-wave potential of the unsubstituted compound [equation (44), Figs. III-20 and III-21]. At a given pH it is possible to estimate the approximate value of the reaction constant $\rho_{\pi,R}$ from the measurement of the half-wave potential of any member of the reaction series. This enables us to estimate the approximate value of the half-wave potential of the unsubstituted compound and, therefore, of any member of the reaction series. This kind of application can be demonstrated by an example.

We suppose that for a hypothetical substance p-CH_3—C_6H_4—R a half-wave potential of -1.05 V was obtained at pH 7. From the graph in Fig. III-20 we can estimate the approximate value for the reaction constant, namely $\rho_{\pi,R} \approx 0.25$ V. Since $\sigma_{p\text{-}CH_3} = -0.17$, $\rho_{\pi,R}$ $\sigma_{p\text{-}CH_3} \approx -0.04$ V and $(E_{1/2})_H \approx -1.01$ V. Then it is possible to estimate approximately the half-wave potentials of, e.g.,

$$p\text{-I-}C_6H_4\text{-R}[E_{1/2} \approx -0.94 \text{ V}(\rho_{\pi,R} \cdot \sigma_{p\text{-I}} = +0.07 \text{ V})];$$

similarly, for p-NH_2—C_6H_4—R we can look for a value near

$$E_{1/2} \approx -1.17 \text{ V}(\rho_{\pi,R} \cdot \sigma_{p\text{-}NH_2} = -0.165 \text{ V});$$

and so on.

It is, thus, possible from one measurement of half-wave potential to estimate the approximate half-wave potentials of 30–50 further derivatives substituted in the *meta* or *para* position.

The values are approximate only, and the treatment can be used only for substances that are reduced by the same mechanism. The accuracy depends on the numerical value of the reaction constant. The probable error to be expected in the estimation of half-wave potentials when $\rho_{\pi,R}$ or ρ'_π is determined by the use of reference compounds is 5 to 20 mV. When not only the half-wave potential, but also $\rho_{\pi,R}$ and E_H are estimated, accuracy is obviously lower. It depends on the value of E_H (and hence of $\rho_{\pi,R}$) and on the nature of the substituent.* Thus, for pH < 3 for reaction series in which $E_H < -0.3$ V the probable error is 5–10 mV, for substances for which $E_H \approx 0.5$ V containing substituents with low substituent constants (e.g., alkyls, p-$NHCOCH_3$, SCH_3) predictions can be made accurately within about 10 mV, while for other substituents the probable error may be 20–30 mV. For substances that are reducible in the medium pH range and have $E_H \approx 1.5$ V the probable error in the case of substituents with high substituent constants [e.g., p-CN, p-$N(CH_3)_2$] may be as great as 50–70 mV.

Ranges of Potentials. For monosubstituted benzene compounds containing various polarographically active groups it is possible to determine [using equation (27a)] ranges of potentials in which the

* This is found to be true even in cases in which the value of $\rho_{\pi,R}$ is determined.

Table III-14
Ranges of Half-Wave Potentials of Reducation Waves of Mono-substituted Benzene Compounds with the Electroactive Group in the Side Chain

Solutions of pH < 3	$E_{1/2}$		E_H
	From	To	
$ArAsO_3H_2$	−1.56	−0.92	−1.27
$ArCOCH_3$	−1.23	−0.72	−1.00
$ArCOAr$	−1.19	−0.33	−0.80
$ArCHO$	−1.12	−0.37	−0.78
$ArC(NOH)CH_3$	−0.97	−0.55	−0.78
$ArC(NOH)Ar$	−0.83	−0.18	−0.54
$ArCSAr$	−0.79	−0.07	−0.46
$ArNO_2$	−0.46	−0.09	−0.29
$Ar\overset{+}{I}Ar$	−0.22	−0.17	−0.20
$ArN=NAr$	−0.09	−0.07	−0.08
Solutions of pH > 11			
$ArCl$	−2.82[a]	−2.40	−2.62[a]
$ArC(NOH)CH_3$	−2.09	−1.49	−1.82
ArI	−1.99	−1.35	−1.68
$ArC(NOH)Ar$	−1.90	−1.23	−1.60
$ArCOCH_3$	−1.81	−1.30	−1.58
$ArCHO$	−1.75	−0.98	−1.40
Solutions of pH 5–8 and unbuffered solutions			
$ArCH=CHAr$	−2.43[a]	−1.89	−2.17
$ArCOOCH_3$	−3.00[a]	−1.03	−2.11
ArI	−1.99	−1.35	−1.68
$ArCOCH_3$	−1.95	−1.14	−1.58
$ArCHO$	−1.73	−1.13	−1.46
$ArCH=CHCOCH_3$	−1.67	−1.02	−1.37
$ArCOAr$	−1.41	−0.81	−1.14
$ArN=NAr$	−0.74	−0.27	−0.51
$ArNO_2$	−0.65	−0.31	−0.50
$ArNO$	−0.28	−0.10	−0.20
$Ar\overset{|}{I}Ar$	−0.22	−0.17	−0.20

[a] Extrapolated.

waves due to the given electroactive group can be expected. The computation was carried out with the use of selected values of $\rho_{\pi,R}$ (Table III-13) and the corresponding values of the half-wave potential of the unsubstituted parent compound (E_H). The limiting values which

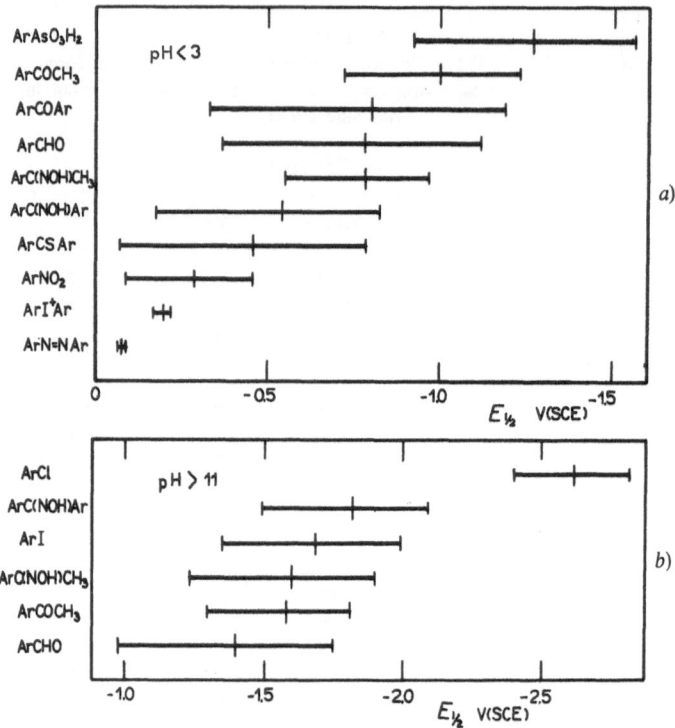

Fig. III-30. Potential ranges in which the half-wave potentials for the reduction of benzene derivatives can be expected for compounds reduced by the same mechanism as the parent unsubstituted compound. a) pH < 3; b) pH > 11.

the half-wave potential reach in the given reaction series (if a change in reaction mechanism is not involved) were calculated from the product $\rho_{\pi,R}\sigma_{\text{X-max}}$. The limiting values of $\sigma_{\text{X-max}}$ were chosen as $\sigma_{p\text{-N(CH}_3)_2} = -1.05$ and $\sigma_{p\text{-NO}_2}^{-} = +1.27$. For compounds like iodobenzenes, for which the application of σ_{X}^{-} is not suitable, the value of $\sigma_{p\text{-N(CH}_3)_3}^{+} = 1.11$ was used. Computed values are summarized in Table III-14 and in Figs. III-30 and III-31. Reaction series are arranged according to the half-wave potential of the unsubstituted compound E_{H}, denoted by a stroke on the abscissa limiting the range of potentials.

It would be possible to calculate similar ranges for di- and poly-substituted benzene derivatives. In the calculation of the limiting values it would then be useful to restrict ourselves to the most frequently encountered substituents.

Structural Studies. From the above it can be easily understood that the nature of a substituent present in a benzene compound containing a known electroactive group might be determined by adopting a procedure opposite to that suggested for the prediction of the half-wave potential. Using the measured half-wave potential and the determined or estimated reaction constant, we may determine the value of the substituent constant numerically or graphically. However, such a problem rarely arises for simple benzene compounds. A more useful application of such a procedure lies in the determination of the position of a substituent in the product of a chemical reaction.

When an electroactive group attached to a benzene ring is to be identified, an examination of the character of polarographic curves and the complex of properties known as polarographic behavior usually affords the best means of identification. Even though identification based on the determination of $\rho_{\pi,R}$ would be time-consuming and inaccurate, nevertheless, linear free energy relations can yield general information in a short time. By the use of the graphs in Figs. III-30 and III-31 it is possible to exclude several types of benzene compounds to which the wave cannot correspond from a single measurement of

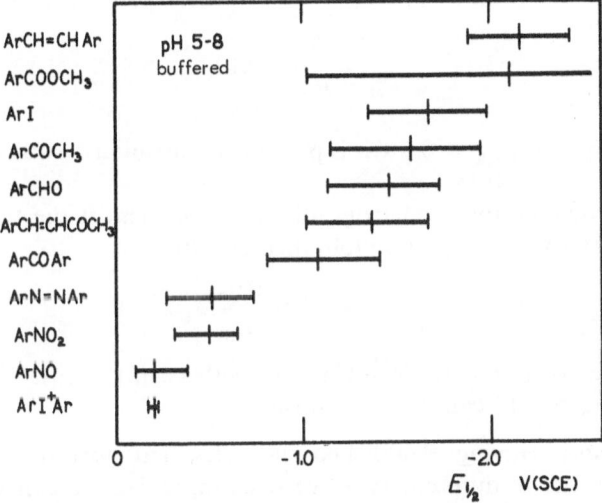

Fig. III-31. Potential ranges in which the half-wave potentials for the reduction of benzene derivatives can be expected for compounds reduced by the same mechanism as the parent unsubstituted compound. For values obtained at pH 5–8 and in unbuffered media (for pH-independent systems).

half-wave potential for a monosubstituted (or unsubstituted) compound. For example, a wave with a half-wave potential of -1.1 V recorded at pH 2.0 cannot correspond to the reduction of any compounds, nitrobenzenes, diphenyliodonium salts, and thiobenzophenones; it is improbable that the substance studied is a benzophenone oxime or acetophenone oxime, and the probability that it is a benzaldehyde or benzenearsonic acid derivative is low.

Conditions which must be met in such applications are that the reaction mechanism must be substantially the same for the substance studied as for the unsubstituted benzene derivative and that no substantial positive or negative *ortho* effects must be involved. Most of the half-wave potentials that are affected by the *ortho* effects of the common substituents lie in the predicted potential range.

Analytical Applications. Finally, linear free energy relations can be used in polarographic analysis for the prediction of the possibility of separating two analogous or isomeric compounds.[26]

The smallest difference in half-wave potential which will allow two waves to be distinguished is about 0.1 V.* To obtain two separated waves of the substances M and N characterized by the substituent constants σ_M and σ_N, respectively, the difference $\sigma_M - \sigma_N$ must fulfill the relation

$$\sigma_M - \sigma_N > \frac{0.10}{\rho_{\pi,R}} \tag{50}$$

For example, for $\rho_{\pi,R} = 0.3$ V it is possible to distinguish substances for which $\sigma_M - \sigma_N > 0.33$.

For oscillographic polarography it is sufficient if the two peaks differ by 0.04 V, and equation (50) then becomes

$$\sigma_M - \sigma_N > \frac{0.04}{\rho_{\pi,R}} \tag{50a}$$

and, hence, for $\rho_{\pi,R} = 0.3$ V substances with peak potentials differing by $\sigma_M - \sigma_N > 0.13$ can be distinguished.

Planning. Having established the rules that govern the effects of substituents in different types of reaction series, we can select by the use of tabulated substituent constants the lowest number of appropriate derivatives which will characterize the reaction series adequately. Substituents should be selected whose constants are

* This quantity depends on the slopes of the two waves concerned.

known and vary over a sufficiently wide range. An understanding of these principles makes it possible to conduct polarographic structural studies more efficiently and economically.

References

[1] F. S. Schultz, *Iowa State Coll. J. Sci.* **26**: 280 (1952).

[2] R. W. Brockman and D. E. Pearson, *J. Am. Chem. Soc.* **74**: 4128 (1952).

[3] E. Imoto and R. Motoyama, *J. Chem. Soc. Japan, Ind. Chem. Sect.* **55**: 384 (1952).

[4] S. Koide and R. Motoyama, *J. Electrochem. Soc. Japan*, **20**: 314 (1952).

[5] H. H. Jaffé, *Chem. Rev.* **53**: 191 (1953).

[6] P. Zuman, *Chem. Listy* **47**: 1234 (1953).

[7] P. Zuman, *Chem. Listy* **48**: 94 (1954).

[8] P. Zuman, *Chem. Zvesti* **8**: 939 (1954).

[9] Z. Grabowski, Roczniki Chem. **28**: 513 (1954).

[10] E. T. Bartel and Z. R. Grabowski, *Prace Konf. Polarog. Warsaw* 1956, 323.

[11] G. Sartori, P. Silvestroni, and C. Calzolari, *Ricerca Sci.* **24**: 1471 (1954).

[12] C. Calzolari and C. Furlani, *Ann. Chim.* **44**: 356 (1954), *Bull. Sci. Fac. Chim. Ind.* (Bologna) **12**: 14 (1959).

[13] G. Giacometti, *Rend. Accad. Naz. Lincei* (8) **18**: 185 (1955).

[14] J. Tirouflet, *Bull. Soc. Chim. France* **1956**, 274.

[15] E. L. Colichman and S. K. Liu, *J. Am. Chem. Soc.* **76**: 913 (1954).

[16] H. E. Bachofner, F. M. Beringer, and L. Meites, *J. Am. Chem. Soc.* **80**: 4274 (1958).

[17] M. Uehara and S. Ono, *Bull. Univ. Osaka Prefect* **9A**: 83 (1960).

[18] T. Iijima and M. Sekido, *J. Soc. Dyers Colourists* **76**: 354 (1960).

[19] M. Kožený and V. Velich, *Collection Czech. Chem. Commun.* **25**: 1031 (1960).

[20] Y. Nagata, I. Tachi, and K. Kitao, *Collection Czech. Chem. Commun.* **25**: 3271 (1960).

[21] G. Klopman, *Hèlv. Chim. Acta* **44**: 1908 (1961).

[22] G. L. K. Hoh, W. E. McEwen, and J. Kleinberg, *J. Am. Chem. Soc.* **83**: 3949 (1961).

[23] J. Boichard, D. Sc. Thesis, Université de Dijon, 1962.

[24] D. Vlachová, R. Zahradnik, K. Antoš, P. Kristián, and A. Hulka, *Collection Czech. Chem. Commun.* **27**: 2826 (1962).

[25] H. Berg, *Chem. Tech.* **8**: 5 (1956).

[26] J. Tirouflet and R. Dabard, *Ricerca Sci.* **29**: Suppl. 211 (1959) (Contributi teor. sper. polarografia, Vol. IV).

[27] H. W. Nürnberg, *Angew. Chem.* **72**: 433 (1960).

[28] Ju. P. Kitajev and G. K. Budnikov, *Uspekhl Khlm.* **31**: 670 (1962).

[29] E. Imoto, *Rev. Polarog.* **9**: 185 (1961).

[30] J. P. Stradinš, "Polarography of Organic Nitro Compounds" (in Russian) Izdat. Akad. Nauk. Latv. SSR, 167 pp.

[31] P. Zuman, D. Sc. Thesis, Czechoslovak Academy of Science, Prague, 1959.

[32] P. Zuman, *Collection Czech. Chem. Commun.* **25**: 3225 (1960).

[33] P. Zuman, *Chem. Listy* **54**: 1244 (1960); *J. Polarog. Soc.* **7**: 66 (1961).

[34] P. Zuman, *Ricerca Sci.* **30**: Suppl. p. 229 (1960). (Contributi teor. sper. polarografia, Vol. V).

[35] P. Zuman, *Chem. Listy* **55**: 261 (1961). *Z. Chem.* (Leipzig) **3**: 161 (1963).

[36] P. Zuman, *Current Trends in the Study of the Influence of Structure in Polarographic Behaviour of Organic Substances* in "Progress in Polarography" (ed. P. Zuman, I. M. Kolthoff) Vol. I. p. 319, Interscience. Publ., John Wiley & Sons, New York, 1962.

[37] P. Zuman, *Transactions of the Conference on Applications of Linear Relationships in Organic Chemistry, Tartu, 1962*, Vol. 2, 3 (1964).

[38] P. Zuman, *Rev. Polarg.* **11**: 102 (1963).

[39] P. Souchay, *Proceedings of the Sixth Meeting C.I.T.C.E. Poitiers 1954*, p. 425, Butterworth, London, 1955.

[40] R. W. Taft Jr., *Separation of Polar, Steric and Resonance Effects in Reactivity* in "Steric Effects in Organic Chemistry," (ed. M. S. Newman), John Wiley & Sons, New York, 1956.

[41] H. C. Brown and Y. Okamoto, *J. Am. Chem. Soc.* **80**: 4979 (1958).

[42] R. W. Taft Jr. and I. C. Lewis, *J. Am. Chem. Soc.* **80**: 2436 (1958).

[43] O. Exner, private communication.

[44] P. J. Bray and R. G. Barnes, *J. Chem. Phys.* **27**: 551 (1957).

[45] A. V. Willi and J. F. Stocker, *Helv. Chim. Acta* **38**: 1279 (1955).

[46] T. Iijima and M. Sekido, private communication.

[47] D. M. Coulson, W. R. Crowell, and S. K. Tendrick, *J. Am. Chem. Soc.* **79**: 1354 (1957).

[48] L. Holleck and H. Marsen, *Z. Elektrochem.* **57**: 944 (1953).

[49] E. Gergely and T. Iredale, *J. Chem. Soc.* **1953**, 3226.

[50] R. M. Powers and R. A. Day Jr., *J. Am. Chem. Soc.* **80**: 808 (1958).

[51] M. Kuna and M. J. Kopac, *Ann. N. Y. Acad. Sci.* **58**: Art. 3, 261 (1954).

[52] C. Calzolari, A. Donda and C. Furlani, *Public Univ. Trieste-Istituto di Merceologia* No. 7, (1957).

[53] G. Sartori and C. Furlani, *Ann. Chim.* **44**: 95 (1945).

[54] H. Sato, *Bull. Natl. Hyg. Lab.* **75**: 58 (1957).

[55] J. Nakaya, H. Inosito, and S. Ono, *J. Chem. Soc. Japan Pure Chem. Sect.* **78**: 935, 940 (1957) *Ref. Zh. Khim.* **1958**, No. 70257.

[56] I. Bergman and J. C. James, *Trans. Faraday Soc.* **48**: 956 (1952).

[57] E. Gergely and T. Iredale, *J. Chem. Soc.* **1951**, 3502.

[58] M. Runner and E. C. Wagner, *J. Am. Chem. Soc.* **74**: 2529 (1952).

[59] J. Tirouflet and J. P. Chané, *Compt. Rend.* **243**: 500 (1956).

[60] J. E. Page, J. W. Smith, and J. C. Waller, *J. Phys. Colloid. Chem.* **53**: 545 (1949).

[61] L. Holleck and R. Schindler, *Z. Elektrochem.* **60**: 1138, 1142 (1956).

[62] L. Moelants and R. Janssen, *Bull. Soc. Chim. Belg.* **66**: 209 (1957).

[63] I. F. Vladimircev and I. Ja. Postovskij, *Dokl. Akad. Nauk SSSR* **83**: 855 (1952).

[64] F. Goulden and F. L. Warren, *Biochem. J.* **42**: 420 (1948).

[65] E. Gergely and T. Iredale, *J. Chem. Soc.* **1951**, 13.

[66] E. S. Levin and Z. I. Fodiman, *Zh. Fiz. Khim.* **28**: 601 (1954).

[67] R. Zahradník, *Collection Chem. Czech. Commun.* **27**: 525 (1962).

[68] B. Breyer, *Ber.* **71**: 163 (1938).

[69] M. Maruyama and T. Furuya, *Bull. Chem. Soc. Japan* **30**: 657 (1957).

[70] J. C. James, *Trans. Faraday Soc.* **47**: 1240 (1951).

[71] K. R. Voronova and A. G. Stromberg, *Zh. Obshch. Khim.* **31**: 2786 (1961).

[72] H. Lund, *Acta Chem. Scand.* **11**: 491 (1957).

[73] R. A. Nash, D. M. Skauen, and W. C. Purdy, *J. Am. Pharm. Assoc. Sci. Ed.* **47**: 433 (1958).

[74] T. Kuwana, D. E. Bublitz, and G. Hoh, *J. Am. Chem. Soc.* **82**: 5811 (1960).

[75] G. Klopman, *Helv. Chim. Acta* **44**: 1908 (1961).

[76] J. K. Kochi, *J. Am. Chem. Soc.* **77**: 3208 (1955).

[77] E. L. Colichman and D. L. Love, *J. Am. Chem. Soc.* **75**: 5736 (1953).

[78] L. P. Hammett, "Physical Organic Chemistry," p. 184, McGraw-Hill, New York, 1940.

[79] J. E. Leffler, *J. Org. Chem.* **20**: 1202 (1955).

[80] O. Exner, *Collection Czech. Chem. Commun.* **29**: 1094 (1964).

[81] H. Sato, *Japan Analyst* **6**: 551 (1957).

[82] V. Bellavita, A. Ricci, and N. Fedi, *Ann. Chim.* **46**: 283 (1956).

[83] R. Ralea and M. A. Petrovanu, *Rev. Chim. Acad. Rep. Populaire Roumaine* **1**: 69 (1956).

[84] K. Schwabe, *Chem. Tech.* (Berlin) **9**: 129 (1957).

[85] L. Holleck and B. Kastening, *Naturwissenschatten* **43**: 298 (1956).

[86] M. Runner, *J. Am. Chem. Soc.* **74**: 3567 (1952).

[87] M. E. Runner, M. L. Kilpatrick, and E. C. Wagner, *J. Am. Chem. Soc.* **69**: 1406 (1947).

[88] H. Sato, *Bull. Natl. Hyg. Lab.* **75**: 52 (1957).

[89] V. Bellavita, N. Fedi, and N. Gagnoli, *Ricerca Sci.* **25**: 504 (1955).

[90] T. Sasaki, *Pharm. Bull.* (Tokyo) **2**: 104 (1954).

[91] Ch. Prévost, P. Souchay, and Ch. Malen, *Bull. Soc. Chim. France 1953*, 78.

[92] R. A. Day Jr. and R. M. Powers, *J. Am. Chem. Soc.* **76**: 3085 (1954).

[93] M. Fields, C. Valle, and M. Kane, *J. Am. Chem. Soc.* **71**: 421 (1949).

[94] H. Schmidt and L. Holleck, *Z. Elektrochem.* **59**: 531 (1955).

[95] J. E. Page, "Proceedings of the First International Polarography Congress," Part I, p. 193, Prague, 1951, Přírodověd. nakl. Prague, 1952.

[96] S. F. Dennis, A. S. Powell, and M. J. Astle, *J. Am. Chem. Soc.* **71**: 1484 (1949).

[97] G. Klopman, *Helv. Chim. Acta* **44**: 1908 (1961).

[98] J. K. Kochi, *J. Am. Chem. Soc.* **77**: 3208 (1955).

[99] S. G. Majranovskij, *J. Electroanal. Chem.* **4**: 166 (1962).

[100] L. Holleck and D. Becher, *J. Electroanal. Chem.* **4**: 321 (1962).

[101] J. Hine, *J. Am. Chem. Soc.* **81**: 1126 (1959).

[102] E. Imoto and R. Motoyama, *J. Chem. Soc. Japan, Ind. Chem. Sect.* **55**: 384 (1952).

[103] C. E. Bennett and P. J. Elving, *Collection Czech. Chem. Commun.* **25**: 3213 (1960).

[104] T. Iredale, *Nature* **177**: 36 (1956).

[105] N. Fuson, M. L. Josien, and E. M. Shelton, *J. Am. Chem. Soc.* **76**: 2526 (1954).

[106] S. V. Gorbačev and S. F. Belevskij, *Zh. Fiz. Khim.* **31**: 1656 (1957).

[107] Z. R. Grabowski, *Zh. Fiz. Khim.* **33**: 728 (1959).

[108] J. Komenda and L. Kišová, *Chem. Zvesti* **16**: 368 (1962).

[109] J. Komenda and J. Tirouflet, *Compt. Rend.* **254**: 3093 (1962).

[110] J. Tirouflet, E. Laviron, R. Dabard, and J. Komenda, *Bull. Soc. Chim. France* **1963**, 857.

[111] L. Holleck and H. J. Exner, *Z. Elektrochem.* **56**: 677 (1952).

IV: Monocyclic Heterocyclic Compounds

1. General

The polarographic reduction of a group in the side chain of a heterocyclic ring usually follows a path analogous to that followed by the corresponding benzene compound. Most deviations from this rule have been observed in the pyridine series (e.g., for aldehydes,[1-9] acids,[10-12] nitriles,[13] and acetyl derivatives[9,14]). In these compounds hydrogenation of the pyridine ring may occur. Moreover, even in cases where the reduction has been proved to occur in the side chain the possibility that the pyridine ring is the first point of electron attack cannot be excluded.

As observed for most benzene compounds, for heterocyclic compounds either the value of the transfer coefficient α does not change markedly in a properly chosen reaction series (Table IV-1), or this coefficient is a linear function† of the substituent constant σ, as was demonstrated (Figs. IV-1 and IV-2) for the half-wave potentials of 3-pyridinecarboxylic acids[16] with a substituent in position 5 and of 5-substituted 2-nitrofurans.[15] Also, in such reaction series the course of the polarographic reduction remains constant. The basic conditions for the application of the general equation (20) thus are fulfilled.

† When substituent constants are not available for a sufficient number of derivatives, $\Delta E_{1/2}$ values (which are related linearly to σ) can also be used for correlation with the transfer coefficient (Fig. IV-2).

Table IV-1

Transfer Coefficients α of Some 5-Nitrofuran Derivatives (according to Ref. 15, expressed as product with the number of electrons transferred n)

Substituent in 5-nitrofuran	αn
H	1.1
2-Br	1.0
2-CHO	1.2
2-CH$_3$CO	1.2
2-COOC$_2$H$_5$	1.0
2-CH$_2$OH	1.0
2-CH$_2$OC$_2$H$_5$	1.0
2-CH=CHCHO	1.1

However, in addition to the influence of the juxtaposition of the reducible group and the substituent, which operates in the benzene series, the positions both of the reducible group and of the substituent with respect to the hetero atom have now to be considered.

Moreover, some heterocyclic rings are attacked by polarographic reduction and can be considered as one polarographically active

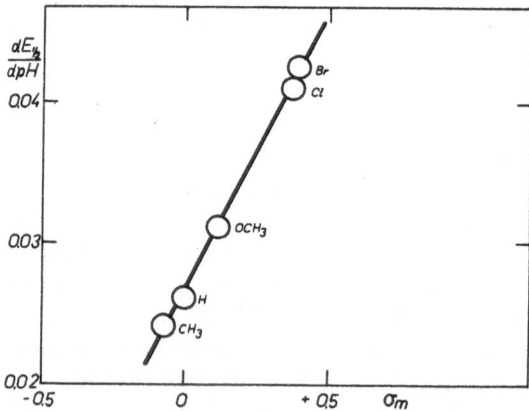

Fig. IV-1. Plot of the slope of the pH-dependence of half-wave potentials ($dE_{1/2}/d$pH) for the reduction of 5-substituted nicotinic acids against Hammett substituent constants $\sigma_{m\text{-}X}$. Reaction Series Nos. 31–35, Table IV-2, measured from graphs in the original paper.[16]

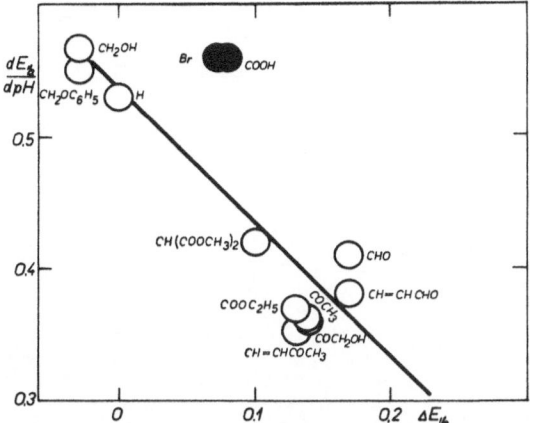

Fig. IV-2. Plot of the slope of the pH-dependence of half-wave potentials ($dE_{1/2}/dpH$) for the reduction of 5-substituted-2-nitrofurans against the shifts in the corresponding half-wave potentials (proportional to σ_X). Reaction Series No. 21. Full point deviates.

group attached to an electroinactive grouping, referred to as the substituent.

This diversity of possibilities expresses itself in a complicated choice of alternative reaction series, and, before discussing particular cases, we shall first suggest how these reactions series can be classified.

The problems met in the polarography of heterocyclic compounds have been treated recently in several review† articles.[7,8,15,21]

2. Classification of Reaction Series

For heterocyclic compounds reaction series can be classified according to whether reduction occurs in the heterocyclic ring (A) or in the side chain (B).

Under (A) the reducible heterocyclic ring is regarded as one polarographically active group, and three types of reaction series are defined:

(Aa) *Het*-X.‡ The reducible heterocycle is substituted in a fixed position by different substituents X.

(Ab) The reducible heterocyclic ring is substituted by the same substituent, but in different positions, as in 3- and 4-methylpyridazines. The reducible group varies (e.g., from 3- to 4-pyridazinyl), and the

† The review by H. H. Jaffé and H. L. Jones (*Advances in Heterocyclic Chemistry* **3**, 209, Academic Press, London 1964) arrived here after the manuscript had been finished.
‡ A reducible heterocyclic ring is denoted by *Het* and an unreducible by Het.

Table IV-2
Reaction Series of Heterocyclic Compounds Following the Linear Free Energy Relations

No.	Reaction series	Solution	Equation	$\rho_{x,R}$, V	$E_{1/2}$ (SCE)	n^a	Substituents Fitting	Substituents Nonfitting	Reference to source of $E_{1/2}$
			(Aa) Electroactive Heterocyclic Ring Carrying Substituents (Het-X)						
1	X–N⊕ / N–O⊖	Borate pH 9.3	29 het	+0.24	-1.51_5^c	4	$CH_3; C_4H_9; CH_2C_6H_5; C_6H_5^b$	C_6H_5	22
2	C_6H_5–N⊕–X / N–O⊖	Borate pH 9.3	29 het	+0.35	-1.36_5^c	4	$CH_3; H; C_6H_5; Cl$	—	23
3	X–N–CH$_2$CH$_2$–N⊕–X / N–O⊖	Borate pH 9.3	29 het	+0.23	-1.37_2^c	3	$CH_3; H; Cl$	—	23
4	X–N–CH$_2$CH$_2$–N–CH–X ⊕ H$_2$N COO⊖ / N–O⊖	Borate pH 9.3	29 het	+0.25	-1.64_7^c	3	$CH_3; H; Cl$	—	23
5	X–C≡CH / OC–O–CO	0.3 M LiCl, methanol-benzene (1:1)	29 het	+0.24	-0.84^c	3	$CH_3; CH_2COOC_4H_9^d; H$	—	24
5a	CH=CH / OC–N–X / OC	BR pH 7.25; 50% ethanol; 1st wave	29 het	+0.09	-0.73^c	9	$CH_3; C_2H_5; C_3H_7; i\text{-}C_3H_7;$ $C_4H_9; C_6H_5; CH_2C_6H_5;$ $CH_2CH_2C_6H_5; cyclo\text{-}C_6H_{11}$	—	24a
5b		BR pH 7.25; 50%; 2nd wave	29 het	+0.24	-1.22^c	9	$CH_3; C_2H_5; C_3H_7; i\text{-}C_3H_7;$ $C_4H_9; CH_2C_6H_5; CH_2CH_2C_6H_5$	$cyclo\text{-}C_6H_{11}$	24a
6	pyridine N⊕–X	0.1 N N(C$_2$H$_5$)$_4^+$ benzenesulfonate	29 het	+0.22	-1.41^c	3	$CH_3; C_2H_5; i\text{-}C_3H_7$	—	25

No.	Structure	Medium	n^a	ρ	n	$\sigma^{\ominus}$	Substituents		Ref.
7	[heterocyclic cation structure]	1 N HCl	38 het	+0.25	7	−0.87^c	[4-CH$_3$,2,6-(C$_2$H$_5$)$_2$]; [4-CH$_3$,2,6-(i-C$_3$H$_7$)$_2$]; [2,6-(CH$_3$)$_2$,6-C$_6$H$_5$]; [2-CH$_3$,4,6-(C$_6$H$_5$)$_2$]; [4-CH$_3$,2,6-(C$_6$H$_5$)$_2$]; 2,4,6-(C$_6$H$_5$)$_3$	2,4,6-(CH$_3$)$_3$; [2,6-(CH$_3$)$_2$, 4-C$_2$H$_5$]	26
8	[X–C(=N)CO–NH–NH structure]	BR buffer pH 8.3	29 het	+0.14	6	−1.41^c	CH$_3$; C$_2$H$_5$; C$_3$H$_7$; C$_5$H$_{11}$; H; C$_6$H$_5$	t-C$_4$H$_9$	27
9	[X–CH=C(CO–NH)(NH–CO) structure]	Borate pH 9.3^m	25 het	+0.40	3	+0.02^c	CH$_3$; H; COO$^-$	—	28
10	[X^1,X^2–C(CO–NH)(CO–NH) structure]	Borate pH 9.2 (i_1)^m	38 het	+0.07	4	+0.13^{c,e}	(CH$_3$, cyclo-C$_6$H$_{11}$); (C$_2$H$_5$)$_2$; (C$_2$H$_5$,C$_5$H$_{11}$); (C$_2$H$_5$,C$_6$H$_5$)	—	29

(Ac) Reducible Ring Carrying a Substituted Phenyl Group (Het-C$_6$H$_4$-X)

No.	Structure	Medium	n	ρ	n	$\sigma^{\ominus}$	Substituents		Ref.
11	[X–C$_6$H$_4$–N$^{\oplus}$(=N–O$^{\ominus}$)O$^{\ominus}$ ring structure]	Borate pH 9.3	27	+0.26	12	−1.23$_5$	H; 3-CH$_3$; 4-CH$_3$; 4-OH; 3-OCH$_3$; 4-OCOCH$_3$; 4-COO$^-$(σ_p); 3-Cl; 4-OCH$_3$; 4-Cl; 3-Br; 4-Br; 3,4-diCl	—	33

^{a} Number of substituted derivatives fitting given equation.
^{b} With a correction for resonance interaction.
^{c} Half-wave potential of the methyl derivative E_{CH_3}.
^{d} Approximate value $\sigma_{CH_2COOC_2H_5} = 0.3$ was usec.
^{e} Extrapolated value.
^{f} On assumption that $\delta_{\pi,R}$ is small.
^{g} These values were obtained with 3-nitrothiophene substituted in position 5. Other values are for 2-nitrothiophenes substituted in position 5.
^{h} $\sigma^{(-)}_{C=NOH}$ and $\sigma^{(-)}_{C(CH_3)=NOH}$ values from Table III-8.
^{j} From the graph in Ref. 16.
^{k} $\sigma_{pyrrole} = +0.17$ from the measurements of diaryl ketones.
^{m} Anodic waves corresponding to a mercury compound formation.

Table IV-2 (continued)

No.	Reaction series	Solution	Equation	$\rho_{x,R}$, V	E_{H_1} (SCE)	n^a	Substituents		Reference to source of $E_{1/2}$
							Fitting	Nonfitting	
12		Borate pH 9.3	27	+0.24	-1.14_5	3	H; CH₃; Cl	—	33
13		Borate pH 9.3	27	+0.19	-1.14_5	3	H; CH₃; Cl	—	33
14		Buffer pH 6.7	27	0.00	-0.35	6	H; NH₂; NHCOCH₃; NHSO₂C₆H₄NH₂; NHSO₂C₆H₄NHCOCH₃; CN	NO₂	34
15		Buffer pH 6.7	27	+0.12	-0.35	6	H; (4-NH₂,3-Cl); (4-NH₂,3-OH); 4-NHCOCH₃; 4-Cl; 4-C₆H₅	4-CH(CH₃)₂; 4-NH₂; 2-Cl^f	34
16		0.1 M LiCl, 0.01% gel.	27	+0.30	-1.02_4	7	H; 3-CH₃; 4-OCH₃; 4-NH₂; 3-NHCOCH₃; 4-NHCOCH₃; 2-naphthyl	3-NH₂	35

(Ba) Nonreducible Ring Carrying a Reducible Group and a Substituent in the Ring (X-Het-R)

No.	Reaction series	Solution	Equation	$\rho_{x,R}$, V	E_{H_1} (SCE)	n^a	Substituents		Reference to source of $E_{1/2}$
							Fitting	Nonfitting	
17		BR 0.1 N KCl; 50% ethanol; 0.01% gel. pH 2.9	27 het'	+0.31	-0.29	4	H; COOC₂H₅(σ_p); COCH₃(σ_p); Bt(σ_p)	—	36

No.	Structure	Conditions	Class	σ^+	σ^-	n	Substituents	(add.)	Ref.
18	X–[thiophene]–NO_2	pH 5.0	27 het'	+0.40	−0.48	4	H; $COOC_2H_5(\sigma_p)$; $COCH_3(\sigma_p)$; $Br(\sigma_p)$	—	36
19	X–[thiophene]–$COOCH_3$	pH 7.5$_5$	27 het'	+0.45	−0.63	4	H; $COOC_2H_5(\sigma_p)$; $COCH_3(\sigma_p)$; $Br(\sigma_p)$	—	36
20	X–[selenophene]–NO_2	Buffers pH 3.1; 10% ethanol	27 het'	+0.16	−0.30[e], −0.40[a,g]	10	$CH_3(\sigma_p)$; $COOH$[a](σ_m); CHO[a](σ_m); $CHO(\sigma_p)$; $COCH_3$[g](σ_m); $COCH_3(\sigma_p)$; CN[a](σ_m); $CN(\sigma_p)$; $I(\sigma_p)$	$COOC_2H_5(\sigma_p)$	38
21	X–[furan]–NO_2	Unbuffered; pH 6.7–7.0; f≠pH, 50% ethanol	27 het'	+0.75	−1.91	4	H; $5\text{-}CH_3(\sigma_p)$; $4\text{-}CH_3(\sigma_m)$; $5\text{-}COOCH_3(\sigma_p)$	—	42
21a		BR pH 0.0[e]	27 het'	+0.13	−0.06[e]	5	H; $COOH(\sigma_p)$; $COCH_3(\sigma_p)$; $CHO(\sigma_p^-)$; $NO_2(\sigma_p^-)$	—	15
22	X–[furan]–CHO	BR pH 0.0[e]	27 het'	+0.15	−0.04[e]	5	H; $COOH(\sigma_p)$; $COOC_2H_5(\sigma_p)$; $COCH_3(\sigma_p^-)$; $CHO(\sigma_p^-)$	Br	15
		BR pH 0.0[e]	27 het'	+0.26	−0.04[e]	5	H; Br; $COOH$; $COOC_2H_5$; $COCH_3$	CHO	15
		Buffers pH 1.0	27 het'	+0.22	−0.87	3	H; CH_3; $COOCH_3(\sigma_p^-)$	—	41
22a		Buffers pH 8.0	27 het'	+0.52	−1.37	3	H; CH_3; $COOCH_3(\sigma_p^-)$	—	41
23	X–[furan]–$COOCH_3$	Unbuffered; pH 6.7–7.0; f≠pH; 50% ethanol	27 het'	+0.67	−1.97	3	H; $COOCH_3(\sigma_p^-)$; $Br(\sigma_p)$	—	42
24	X–[pyrrole NH]–NO_2	Prideaux–Ward pH 4.2; 10% ethanol	27 het'	+0.29	−0.55	8	H; $CN(\sigma_p)$; $COOH(\sigma_p)$; $CH{=}NOH$[b]; $C(CH_3){=}NOH$[b]; $COCH_3(\sigma_p^-)$; $CHO(\sigma_p^-)$; $NO_2(\sigma_p^-)$	—	40
25	X–[pyrrole NH]–NO_2	0.1 N NaOH; 10% ethanol	27 het'	+0.21	−1.01	5	H; $COO^-(\sigma_p^-)$; $CH{=}NOH$[b]; $C(CH_3){=}NOH$[b]; $CHO(\sigma_p^-)$	CN	40
26	X–[pyrrole NH]–NO_2	Prideaux–Ward pH 4.2; 10% ethanol	27 het'	+0.44	−0.69	5	H; $COOH(\sigma_m)$; $CHO(\sigma_m)$; $COCH_3(\sigma_m)$; $C(CH_3){=}NOH$[b]	CN; $CH{=}NOH$; NO_2	40
27	X–[pyrrole NH]	0.1 N NaOH; 1C% ethanol	27 het'	+0.42	−1.06	5	H; $COOH(\sigma_m)$; $CHO(\sigma_m)$; $COCH_3(\sigma_m)$; $NO_2(\sigma_m)$	CN	40

Table IV-2 (*continued*)

No.	Reaction series	Solution	Equation	$\rho_{\pi,R}$, V	$E_{\frac{1}{2}}$, (SCE)	n^a	Substituents Fitting	Substituents Nonfitting	Reference to source of $E_{1/2}$
28	[structure: CH3, CHO, X, H3C, N–H]	BR, N(CH3)4+; 50% dioxane; pH 7.1	27 het'	+0.38	−1.96	4	H; CH3(σ_m); C2H5(σ_m); COOC2H5(σ_m)	CHO	43
29	[structure: CH3, CHO, X1, X2, N–H]	BR, N(CH3)4+; 50% dioxane; pH 7.1	27 het'	+0.34	−1.87^e	3	(4,5-diCH3); (4-CH3,5-COOC2H5); (4-COOC2H5,5-CH3) ($\sigma_m = \sigma_p$)	—	43
30	[structure: X2, Br, N, S, X1]	Alkaline; $f \neq$ pH	27 het'	+1.10	−1.47	4	H; 2-NH2(σ_p); 2-Cl(σ_p); (2-NH2(σ_p),4-CH3(σ_m))	—	44
31	[structure: X, COOH, N, S]	Buffers pH 1	27 het'	+0.43	−1.07^j	5	H; CH3(σ_m); OCH3(σ_m); Cl(σ_m); Br(σ_m)	—	16
32		pH 2	27 het'	+0.38^j	−1.11^j	5	H; CH3(σ_m); OCH3(σ_m); Cl(σ_m); Br(σ_m)	—	16
33		pH 3	27 het'	+0.29^j	−1.16^j	5	H; CH3(σ_m); OCH3(σ_m); Cl(σ_m); Br(σ_m)	—	16
34		pH 7	27 het'	+0.19	−1.21	5	H; CH3(σ_m); OCH3(σ_m); Cl(σ_m); Br(σ_m)	—	16
35		pH 8	27 het'	+0.13	−1.26	5	H; CH3(σ_m); OCH3(σ_m); Cl(σ_m); Br(σ_m)	—	16
(Bc)	Nonreducible Ring Carrying a Reducible Group and a Substituent in the Same Side Chain (Het-R-X)								
36	[structure: COOR, O]	Unbuffered; pH 6.7–7.0; $f \neq$ pH	29 het	+0.18	−1.97^e	3	CH3; C2H5; C6H5	—	42
37	[structure: COOR, S]	Unbuffered; pH 6.7–7.0; $f \neq$ pH	29 het	+0.17	−1.91^e	3	CH3; C2H5; C6H5	—	42
38	[structure: COR, N–H]	BR, N(CH3)4+; 50% dioxane; pH 7.4$_5$	29 het	+0.54	−1.58^e	3	CH3; C6H5; pyrrole^k	—	45

conditions for the application of the general equation (20) are not fulfilled. This case is, therefore, only mentioned briefly.

(Ac) Het-C_6H_4-$X(m,p)$. The reducible heterocyclic ring is substituted in a fixed position by a phenyl group with a substituent in the *meta* or *para* position with respect to the heterocyclic ring.

Under (B) (electroactive group in the side chain) three types of reaction series are considered:

(Ba) X-Het-R. The aromatic heterocyclic ring carries both a reducible group R in a fixed position relative to the hetero atom, and a polarographically inactive (in a given potential range) substituent X.

(Bb) Het-R. The unreducible heterocyclic ring carries only a reducible group R in a given position.

(Bc) Het-R-X. The aromatic ring carries one side chain containing both the reducible group R and the substituent X, which influences the reducibility of the group R.

The reaction series covered by special forms of the general equation (20) are given in Table IV-2.

3. (A) Reducible Heterocyclic Rings

(Aa) *The Role of Substituents in a Reducible Heterocyclic Ring*. In a system of type (Aa) (i.e., *Het*-X) the reducible group can be in a direct interaction with the substituent X. If bulkier substituents (with probable steric interaction) and substituents forming conjugated systems (with probable mesomeric interaction) are ignored for the moment, conditions are fulfilled for the shift in half-wave potential $(\Delta E_{1/2})$ to obey the modified Taft equation (25 het) or (29 het).

$$\Delta E_{1/2} = \rho^I_{\pi,\text{het}} \, \sigma^I_X \qquad (25 \text{ het})$$

$$\Delta E_{1/2} = \rho^*_{\pi,\text{het}} \, \sigma^*_{\text{CH}_2X} \qquad (29 \text{ het})$$

Certain complications can arise with polysubstituted compounds. When the positions into which the substituents are introduced are not equivalent it is necessary to determine the value of the reaction constant $\rho^*_{\pi,\text{het}}$ for each position separately. Hence, for polysubstituted compounds the shift in half-wave potential is given by

$$\Delta E_{1/2} = \sum \rho^{*i}_{\pi,\text{het}} \, \sigma^*_X \qquad (39 \text{ het})$$

Only for substituents in equivalent positions can the simpler equation (38 het) be used.

$$\Delta E_{1/2} = \rho^*_{\pi,\text{het}} \sum \sigma^*_X \qquad (38 \text{ het})$$

If mesomeric interaction between the heterocyclic ring *Het* and the substituent X takes place, the shift in half-wave potential is given by

$$\Delta E_{1/2} = \rho^*_{\pi,\text{het}} \, \sigma^*_{\text{CH}_2\text{X}} + M_\pi \qquad\qquad (33 \text{ het})$$

or by the analogous equation with the constants σ^i_X. Steric interactions could be treated quantitatively in a similar way, but in the absence of any such example at present, the corresponding equation is not cited.

In the above equations σ^i_X is the inductive substituent constant and $\sigma^*_{\text{CH}_2\text{X}}$ is the polar substituent constant, related to σ^i_X by the relation $\sigma^i_X = 0.45\sigma^*_{\text{CH}_2\text{X}}$ (Table III-1). The $\rho_{\pi,\text{het}}$ with an upper index is the reaction constant, expressing the susceptibility of the reduction of the heterocyclic ring to the effects of substituents in a given position. Its value depends on the ring type, on the position of the substituent, and on reaction conditions, but it is independent of the nature of the substituents.

Examples of the application of equation (25 het), (29 het), and (38 het) are given in Table IV-2. The shift, due to a phenyl group attached to a sydnone ring in position 3, was calculated by equation (33 het) to be $M_\pi = +0.10$ V toward more positive values (Fig. IV-3) and attributed to a mesomeric effect.[30] This explanation was supported by the effect of steric hindrance to coplanarity[30] [see Part (Ac) of this section]. Thus, whereas, the phenyl group in position 3 excites mesomeric interaction with the sydnone ring (seemingly due to the polarizibility of the nitrogen), the same group in position 4 influences the reduction of the N-O bond of the sydnone ring by a predominantly polar effect.

Deviations for the t-butyl derivatives in Reaction Series No. 8 (Fig. IV-4) are accompanied by change in the slopes of the logarithmic analysis and pH-dependence curves (Table IV-3).

The shift of about 30 mV toward more negative potentials and the deviations in α-value can be explained either by a steric effect on the reduction process or by a sterically influenced change in reduction mechanism.

The validity of the linear free energy relations has been also extended to anodic waves, corresponding to mercury salt formation.

Plotting of the regression line and determination of the value of the reaction constant $\rho^*_{\pi,\text{het}}$ usually present no difficulty. An exception is Reaction Series No. 7 (Fig. IV-5), containing polysubstituted pyrylium salts, the half-wave potentials of which were correlated using equation (38 het). The simple objection, namely that $\rho^i_{\pi,\text{het}}$ values differ in positions 2 and 4 and that equation (39 het) should be used instead

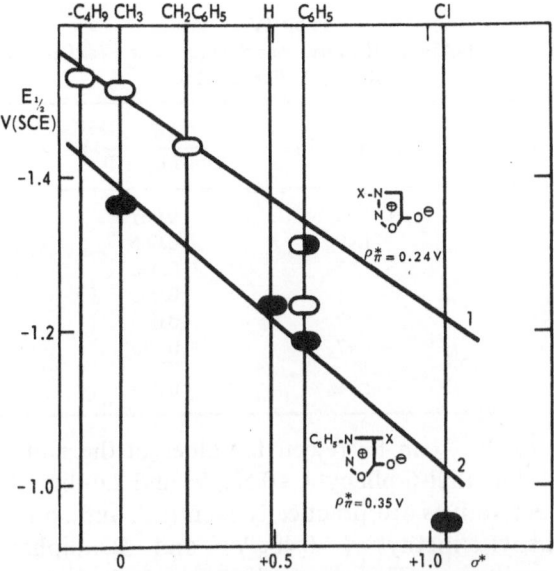

Fig. IV-3. Relation of half-wave potentials of substituted sydnones to Taft polar sub-stituent constants σ_X^*: 1) (circles) sydnones substituted in position 3 (Reaction Series No. 1, Table IV-2); 2) (full points) 3-phenylsydnones substituted in position 4 (Reaction Series No. 2). Halved point: Half-wave potential of 3-phenylsydnone, corrected for the contribution of mesomeric interaction determined from the study of the steric hindrance to coplanarity.

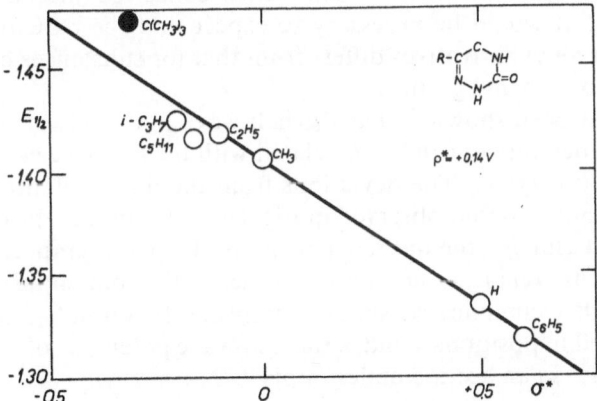

Fig. IV-4. Relation of half-wave potentials of 6-azauracil derivatives substituted in position 5 to Taft polar substituent constants σ_X^*. Reaction Series No. 8, Table IV-2. Full point deviates.

Table IV-3
Characteristics of 6-Azauracil Derivatives Substituted in Position 5 in a Britton–Robinson Buffer of pH 8.3

Substituent	$E_{1/2}$, V (SCE)	$\dfrac{\Delta E}{\Delta \log i/(i_d - i)}$, V	$\dfrac{dE_{1/2}}{d\mathrm{pH}}$, V
H	-1.33_6	0.067	0.051
CH_3	-1.40_8	0.066	0.050
C_2H_5	-1.41_9	0.066	0.047
C_5H_{11}	-1.41_7	0.068	0.050
$i\text{-}C_3H_7$	-1.42_6	0.065	0.050
$t\text{-}C_4H_9$	-1.47_5	0.080	0.037
C_6H_5	-1.31_9	0.069	0.050

of equation (38 het) can be rejected. Values of the half-wave potentials[26] of 2,4-dimethyl-6-phenyl (-0.58_0 V) and 2,6-dimethyl-4-phenyl (-0.57_7 V) derivatives are practically identical, and so are the values of 2,4-diphenyl-6-methyl (-0.40_8 V) and 2,6-diphenyl-4-methyl (-0.39_4 V) derivatives. This shows that $\rho_{\pi,\mathrm{het}}^2$ and $\rho_{\pi,\mathrm{het}}^4$ are practically equal and that the use of one value $\rho_{\pi,\mathrm{het}}$ in equation (38 het) provides a good approximation. Values for the deviating 2,4,6-trimethyl and 2,6-dimethyl-4-ethyl derivatives were unfortunately recorded under conditions which differed slightly from those used for the other compounds. If the half-wave potentials are comparable, the possibility cannot be excluded that in this series concerted mechanisms are operating. It would be necessary to expect that the reduction mechanism for alkyl derivatives differs from that for substances containing one or more phenyl groups.

It has been shown[26] that the half-wave potentials of substituted pyrylium derivatives can be correlated with the frequencies of infrared absorption maxima. The deviations from the linear relations[26] show a trend similar to that observed in Fig. IV-5. It thus, can, be concluded that the reason for the discrepancy lies in the polarographic data and that such discrepancies are not properties of the compounds.

Another complicated system comprises 1,2-dithiole-3-thiones (I) substituted in positions 4 and 5, the half-wave potentials of which were measured[31] in an acetate buffer of pH 4.7.

The half-wave potentials of the second waves of these substances show rather poor correlation when equation (38 het) is applied (Fig. IV-6). It was calculated that for 4-methyl derivatives having different substituents in

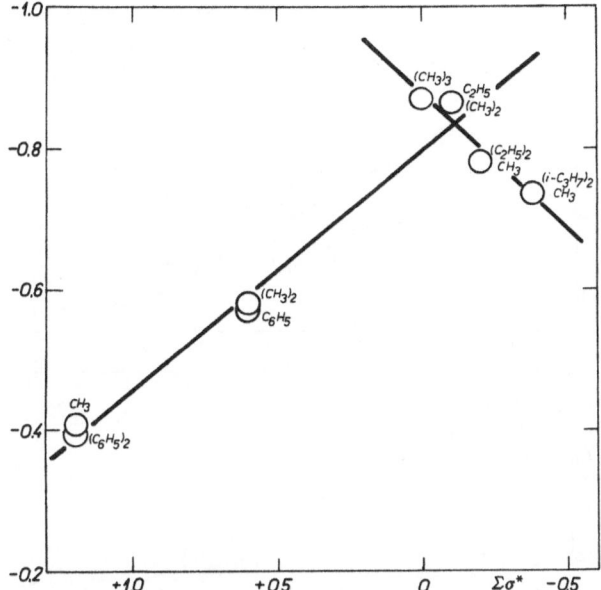

Fig. IV-5. Relation of half-wave potentials of 2,4,6-trisubstituted pyrylium salts to Taft polar substituent constants σ_X^*. Half-wave potentials from Ref. 26. Two points for dimethyl phenyl and methyl diphenyl derivatives correspond to possible position isomers. 2,4,6-Triphenylpyrylium also fits the line for other phenyl derivatives.

position 5, the half-wave potentials can be correlated using $\rho_{\pi,\text{het}}^5 = +0.39$ V, and for 5-phenyl derivatives having different substituents in position 4 they can be correlated using $\rho_{\pi,\text{het}}^4 = +0.09$. From equation (39 het) it was possible to calculate for polysubstituted compounds the shift

$$\Delta E_{1/2} = \rho_{\pi,\text{het}}^4 \, \sigma_{4\text{-X}} + \rho_{\pi,\text{het}}^5 \, \sigma_{5\text{-X}}.$$

This shift shows a good correlation with measured half-wave potentials (Fig. IV-6) with practically unit slope. The significance of this correlation is diminished by the fact that of six values in this graph five were used in the computation of $\rho_{\pi,\text{het}}^4$ and $\rho_{\pi,\text{het}}^5$.

Differences in the value of reaction constant $\rho_{\pi,\text{het}}^*$ for two successive electrode processes have been observed for N-substituted maleimides[24a] (Fig. IV-7): for the first wave the value $\rho_{\pi,\text{het}}^* = 0.09$ V, and for the second wave the value $\rho_{\pi,\text{het}}^* = 0.24$ V have been found (No. 5a and 5b, Table IV-2).

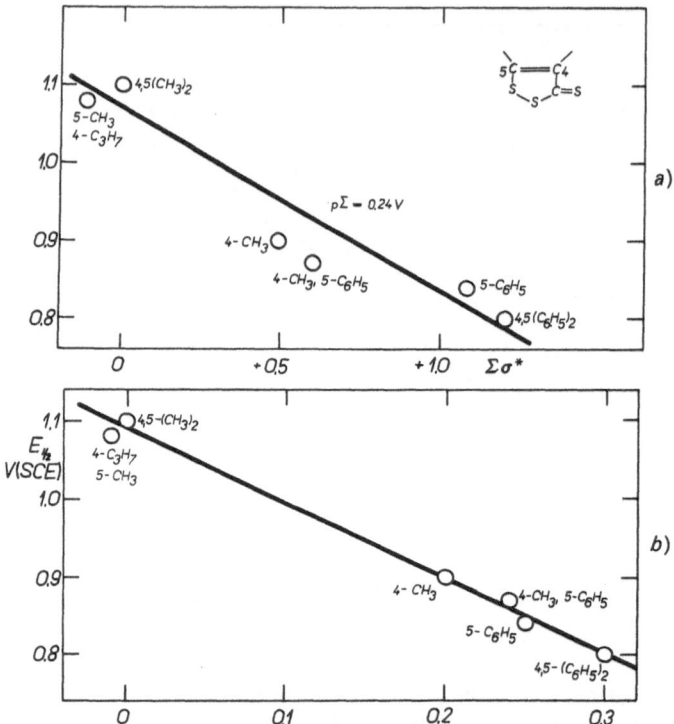

Fig. IV-6. Substituent effects on the second reduction waves of 1,2-dithiole-3-thiones. *a*) Relation of half-wave potentials to the sum of Taft polar substituent constants for substituents in both positions 4 and 5; *b*) relation of measured half-wave potentials to half-wave potential shifts calculated according to the equation $\Delta E_{1/2} = \rho^4_{\pi,\text{het}}\sigma^*_{X_4} + \rho^5_{\pi,\text{het}}$ $\sigma^*_{X_5}$ for $\rho^4_{\pi,\text{het}} = 0.09$ V and $\rho^5_{\pi,\text{het}} = 0.39$ V. Half-wave potentials from Ref. 31.

Ferrocenes, which were treated in Chapter III as aromatic systems carrying a reactive group (iron) separated by a rigid aromatic system from the substituent in the cyclopentadienyl ring, also can be treated as one single (heterocyclic) electroactive group with a directly attached substituent. Good correlation between $E_{1/4}$ values and σ^*_X constants for 20 substituents with $\rho^*_\pi = 0.10_4$ V was found.[32]

(Ac) *Effects of Substituents in a Phenyl Group Attached to a Reducible Heterocyclic Ring.* When a reducible heterocyclic ring carries a phenyl group, then *meta* and *para* substituents in the phenyl ring exert polar effects on the reduction of the heterocyclic ring comparable to those exerted on any other reducible group in this position of the phenyl ring. The modified Hammett equation thus can be used in the form

$$\Delta E_{1/2} = \rho_{\pi,R}\sigma_X \tag{27}$$

where R represents the whole reducible heterocyclic ring. The significance of the total polar substituent constant σ_X and the reaction constant $\rho_{\pi,R}$ (expressed in volts) is the same as for benzene compounds (Chapter III). The value of the reaction constant $\rho_{\pi,R}$ also depends on the position of attachment between the phenyl and heterocyclic rings.

Instances of the successful application of equation (27) are given in Table IV-2. It is noteworthy that in sydnones the reducible bond N-O is in the β-position to N-3 and in the γ-position to C-4, which

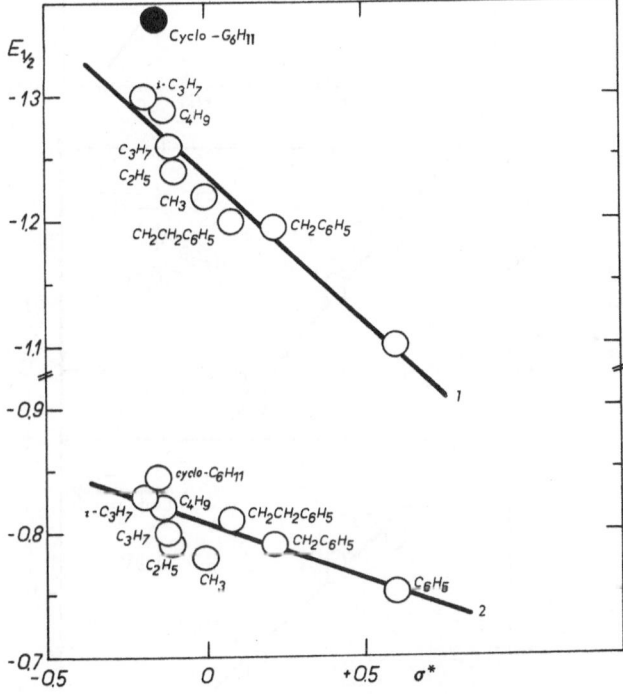

Fig. IV-7. Relation of half-wave potentials of N-substituted maleimides to Taft polar substituent constants σ_X^*. 1) First wave ($\rho_{\pi,het}^* = 0.09$ V); 2) second wave ($\rho_{\pi,het}^* = 0.24$ V). Reaction Series No. 5a and 5b, Table IV-2. Full point deviates.

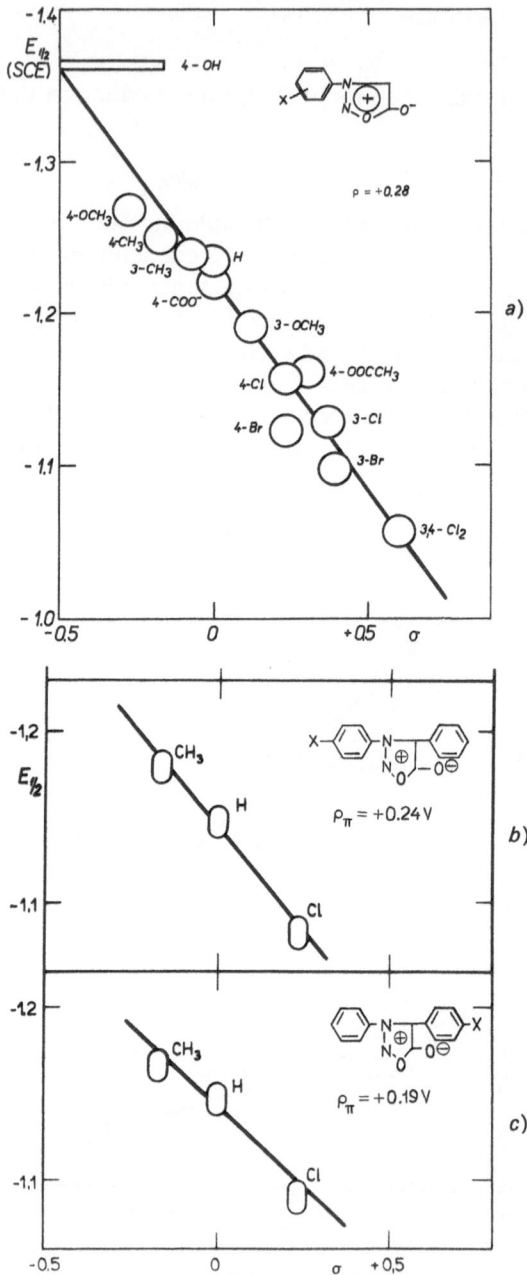

Fig. IV-8. Relation of half-wave potentials for the reduction of substituted phenylsyd-nones to Hammett substituent constants σ_X. a) 3-Phenylsydnones, Reaction Series No. 11, Table IV-2; b) 3,4-diphenylsydnones substituted in phenyl in position 3, Reaction Series No. 12; c) 3,4-diphenylsydnones substituted in phenyl in position 4, Reaction Series No. 13.

carries the phenyl group, and the long distance over which the polar effects are transmitted shows the mobility of π-electrons in the sydnone ring. The small differences in the values of reaction constants in Reaction Series No. 11–13 are too insignificant to allow discussion. The correlations are shown in Fig. IV-8.

The 3-phenylsydnone derivatives containing alkyl groups in *ortho* positions shows a further shift toward more negative potential, in addition to the shift caused by the polar effects of the alkyl groups. The shift of approx. $+0.06_5$ V was attributed to the loss of mesomeric interaction due to the steric hindrance to coplanarity[30] and is comparable with the shift, expressed as M_π, calculated using equation (33 het) [see Part (Aa) of this section].

Tetrazolium salts (No. 14 and 15, Table IV-2) show a marked difference between the value of $\rho_{\pi,R}$ for C-phenyl derivatives ($\rho_{\pi,R} \approx 0.00$ V) and the value for N-phenyl derivatives ($\rho_{\pi,R} = +0.12$ V). If the bond between the nitrogen atoms carrying aryls is reduced first, as suggested by Campbell and Kane,[34] this difference can be explained by the greater distance of the point of substitution from the reaction

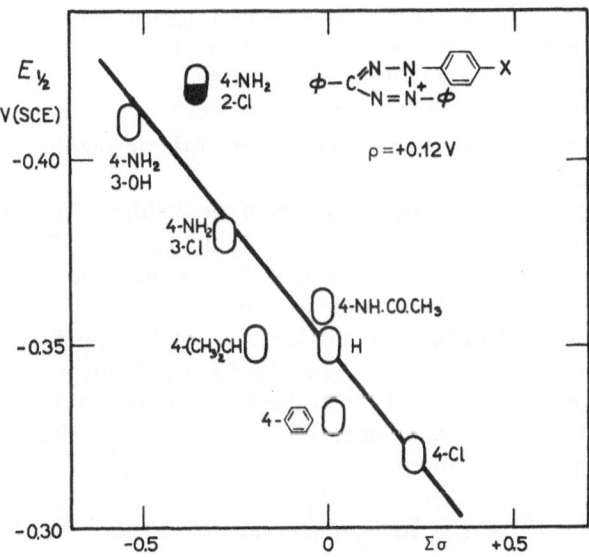

Fig. IV-9. Relation of half-wave potentials for the reduction of phenyl-substituted triphenyltetrazolium salts to Hammett substituent constants σ_X. Reaction Series No. 15, Table IV-2. Halved point: *ortho* derivative.

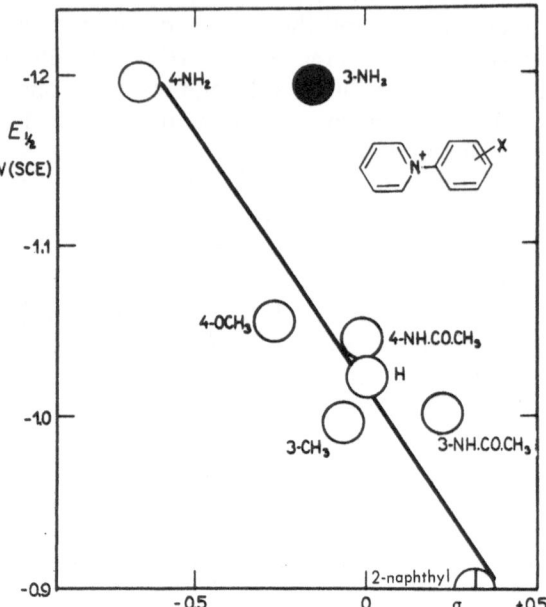

Fig. IV-10. Relation of half-wave potentials for the reduction of substituted 1-phenyl-pyridinium ions to Hammett substituent constants σ_X. Reaction Series No. 16, Table IV-2. Full point deviates, halved circle: $\sigma_{naphthyl}^{\pi}$ was used.

center and by the nonaromatic character of the tetrazolium ring with its rather immobile electron system.

In Reaction Series No. 15, in which the 2-chloro-4-amino deriva-tive contains an *ortho* substituent, the value $\sigma_{o-Cl} = 0.20$ was used as a first approximation and the steric effect was neglected. Figure IV-9 shows that this approximation is plausible.

The deviation for the 3-amino derivative of the 1-phenylpyridinium ion (Reaction Series No. 16, see Fig. IV-10) is difficult to explain. The identical half-wave potentials[35] of the 3- and 4-amino derivatives are unexpected, and the possibility of the mislabeling of the 3-amino derivative specimen cannot be excluded.

4. (B) Reduction in the Side Chain

(Ba) *The Effect of Substituents in Heterocyclic Rings on the Reduc-tion of a Group in Another Position in the Ring.* If the substituent in the heterocyclic ring is not in a position adjacent to the polarographically active group, conditions are fulfilled for the application of the modified Hammett equation (27) in the form (27 het). Both the substituent and the polarographically active group are separated by a rigid cyclic system, preventing steric interactions, but permitting the transfer of polar effects.

$$\Delta E_{1/2} = \rho_{\pi,het,R}\, \sigma_{het,X} \qquad\qquad (27\ \text{het})$$

In this equation $\sigma_{het,X}$ is a substituent constant, dependent on the kind of the substituent and its position relative to the polarographically active group. The value of $\sigma_{het,X}$ will generally be dependent on the type of heterocyclic system and on the position of the substituent relative to the hetero atom.

The reaction constant $\rho_{\pi,het,R}$ expresses the susceptibility of reduction in the side chain R to the effects of substituents in certain positions. The value of this constant depends on the kind of the polarographically active group R, on its position relative to the hetero atom, on the type of the heterocyclic ring, and on reaction conditions, but is independent of the kind and position of the substituent X.

It would be most logical to define the values of the substituent constants $\sigma_{het,X}$ by the difference of the logarithms of the dissociation constants of the corresponding carboxylic acids, where R is replaced by the carboxy groups, but these constants are known for a limited number of compounds only.

Imoto and co-workers,[36,37] and independently Tirouflet and Chané,[7,38,39] have shown from polarographic and kinetic studies that for a thiophene ring carrying the polarographically active group in position 2 and the substituent in position 5 it is possible to use, as a first approximation, $\sigma_{p\text{-}X}$ (derived for benzene derivatives) instead of $\sigma_{het,X}$. For the pair benzene–thiophene the correlation of σ_X and $\sigma_{thiophene,X}$ is clearly demonstrated in Fig. IV-11, where the half-wave potentials of substituted nitrothiophenes show linear correlation of unit slope with the half-wave potentials of substituted nitrobenzenes. In addition to $\sigma_{p\text{-}X}$ for 2,5-thiophene derivatives, it is possible to use $\sigma_{m\text{-}X}$ values for thiophenes, carrying the polarographically active group in position 2 and the substituent in position 4 or, reversely, the electroactive group in position 4 and the substituent in position 2.

$$\sigma_{p\text{-}X} \qquad\qquad \sigma_{m\text{-}X} \qquad\qquad \sigma_{m\text{-}X}$$

Hence, equation (27 het) can be simplified, until better values of $\sigma_{het,X}$ are available, and used in the form

$$\Delta E_{1/2} = \rho_{\pi,het,R}\sigma_X \qquad\qquad (27\ het')$$

The unit slope of the graph in Fig. IV-11 shows that the ratio of the reaction constant for thiophene derivatives $\rho_{\pi,het,R}$ to that for benzene derivatives $\rho_{\pi,R}$ is approximately unity ($\rho_{\pi,het,R}/\rho_{\pi,R} = 1.03$). This means that the ability of the electronic system of the thiophene ring to transfer effects from the substituent (e.g., in position 5) to the electroactive group (e.g., in position 2) differs little from the analogous property of the benzene electronic system for transfer (between positions 1 and 4). With increasing pH an increase in the ratio $\rho_{\pi,het,R}/\rho_{\pi,R}$ can be observed (Fig. IV-12).

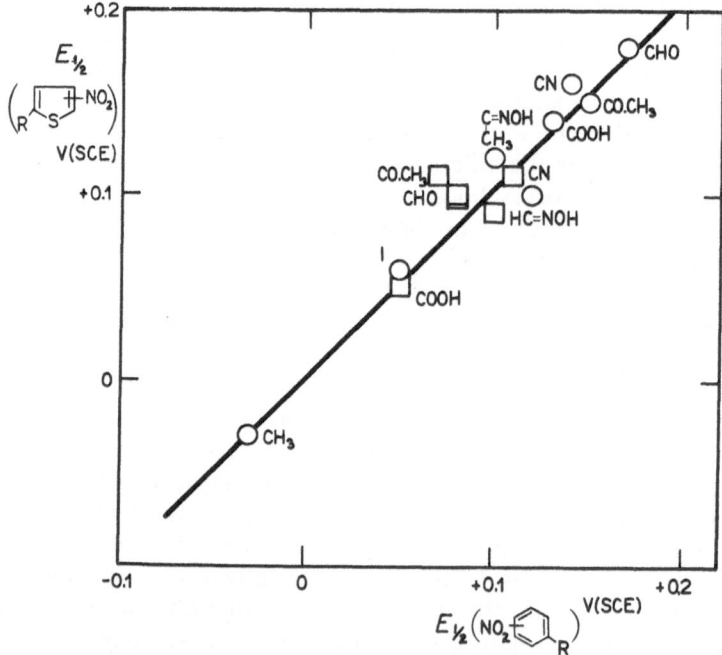

Fig. IV-11. Relation of half-wave potentials for the reduction of substituted nitrothiophenes to half-wave potentials for the reduction of substituted nitrobenzenes. Reaction Series No. 17, Table IV-2 and No. 40, Table III-4. Circles: 2-nitrothiophenes substituted in position 5 were compared with *p*-nitrobenzenes. Squares: 3-nitrothiophenes substituted in position 5 were compared with *m*-nitrobenzenes.

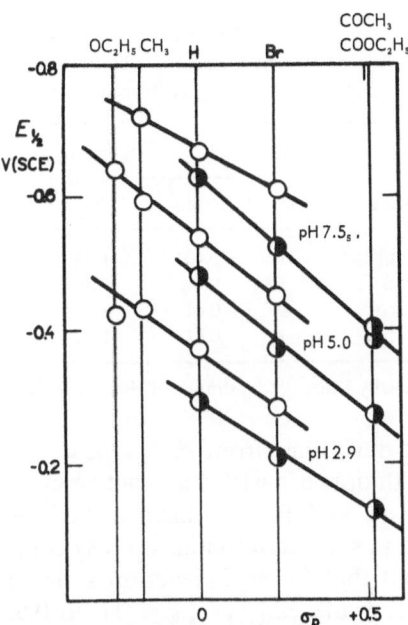

Fig. IV-12. Relation of half-wave potentials for the reduction of nitro compounds to Hammett substituent constants σ_X. Circles: *para*-substituted nitrobenzenes, Reaction Series No. 40 and 41, Table III-4. Halved points: 2-nitrothiophenes substituted in position 2, Reaction Series No. 17, Table IV-2.

The above treatment was extended[16,17,21,40] to other heterocyclic compounds. Generally, it has been demonstrated that equation (27 het') gives good linear correlation between $\Delta E_{1/2}$ and σ_X. Hence, it can be assumed that a linear relation between σ_X and $\sigma_{het,X}$ exists, i.e., that substituents affect the electroactive center in the same order and in the same relative degree. On the other hand, the susceptibility of the system to substituent effects, expressed in the value of the reaction constant, can differ—as could be predicted—for the heterocyclic and benzene rings. Thus, for the shifts in half-wave potential in the furan series the ratio $\rho_{\pi,het,R}/\rho_{\pi,R}$ was found to be 0.88,[16] whereas for the pyrrole series[17,40] the value of this ratio is between 1.7[17] and 2.1.[40]

It has been recently demonstrated[16] that such correlations between the benzene and heterocyclic series are not restricted to polarographic data, but can be found for equilibrium constants and kinetic data as well. The results obtained for 25 reaction series[16] are summarized in Table IV-4.

Table IV-4

Comparison of Reaction Constants in Benzene $(\rho_{\pi,R})$ and
Heterocyclic $(\rho_{\pi,\text{het},R})$ Reaction Series[a]

| Ring | n^b | $\rho_{\pi,\text{het},R}/\rho_{\pi,R}$ | | |
		From	To	Mean
Thiophene	11	0.63	1.34	0.98_5
Furan	3	0.88	1.39	1.15
Thiazole	6	0.61	2.35	1.08
Pyridine	4	0.61	1.44	1.02

[a] Values by Imoto[16]; [b] number of reaction series.

The greatest deviations from $\rho_{\pi,\text{het},R}/\rho_{\pi,R} = 1.0$ were observed[16] for 2-substituted thiophenes with the reactive group in position 4, for thiazole derivatives, and for 2,5-disubstituted pyridine derivatives. It is deduced that in such compounds a cross-conjugation takes place and it is suggested that for such reaction series in which the cross-conjugation plays a role, $\rho_{\pi,\text{het},R}/\rho_{\pi,R} \approx 0.6$ to 0.65 should rather be expected. Nevertheless, as mentioned above, the polarographic data for 4-nitro-2-substituted thiophenes give a good fit to the equation used for the other thiophene derivatives.

The applicability of equation (27 het') to five-membered heterocyclic compounds and some pyridine derivatives has been proved by the linear correlation found for the examples cited in Table IV-2.

For 2-nitrothiophene derivatives substituted in position 5 the application of the substituent constants σ_{p-X}^- proved to be useful. This is taken as evidence for mesomeric interaction between the nitro group and the substituent via the thiophene nucleus. Similarly, σ_{p-X}^- fitted the linear correlation better than σ_{p-X} for analogous nitroselenophenes, and nitropyrroles. Even in 2-furaldehydes substituted in position 5 (Reaction Series No. 22 and 22a) mesomeric interaction seems to be operating. In the case of 2-nitrofurans substituted in position 5, the limited number of values of half-wave potential available makes it difficult to decide which kind of constants should be preferred. When σ_{p-X}^- values are used (Reaction Series No. 21) the value for the 5-bromo derivative deviates (the reduction of C-Br before that of NO_2 cannot be excluded); when σ_{p-X} constants are used, the half-wave potential of 5-nitro-2-furaldehyde shows a marked deviation (Reaction Series No. 21a). It is a pity that numerous data on nitrofurans given by Sasaki[46] cannot be used for linear free energy relations, as either the $E_{1/2}$ versus pH plots for nitrofurans carrying various substituents vary so much that it is impossible to find a pH range suitable for correlation, or substituent constants for the substituents studied are not known. For nitropyrroles the CN-derivative either deviates or can be correlated by using σ_{p-x}; certainty concerning this group resembles that observed in the benzene series. In Reaction Series No. 29 and 30 additivity of the substituent effects seems to be proved.

For 2-furoic esters with substituents in position 5 (Series No. 23, Table IV-2) it has been remarked[42] that the 5-methyl derivative is not reducible in the potential range available. This now can be easily understood, for by the use of equation (27 het′) the half-wave potential was calculated to be -2.15 V, which is too negative to be observed with conventional buffers.

For substituted thiophene aldehydes no correlation was found.[17] The few available half-wave potentials of 2-bromo-5-substituted thiophenes[41] and furans[41] seems to indicate a great sensitivity toward substituent effects ($\rho_{\pi,\text{het,Br}} \approx 0.66$ V at pH 1 for 2-bromothiophenes and $\rho_{\pi,\text{het,Br}} \approx 0.80$ V at pH 1 and 0.77 V at pH 8 for 2-bromofurans).

The values for all the reaction constants collected so far (Table IV-2, No. 17–35) are positive, from which it is assumed that all these reductions follow a nucleophilic mechanism. The great variability and limited number of related data available prevent discussion of the structural effects of the electroactive group, and the type of ring on the value of the reaction constant. Nevertheless, in a limited series of

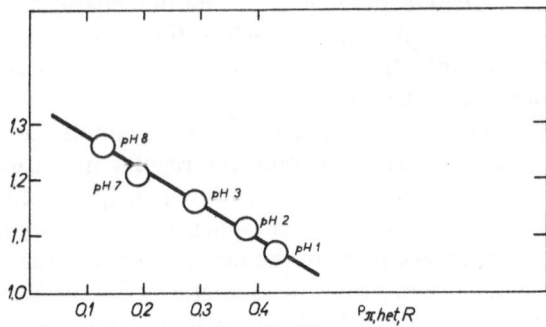

Fig. IV-13. Reduction of 5-substituted nicotinic acids. Relation of half-wave potentials for the parent unsubstituted compound E_H at a given pH value to the value of the reaction constant $\rho_{\pi,\text{het,R}}$ obtained at the same pH value. Reaction Series No. 31–35, Table IV-2.

TABLE IV-5

Percentage of 4-Nitro-2-Substituted Isomer Formed in the Nitration of 2-Substituted Thiophenes and Pyrroles[47]

X—[structure with NO₂, Het]	X = CHO	COCH$_3$	H	COOH	CH(OCOCH$_3$)$_2$	NO$_2$
Het = S	75	52	43	31	14	80
NH	58	56	44	52	—	54

reaction constants given by Imoto[16] for 5-substituted nicotinic acids measured at various pH values, the relationship between E_H and $\rho_{\pi,het,R}$ [equation (44)] was again demonstrated (Fig. IV-13).

In the reaction series belonging to this group three kinds of *ortho* effects can be assumed: the interaction of the substituent X with the reducible group R (when present in an adjacent position), the interaction of the substituent X in the α-position (relative to the hetero atom) with the hetero atom [which would express itself as a deviation from equation (27 het')], and the interaction of the polarographically active group R with the hetero atom (expressed in the value of the half-wave potential of the unsubstituted parent compound and possibly in the value of $\rho_{\pi,het,R}$); however, available data do not permit such analysis.

Finally, the analytical aspect should be mentioned briefly. When a thiophene or pyrrole substituted in position 2 is nitrated, either a 4- or 5-nitro derivative can result. The difference in the half-wave of these two isomers can be computed using the equation $\Delta E_{1/2} = \rho_{\pi,het,R}(\sigma_{p-X} - \sigma_{m-X})$. Hence, it can be predicted whether the separation of waves (or peaks in the single-sweep method) will be sufficient to make the analysis of the mixture possible. By the use of such a method, analyses of mixtures of isomers were carried out (Table IV-5).

Some complications may arise[47]: for example, 5-nitro-2-thiophenecarboxaldehyde gives not only the regular first wave, but also another wave at more negative potentials which coincides with the four-electron wave of 4-nitro-2-thiophenecarboxaldehyde. This, of course, decreases the accuracy of the measurement of the wave height for the latter substance in a mixture. Sometimes the half-wave potential can be affect by the presence of the second isomer in a mixture or can be concentration-dependent. Hence, information and predictions derived from treatments such as the above are again to be treated as tentative.

(Bb) *Type of Heterocyclic Ring and Reduction in the Side Chain.* Reduction in the side chain can in principle be treated in three ways: (*i*) the introduction of a hetero atom into the aromatic cyclic system is considered as a defect comparable to the introduction of a substituent; this is equivalent to the treatment used by Jaffé[48] for pyridine rings; (*ii*) the whole heterocyclic ring is considered as a substituent; (*iii*) correlations are restricted to the polarographic data under consideration on the assumption that the mesomeric and steric effects of a heterocyclic ring will remain substantially constant when different reducible groups are attached.

(*i*) When the hetero atom is considered as a substituent, it is impossible to use equation (27) in its simple form as there is the possibility of a direct mesomeric interaction between the reducible group R and the hetero atom, for the p-electrons of the hetero atom are included in the aromatic sextet of the ring.

The mesomeric interaction in a nucleophilic reduction is formally considered in

$$\Delta E_{1/2} = \rho_h \sigma_F^-$$ (51)

where the reaction constant ρ_h is defined in the same way as the reaction constants above. The substituent constant σ_F^- (which takes account of the mesomeric interaction) expresses the effect of the hetero atom, its position relative to the reducible group, and possibly also the size of the ring. It is defined relative to phenyl. The values of this constant should be obtained using dissociation constants of heterocyclic acids, as indicated for $\sigma_{het,X}$ in Part (Ba), with the carboxy group in place of the reducible group; as standard substance benzoic acid is most suitable.

When dissociation constants were compared with the rates of the alkaline hydrolysis of the esters Het-COOC$_2$H$_5$, it was found[49] that the relation is expressed by two straight lines: β-carboxylic esters fall on the same line as *meta* and *para*-substituted benzoic esters, whereas α-carboxylic esters fall on a line which is followed also by some *ortho*-substituted benzoic acid derivatives (Fig. IV-14). This second line has the same slope as the first, but is shifted by a constant increment. This shift is attributed[16,49] to "the adjacent hetero atom effect." The effect of ethanol on dissociation constants has proved that this effect is not due primarily to hydrogen bonding.

On the basis of these observations σ_a constants, called "aryl values" were defined for β-carboxylic acids as

$$\sigma_a = \log K/K_0$$ (52)

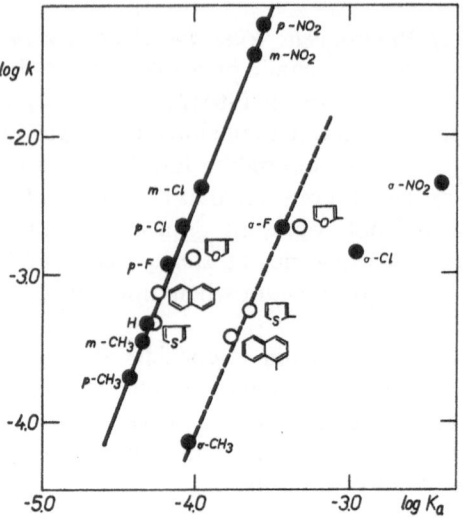

Fig. IV-14. Relation of logarithms of rate constants for alkaline ester hydrolysis to logarithms of dissociation constants. Full points: benzoic acid derivatives; circles: heterocyclic derivatives. According to Ref. 49.

and for α-carboxylic acids as

$$\sigma_a = \log K/K_0 - 0.41 \tag{53}$$

The term 0.41 is a quantum-mechanically derived correction for the steric effect. The values of "aryl values" σ_a obtained in this way are given in Table IV-6. The aryl value is, in fact, analogous to σ_F^-, but expressed for the hetero atom or hetero atoms and condensed rings.

These constants have been successfully applied for correlations in some reaction series (Fig. IV-15). On the other hand, as 2-furyl derivatives are, in most instances, reduced at more negative potentials than

Table IV-6
Aryl Values Determined by Otsuji, Kubo, and Imoto[49]

Ring	σ_a	Ring	σ_a
Phenyl	0.00	2-Thienyl	0.30
2-Furyl	0.67	3-Thienyl	0.12
3-Furyl	0.25	3-Pyridyl	0.6
		4-Pyridyl	0.5

2-thienyl derivatives (which would correspond to $\sigma_{\text{a-2-furyl}} < \sigma_{\text{a-2-thienyl}}$, the values obtained by the use of $\sigma_{\text{a-2-furyl}} = 0.67$ cannot be regarded as useful for polarographic data. Also, the use of the value for 4-pyridyl cannot be recommended, since changes in the mechanism of the electrode process often occur for these derivatives.

No reaction series was available which contained enough α- and β-derivatives simultaneously to allow us to distinguish whether the correction for steric factors for α-derivatives is preferable, or whether simple correlation with pK_a values is sufficient. As nitro compounds fitted the $E_{1/2}-\sigma_a$ correlation, at least as good as compounds containing other electroactive groups, no significant steric hindrance to coplanarity is operating.

The confidence in the applicability of σ_a values for linear free energy correlations is shaken by Japanese authors,[49] who showed that split

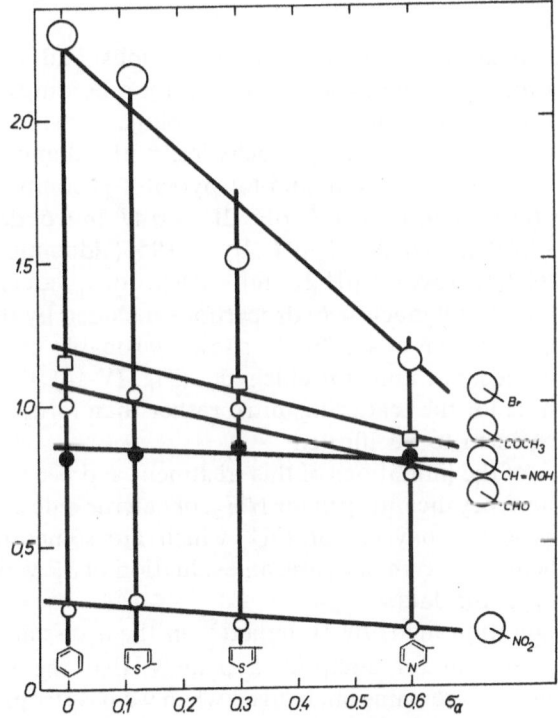

Fig. IV-15. Relation of half-wave potentials for reduction of various groups on heterocyclic rings to aryl values σ_a derived by Otsuji, Kubo, and Imoto.[49]

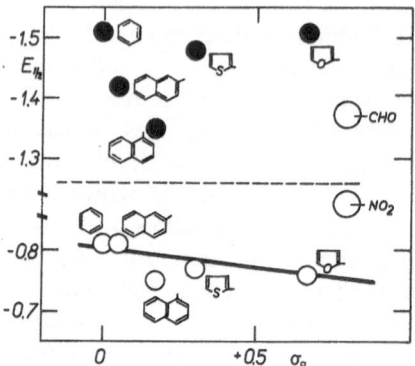

Fig. IV-16. Relation of half-wave potentials for reduction of aldehydes and nitro derivatives derived from heterocyclic and polycyclic systems to aryl values σ_a derived by Otsuji, Kubo, and Imoto.[49] Circles: nitro compounds; full points: aldehydes. Half-wave potentials at pH 10.3 from Ref. 36.

log K–σ_a plots are obtained for several reactions, and is shaken also by the variation in the values of σ_a constants for other rings, determined by inter- and extra-polation of $E_{1/2} - \sigma_a$ plots. Thus, for 2-thiazolyl compounds values of $\sigma_a = 1.35$ (aldehydes,[18] pH 2) and 0.85 (bromides,[18] pH 2) have been found, and for pyrrol-2-yl compounds values of -0.82_5 (nitro derivatives,[18] pH 2), -0.47 (nitro derivatives,[40] pH 4.2); -0.20 (aldehydes,[18] pH 2); -0.95 (aldoximes,[18] pH 2); -0.38 (acetyl derivatives,[18] pH 2), and -0.9 (from σ_n values, see p. 160) have been found. Polynuclear hydrocarbons included by the Japanese authors[49] in this type of treatment show reasonable correlation for nitro compounds, but none for aldehydes (Fig. IV-16). If steric effects are responsible for the scattering, nitro rather than formyl derivatives would be expected to be affected.

Because of the limitations of this treatment and because the interchanges of heterocyclic nitrogen for NH_2, of heterocyclic sulfur for SH, and of heterocyclic oxygen for OH, which are sometimes used in quantum chemistry, seem insecure, an evaluation of σ_F^- was attempted from polarographic data.

An assumption made by Hammett[50] in the assessment of the σ_X substituent constant for 2-naphthyl, namely, that the value of the reaction constant ρ remains unchanged when we pass from substituted benzene derivatives to compounds with different aromatic rings (see Fig. IV-14), was adopted here. For a given reducible group $\rho_{\pi,R} \approx \rho_h$

is adopted as a first approximation. This implies that the value of the reaction constant $\rho_{\pi,\text{CHO}}$, determined for substituted benzaldehydes, remains unchanged for the series: benzaldehyde, 2-furaldehyde, 2-thiophenecarboxaldehyde, pyrrole-2-carboxaldehyde, etc. The values given in Table IV-7 were computed on this basis.

Below we write the calculated values of σ_F^- in ascending order, the most probable range (underlined) to be selected, and give the arithmetic mean over this range; in this way the number and kind of the deviations are revealed more clearly than by a simple statistical treatment:

(structure)						σ_F^- (aver.)
[S]	−0.17	0.0	0.0	+0.03	+0.04	
	+0.05	+0.06	+0.09	+0.09	+0.09	
	+0.17	+0.18	+0.18	+0.22	+0.22	+0.24
	+0.23	+0.23	+0.26	+0.31	+0.35	
	+0.36	$+0.37_5$	+0.40	+0.42	+0.44	
	+0.50	+1.08	+1.2	+1.4	+1.7	
[S]	−0.19	-0.12_5	−0.12	−0.07	−0.07	−0.09
	−0.06	+0.22				
[O]	0.0	+0.01	+0.07	+0.18	+0.20	+0.18
	+0.21	+0.21	+0.21	+0.31	+0.42	
[NH]	−1.14	−1.07	−1.06	−1.00	−0.96	
	−0.94	−0.93	−0.73	−0.72	−0.69	−0.70
	−0.67	−0.60	−0.58	−0.53	−0.47	
	−0.43	−0.39	−0.38	−0.27	−0.07	
[NH]	−2.00	−1.78	−1.27			−1.68
[N,S]	+0.89	+1.87	+2.16			+1.64
[N]	+0.72	+1.00	+1.68	+1.81	+2.16	+1.5
[N]	+0.06	+0.50	+0.78	+0.88	+1.57	+0.72
[N]	+0.94	+1.78	+1.91	+2.81		+1.86

Table IV-8
Nucleus Constants σ_n Derived by Nakaya and Co-Workers[42]

Ring	σ_n
Pyrrolyl	0.6
Phenyl	1.5
2-Furyl	2.27
2-Thienyl	2.55

The number and magnitude of the deviations are greater than is usual for the substituent constants σ. The difference between the values of σ_F^- for different heterocyclic rings computed from half-wave potentials obtained under identical conditions is somewhat more constant. For example, $\sigma_{\text{thiophene}}^- - \sigma_{\text{furan}}^- \approx +0.08$ units. Generally speaking, this kind of treatment offers only an indication and a very rough approximation to the values of the σ_F^- constants.

One reason for these discrepancies lies in the fact that in the computation of σ_F^- values in Table IV-7 selected values of $\rho_{\pi,R}$ (given in Table III-13, p. 102) were used instead of the $\rho_{\pi,R}$ values obtained under reaction conditions identical to those employed for the heterocyclic compounds. Errors from this source account for the fact that in some reaction series the σ_F^- values are considerably lower (e.g., for aldoximes), or higher (e.g., for bromides and iodides) than in other series. A more important reason for the observed departures from the approximation $\rho_{\pi,R} \approx \rho_h$ may be that the mesomeric interaction between the heterocyclic ring and the reducible group differs from one reaction series to another. If this is so, equation (51) cannot be valid. However, we demonstrate in Section 3 below that parallelism in mesomeric interaction can be found among some, at least, of the reaction series.

Another series of substituent constants called nucleus constants (σ_n), expressing the effect of the heterocyclic ring, has been derived† by Nakaya and co-workers[42] (Table IV-8).

In addition to carbonyl derivatives, from which these values were derived, these constants have proved satisfactory for other derivatives[18] (Fig. IV-17).

(ii) When the whole heterocyclic ring is treated as a substituent, equation (33 het) can be used.

$$\Delta E_{1/2} = \rho_h^* \sigma_h^* + M_\pi \qquad \text{(33 het)}$$

† Due to the language difficulties, the present author was unable to find out how these values were derived and defined.

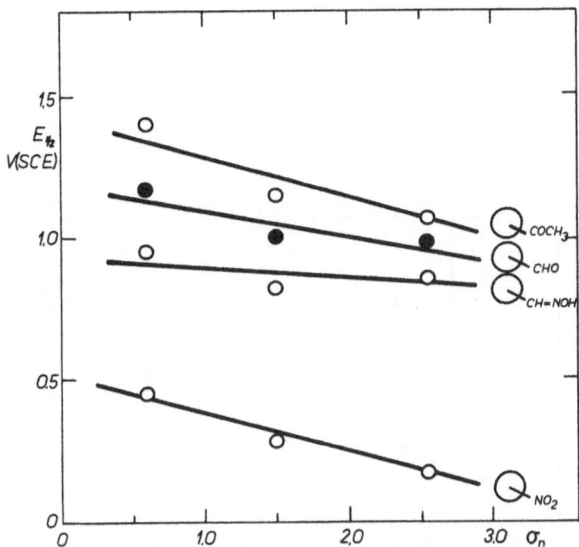

Fig. IV-17. Relation of half-wave potentials for reduction of various groups on hetero-cyclic rings to nucleus constants σ_n derived by Nakaya, Inosito, and Ono.[42]

Here gain, ρ_h^* is defined as above, whereas M_π is the contribution due to mesomeric interaction and the substituent constants σ_h^* express the polar effect of the heterocyclic ring on the reducible group R in side chain. The value of σ_h^* is dependent on the kind of heterocyclic ring and the position of R relative to the hetero atom; it is independent of the nature of R.

Although the values of the constants σ_h^* should be determined from the acid- and base-catalyzed ester hydrolysis of the corresponding acids, as for aliphatic acids,[53] such treatment has not been carried out yet. Polarographic data do not allow a computation such as that in Section (*i*), for a comparison with alkyl derivatives is precluded because of the differences in mechanism in the reduction of aliphatic and aromatic compounds.

(*iii*) In addition to the above treatments, we can rely on polaro-graphic data alone. The heterocyclic aromatic aldehydes with five-membered rings were chosen as standard comparison substances, for most experimental data were available for this reaction series, and it seems that these compounds are reduced by an identical mechanism.

TABLE IV-7

Values of σ_F^- Computed by the Use of Equation (51) for $\rho_h = \rho_{\pi,R}$; $\sigma_F^- = \Delta E_{1/2}/\rho_{\pi,R}$

Reaction series	pH	$\rho_{\pi,R}$	σ_F^- thiophene (S)	σ_F^- thiophene (S)	σ_F^- furan (O)	σ_F^- pyrrole (NH)	σ_F^- pyrrole (NH)	σ_F^- thiazole (N,S)	σ_F^- pyridine 2- (N)	σ_F^- pyridine 3- (N)	σ_F^- pyridine 4- (N)	Ferrocene	$E_{1/2}$ from Ref.
Het-NO₂	2	+0.16	+0.44	-0.12	—	-1.00	-2.00	—	—	—	—	—	3;7
	7.9	+0.14	+1.2	-0.07	—	-1.07	-1.78	—	—	—	—	—	3;7
	10.3	+0.24	+0.17	—	+0.21	—	—	—	—	—	—	—	36
	2	+0.16	$+0.37_5$	-0.19	—	-1.06	—	—	+1.00	+0.50	—	—	18
	4.2	+0.15	+0.40	-0.07	—	-0.93	-1.27	—	—	—	—	—	40
Het-CHO	10.3	+0.33	+0.09	—	0.0	-0.27	—	—	—	—	—	—	36
	1.0	+0.32	+0.31	—	+0.20	-0.60	—	—	—	—	—	—	42
	4.0	+0.26	—	—	—	-0.58	—	—	—	—	—	—	42
	8.0	+0.26	+0.42	—	+0.31	-0.96	—	—	—	—	—	—	42
	7.5	+0.26	+0.50	—	+0.42	-0.47	—	—	—	—	—	—	43
	10.6	+0.33	+0.26	—	+0.21	—	—	—	—	—	—	—	43
	2.0	+0.32	+0.03	—	—	-1.04	—	—	—	—	—	—	3
	7.9	+0.26	0.0	-0.12	—	-0.53	—	—	—	—	—	—	3;7
	2.0	+0.32	+0.06	—	—	—	—	+1.87	+1.68	+0.78	—	—	18
	4.2	+0.26	+0.04	—	—	-0.73	—	—	—	—	+1.78	—	40
Het-CH=NOH	2.0	+0.18	—	—	—	-0.94	—	+0.89	+0.72	+0.06	+0.94	0.0	7
	2.0	+0.18	-0.17	-0.06	—	-0.72	—	—	—	—	—	—	18
	4.2	+0.18	0.0	—	—	-0.44	—	—	—	—	—	—	40

Het-COCH₃	1.0	+0.22	+0.35	—	+0.21	—	—	—	—	—	—	42
	4.0	+0.35	+0.18	—	+0.07	—	—	—	—	—	—	42
	8.0	+0.35	+0.18	—	—	−0.67	—	—	—	—	—	42
	2.0	+0.22	+0.23	—	—	—	—	—	—	—	—	3
	7.9	+0.35	+0.09	—	—	—	—	—	—	—	—	3
	2.0	+0.22	+0.36	—	—	−1.14	—	+2.73	+0.88	+2.81	−0.86	18
	4.2	+0.35	+0.09	—	—	−0.69	—	—	—	—	—	40
Het CCH₃ =NOH	2.0	+0.18	+0.22	—	—	—	—	—	—	—	—	3
	4.2	+0.18	+0.22	—	—	−0.39	—	—	—	—	—	40
Het CH=CHCOCH₃	4.0	+0.28	+0.05	—	+0.01	−0.38	—	—	—	—	—	42
Het-COOCH₃	unbuffered	+0.85	+0.23	—	+0.18	—	—	—	—	—	—	42
Het-Br	unbuffered	+0.74	+1.08	+0.22	—	—	—	+2.16	+1.81	+1.57	+1.91	18
Het-I	unbuffered	+0.30	+1.7	—	—	—	—	—	—	—	—	51
	unbuffered	+0.30	—	—	—	−0.07	—	—	—	—	—	52
	unbuffered	+0.30	+1.4	—	—	—	—	—	—	—	—	7

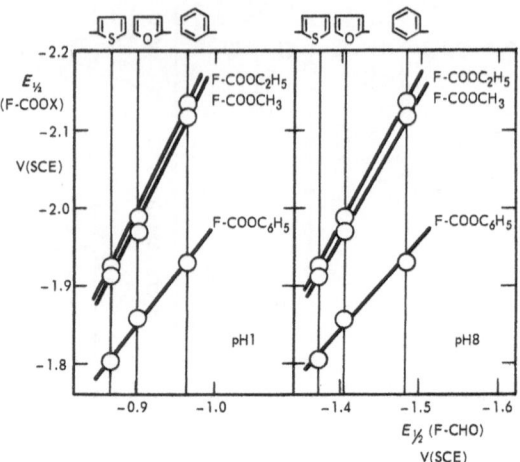

Fig. IV-18. Relation of half-wave potentials for the reduction of aromatic carboxylic esters (F-COOX) to half-wave potentials for the reduction of the corresponding aromatic aldehydes (F-CHO) for various groups X. Half-wave potentials at pH 1 (left) and pH 8 (right) from Ref. 42.

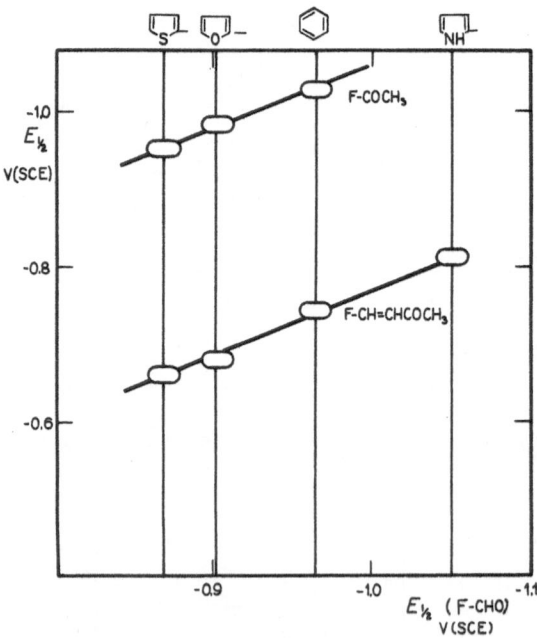

Fig. IV-19. Relation of half-wave potentials for the reduction of aryl–alkyl derivatives of the type F $(CH=CH)_n COCH_3$ to half-wave potentials of the corresponding aromatic aldehydes (F—CHO) at pH 1.0. Half-wave potentials from Ref. 42.

Pyridine aldehydes could not be included because of complications due to hydration.[1-9]

To avoid any assumptions concerning the value of the reaction constant, the half-wave potentials were compared directly using

$$(E_{1/2})_{R^1,het} - (E_{1/2})_{R^1,F} = \rho'_R[(E_{1/2})_{R^2,het} - (E_{1/2})_{R^2,F}] \qquad (54)$$

where the index F denotes the benzene derivative, and the index Het denotes derivative with the reducible group R in a heterocyclic ring. For $(E_{1/2})_{R^1}$ the half-wave potentials of the aldehydes were used, and for $(E_{1/2})_{R^2}$ the half-wave potentials for corresponding heterocyclic compounds containing the reducible group R^2 were used.

The validity of equation (54) has been proved for esters of the type $Het\text{-}COOCH_3$, $Het\text{-}COOC_2H_5$, and $Het\text{-}COOC_6H_5$ (Fig. IV-18), for

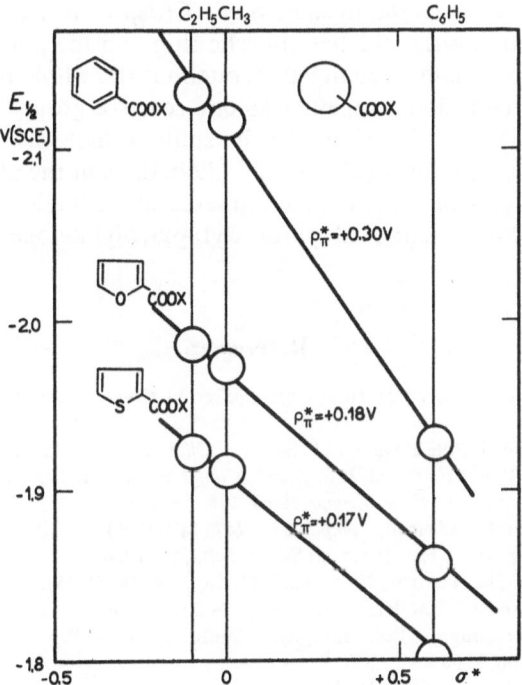

Fig. IV-20. Relation of half-wave potentials for the reduction of aromatic carboxylic esters to Taft polar substituent constants σ_X^*. Reaction Series No. 23, Table V-2, and No. 36 and 37, Table IV-2.

Het-COCH$_3$, Het-CH = CHCOCH$_3$ (Fig. IV-19), and Het-CH = CHCOOCH$_3$ (the half-wave potentials for which were measured by Nakaya and Inosito and Ono[42]), and also for two series of values for the aldehydes Het-CHO obtained under different conditions. These results show that there is a parallelism in mesomeric interaction between the reducible group and the aromatic heterocyclic ring among these series.

Poor agreement with equation (54) resulted when the half-wave potentials of aldehydes were compared with those of nitro compounds (see Ref. 36). It appears that for the nitro group the relation between the extent of mesomeric interaction with the heterocyclic ring and the structure of the latter differs from the relation holding for the other groups mentioned above.

(Bc) *Effect of a Substituent in the Side Chain of a Nonreducible Heterocyclic Ring on the Reduction of a Group in the Same Side Chain.* In this reaction series the heterocyclic ring and the polarographically active group remain unchanged throughout the whole reaction series, and can be considered together as one reactive group. The effects of substituents in the side chain on the shifts in half-wave potential are expressed by equations (25 het) and (29 het), as in the aliphatic series.

Examples of this type of reaction series are reductions of carboxylic esters (Fig. IV-20) and of alkyl (or aryl) pyrrolyl ketones (Table IV-2).

References

[1] J. Volke, *Chem. Listy* **52**: 16 (1958); *Collection Czech. Chem. Commun.* **23**: 1486 (1958).

[2] V. Preininger, H. Potěšilová, and F. Šantavý, *Chem. Listy* **52**: 25 (1958).

[3] J. Tirouflet, P. Fournari, and J. P. Chané, *Compt. Rend.* **242**: 1799 (1956).

[4] J. Tirouflet and E. Laviron, *Compt. Rend.* **246**: 397 (1958).

[5] J. Tirouflet and E. Laviron, *Compt. Rend.* **247**: 217 (1958).

[6] J. Volke, *Z. physik Chem.* (Leipzig) Sonderheft, **268** (1958).

[7] J. Tirouflet and E. Laviron, *Ricerca Sci.* **29**: Suppl. A, 189 (1959).

[8] E. Laviron and J. Tirouflet, "Advances in Polarography, Proceedings of the Second International Congress Polarography," Cambridge, 1959. Vol. II, p. 727, Pergamon Press, London, 1960.

[9] E. Laviron, Thesis, Dijon, 1961.

[10] F. Šorm, *Chem. Obzor* **18**: 213 (1943).

[11] J. Volke and V. Volková, *Collection Czech. Commun.* **20**: 1332 (1955); *Chem. Listy* **49**: 490 (1955).

[12] J. Volke, *Collection Czech. Chem. Commun.* **22**: 1777 (1957); *Chem. Listy* **51**: 414 (1957).

[13] J. Volke, R. Kubíček, and F. Šantavý, *Collection Czech. Chem. Commun.* **25**: 1510 (1960).

[14] J. Volke, *Collection Czech. Chem. Commun.* **25**: 3397 (1960).

[15] J. Stradiņš, "Poljarografija organičeskich nitrosoedinenij" ("Polarography of Organic Nitrocompounds"). 163 pp. Izdat. Akad. Nauk. Latv. SSR. Riga, 1960.

[16] E. Imoto, *Rev. Polarog.* **9**: 185 (1961).

[17] J. Tirouflet and R. Dabard, *Ricerca Sci.* **29**: Suppl. A, 189 (1959).

[18] J. Tirouflet and M. Person, *Ricerca Sci.* **30**: Suppl. A, 269 (1960).

[19] J. Volke, *Chem. Listy* **55**: 26 (1961).

[20] J. Volke, *Electrochemical Properties in Solutions* in (ed. Katritzky): "Physical Methods in Heterocyclic Chemistry," Vol. 1, p. 217 New York, 1963.

[21] P. Zuman, *Collection Czech. Chem. Commun.* **27**: 630 (1962).

[22] P. Zuman, *Z. Physik. Chem.*, Sonderheft *1958*, 246.

[23] P. Zuman, Unpublished results.

[24] C. Ricciuti, C. O. Willits, H. B. Knight, and D. Swern, *Anal. Chem.* **25**: 933 (1953).

[24a] A. Němečková, M. Maturová, M. Pergál, and F. Šantavý, *Collection Czech. Chem. Commun.* **26**: 2749 (1961).

[25] S. G. Majranovskij, *Dokl. Akad. Nauk SSSR*, **110**: 593 (1956).

[26] E. Gird and A. T. Balaban, *J. Electroanal. Chem.* **4**: 48 (1962).

[27] J. Krupička and J. Gut, *Collection Czech. Chem. Commun.* **27**: 546 (1962).

[28] O. Manoušek and P. Zuman, *Collection Czech. Chem. Commun.* **20**: 1340 (1955); *Chem. Listy* **49**: 668 (1955).

[29] K. R. Voronova and A. G. Stromberg, *Zh. Obshch. Khim.* **31**: 2786 (1961).

[30] P. Zuman and D. Voaden, *Tetrahedron* **16**: 130 (1961).

[31] L. Stárka and L. Jirousek, *Pharmazie* **14**: 473 (1959).

[32] L. K. Hoh, W. E. McEwen, and J. Kleinberg, *J. Am. Chem. Soc.* **83**: 3949 (1961).

[33] P. Zuman, *Collection Czech. Chem. Commun.* **25**: 3265 (1960).

[34] H. Campbell and P. O. Kane, *J. Chem. Soc.* **1956**, 3130.

[35] K. Schwabe, *Chem. Tech.* (Berlin) **9**: 129 (1957).

[36] E. Imoto, R. Motoyama, and H. Kakiuchi, *Bull. Naniwa Univ.* **3A**: 203 (1955).

[37] E. Imoto, Y. Otsuji, and T. Hirai, *J. Chem. Soc. Japan* **1956**, 804.

[38] J. Tirouflet and J. P. Chané, *Compt. Rend.* **243**: 500 (1956).

[39] J. Tirouflet and J. P. Chané, *Compt. Rend.* **245**: 80 (1957).

[40] P. Fournari, Thesis, Dijon, 1961.

[41] J. Nakaya, K. Mori, H. Kinoshita, and S. Ono, *Nippon Kagaku Zasshi* **80**: 1212 (1959).

[42] J. Nakaya, H. Inosito, and S. Ono, *J. Chem. Soc. Japan, Pure Chem. Sect.* **78**: 935, 940 (1957): *Ref. Zh. Khim.* **1958**, No. 70257.

[43] F. Cappellina and V. Lorenzelli, *Ann. Chim.* **48**: 893 (1958).

[44] J. Tirouflet and E. Laviron, *Compt. Rend.* **244**: 2063 (1957).

[45] F. Cappellina and V. Lorenzelli, *Atti Accad. Sci Ist. Bologna* 9 **5**: 131 (1958).

[46] T. Sasaki, *Pharm. Bull.* **2**: 104 (1954).

[47] J. Tirouflet, "Advances in Polarography, Proceedings of the Second International Congress Polarography," Cambridge, 1959, Vol. II, p. 740, Pergamon Press, London, 1960.

[48] H. H. Jaffé, *Chem. Rev.* **53**: 191 (1953).

[49] Y. Otsuji, M. Kubo, and E. Imoto, *Bull. Osaka Prefect.* **7A**: 61 (1959).

[50] L. P. Hammett, "Physical Organic Chemistry," p. 184, McGraw Hill, New York, 1940.

[51] F. M. Beringer, H. E. Bachofner, R. A. Falk, and M. Leef, *J. Am. Chem. Soc.* **80**: 4279 (1958).

[52] E. Gergely and T. Iredale, *J. Chem. Soc.* **1951**: 3502.

[53] R. W. Taft Jr., *Separation of Polar, Steric and Resonance Effects on Reaction*, in "Steric Effects in Organic Chemistry" (ed. M. S. Newman) John Wiley & Sons, New York, 1956.

V: Reaction Series in which the Electro-active Group is Directly Attached to an Alkyl or Aryl Substituent

1. General

In substances in which the electroactive group is attached directly to a substituent, a direct interaction between the reactive group and the substituent can take place. The shifts in the half-wave potentials of compounds of this kind, in which the polarographically active group is not cyclic system (as in some heterocyclic compounds treated in Chapter IV), are discussed in this chapter. Substituents of the type CH_2X and phenyl, which exert predominantly polar effects, cause shifts in half-wave potential which follow

$$\Delta E_{1/2} = \rho^*_{\pi,R}\sigma^*_X \qquad (29)$$

Systems in which inductive effects are propagated from the substituent X to the reactive center R show shifts in half-wave potential which follow the equivalent equation

$$\Delta E_{1/2} = \rho^I_{\pi,R}\sigma^I_X \qquad (25)$$

In the presence of several exchangeable substituents in a molecule containing one single electroactive group, the half-wave potentials are shifted according to equations (38a) and (38b), respectively,

$$\Delta E_{1/2} = \rho^I_{\pi,R}\Sigma\sigma^I_X \qquad (38a)$$

$$\Delta E_{1/2} = \rho^*_{\pi,R}\Sigma\sigma^*_X \qquad (38b)$$

169

In molecules in which polar and mesomeric interaction between the substituent and the electroactive group takes place, equation (33) is applied, whereas, when polar and steric effects participate equation (36) can be used:

$$\Delta E_{1/2} = \rho^*_{\pi,R}\sigma^*_X + M_\pi \tag{33}$$

$$\Delta E_{1/2} = \rho^*_{\pi,R}\sigma^*_X + \delta_{\pi,R}(E_S)_X \tag{36}$$

The possibility of the application of polarographic data in equation (29) was recognized early.[1] Later, such a treatment was applied to a few selected reaction series,[2,3] and its general applicability was demonstrated.[4–14]

The choice and treatment of half-wave potentials in this chapter follow the same principles as for benzene derivatives (Chapter III).

The values of polar substituent constants σ^*_X and of steric substituent constants $(E_S)_X$ used in correlations are given in Table V-1. Most of the values are taken from Ref. 1; the sources of the other values are stated. Values of the inductive substituent constants σ^I_X are to be found in Table III-1.

Table V-1
Values of Polar and Steric Substituent Constants Used in Correlations

Substituent	σ^*_X	E_S	Substituent	σ^*_X	E_S
CH_3	0	0	$n\text{-}C_7H_{15}$	-0.17^a	—
C_2H_5	-0.10	-0.07	$n\text{-}C_8H_{17}$	-0.15^a	-0.33
$n\text{-}C_3H_7$	-0.11_5	-0.36	cyclo-C_5H_9	-0.20	-0.51
$i\text{-}C_3H_7$	-0.19	-0.47	cyclo-C_6H_{11}	-0.15	-0.79
$n\text{-}C_4H_9$	-0.13	-0.39	H	$+0.49$	$+1.24$
$i\text{-}C_4H_9$	-0.12_5	-0.93	$C_6H_5CH_2$	$+0.21_5$	-0.38
$s\text{-}C_4H_9$	-0.21	-1.13	C_6H_5	$+0.60$	—
$t\text{-}C_4H_9$	-0.30	-1.54	$C_6H_5C_6H_4$	$+1.2$	—
$n\text{-}C_5H_{11}$	-0.16^a	-0.40	COOH	$+1.1^b$	—
$i\text{-}C_5H_{11}$	-0.16^a	-0.35	$C_6H_5OCH_2$	$+0.85$	—
$(C_2H_5)_2CH$	-0.22_5	-1.98	CH_3COCH_2	$+0.60$	—
$(CH_3)_3CCH_2$	-0.16_5	-1.74	$CH_3CH{=}CH$	$+0.36$	—
$n\text{-}C_6H_{13}$	-0.16^a	—	$C_6H_5CH{=}CH$	$+0.41$	—

[a] From Hoefelmeyer et al.[15]
[b] See p. 173.

2. Reactions for which Equations (25), (29), and (36) Hold

Ninety-seven reaction series, with half-wave potentials for 389 compounds that fit one of the forms of equation (20), are summarized in Table V-2. Section A contains reaction series in which $\rho > 0$ and for which a nucleophilic mechanism of the electrode process is expected; in Section B those in which $\rho < 0$ are brought together. If Hammett's[69] assumption is valid, an electrophilic process should operate in the latter case. In Section C reaction series in which the electroactive group is bound to several different substituents, and also those in which

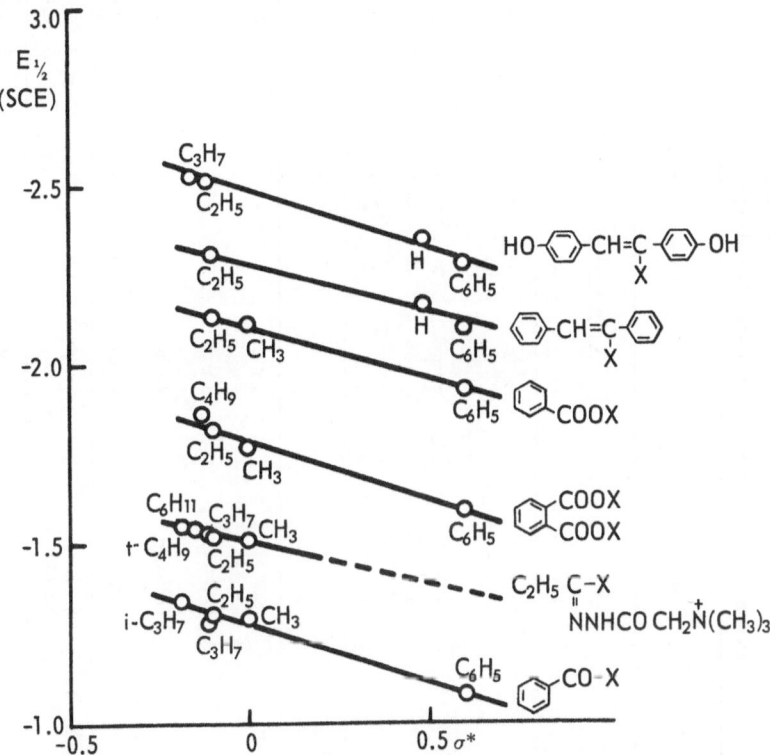

Fig. V-1. Relation of half-wave potentials for reductions of alkyl- and phenyl-substituted compounds to Taft polar substituent constants σ_X^*. Aryl alkyl ketones, Reaction Series No. 2, Table IV-2; ethyl alkyl ketone betainylhydrazones, Reaction Series No. 17; phthalic diesters, Reaction Series No. 26; benzoic esters, Reaction Series No. 23; substituted stilbenes, Reaction Series No. 35 and 36.

Table V-2

Reaction Series of Compounds For Which the Validity of Equations (25) and (29) Has Been Proved

No.	Reaction series	Solution	Note	Equation	$\rho^*_{z,R}$, V	E_{CH_3}, (SCE)	n	Substituents Fitting linear relation	Substituents Deviating	Ref. to source of $E_{1/2}$
						A. Reduction by Nucleophilic Mechanism				
1	OHC—Z	0.1 N LiOH		29	+0.35	−1.87	4	CH_3; C_2H_5; C_3H_7; $i\text{-}C_3H_7$	—	16
2	C_6H_5CO—Z	Acetate, pH 5		29	+0.31	−1.29	5	CH_3; C_2H_5; $n\text{-}C_3H_7$; $i\text{-}C_3H_7$; C_6H_5	—	2
3		NH_3, NH_4Cl, pH 9		29	+0.39	−1.48	6	CH_3; C_2H_5; $n\text{-}C_3H_7$; $i\text{-}C_3H_7$; $t\text{-}C_4H_9$; C_6H_5	—	2
4		Buffers, pH 0	extrapol.	29	+0.39	−1.02	3	CH_3; $CH_2C_6H_5$; C_6H_5	—	17
5		Buffers, alkal. 30% i-PrOH	$f \neq$ pH	29	+0.42	−1.58	3	CH_3; $CH_2C_6H_5$; C_6H_5	—	17
6		0.1 M LiOH		29	+0.16	−1.61	3	CH_3; $CH_2C_6H_5$; $C_6H_4.C_6H_5$	C_6H_5	18
7	$CH_3C(NH)$—Z	0.1 M NH_3, 1 M NH_4Cl	pH 8.2	29	+0.31	−1.44	3	CH_3; C_2H_5; CH_2COCH_3	—	19
8	$C_6H_5C(NOH)$—Z	Buffers, alkal. NH_4^+; 3% i-PrOH	$f \neq$ pH	29	+0.48	−1.77	3	CH_3; $CH_2C_6H_5$; C_6H_5	—	18
9		Buffers, alkal.	$f \neq$ pH	29	+0.31	−1.82	4	CH_3; C_6H_5; $C_6H_4.C_6H_5$ (*syn*); $C_6H_4.C_6H_5$ (*anti*)	$CH_2C_6H_5$	17
10		Buffers, pH 0	extrapol.	29	+0.31	−0.85	4	CH_3; $CH_2C_6H_5$; $C_6H_4.C_6H_5$ (*sym*); $C_6H_4.C_6H_5$ (*anti*)	C_6H_5	17
11		Buffers, pH 0	extrapol.	29	+0.36	−0.77	3	CH_3; H; C_6H_5	$CH(OH)C_6H_5$	20
12	$C_6H_5C_6H_4C(NOH)$—Z	Buffer, pH 0 30% i-PrOH	extrapol.	29	+0.22	−0.70	4	CH_3; C_6H_5 (*syn*); C_6H_5 (*anti*); $C_6H_4.C_6H_5$	—	17
13	$C_6H_5C(NOH)CO$—Z	Buffer, pH 0	extrapol.	29	+0.18	−0.49	3	CH_3; $CH_2C_6H_5$; $C_6H_5(\alpha)$	$C_6H_5(\beta)$	20

No.	Compound	n				Substituents R		Ref.	Conditions
14	$HC{-}Z$ ‖ $NNHCONH_2$	29	+0.21	−1.16	3	CH_3; H; $CH_3CH{=}CH$	C_4H_9	21	Acetate, pH 4.7 0.0016% Triton X
15	$CH_3C{-}Z$ ‖ $NNHCONH_2$	29	+0.31	−1.15	7	CH_3; C_6H_5; $i\text{-}C_3H_7$; $t\text{-}C_4H_9$; $CH_2C_6H_5$; C_6H_5; $CH_2OC_6H_5$	C_6H_{13}[a]; $(CH_2)_2C_6H_5$	22	Acetate, pH 5.3
16	$CH_3C{-}Z$ ‖ $NNHCOCH_2N(CH_3)_3Cl$	29	+0.15	−1.50	7	CH_3; C_2H_5; $n\text{-}C_3H_7$; $i\text{-}C_3H_7$; $n\text{-}C_4H_9$; $t\text{-}C_4H_9$; cyclo-C_6H_{11}	$i\text{-}C_4H_9$	23	BR Buffer, 0.2 M KCl pH 8, 50% EtOH
17	$C_2H_5C{-}Z$ ‖ $NNHCOCH_2N(CH_3)_3Cl$	29	+0.24	−1.51	5	CH_3; C_2H_5; C_3H_7; $i\text{-}C_3H_7$; cyclo-C_6H_{11}	$t\text{-}C_4H_9$	23	BR Buffer, 0.2 M KCl pH 8, 50% EtOH
18	$C_3H_7C{-}Z$ ‖ $NNHCOCH_2N(CH_3)_3Cl$	29	+0.27	−1.51	5	CH_3; C_2H_5; C_3H_7; $i\text{-}C_3H_7$; cyclo-C_6H_{11}	—	23	BR Buffer, 0.2 M KCl pH 8, 50% EtOH
19	$i\text{-}C_3H_7C{-}Z$ ‖ $NNHCOCH_2N(CH_3)_3Cl$	29	+0.27	−1.54	4	CH_3; C_2H_5; C_3H_7; cyclo-C_6H_{11}	—	23	BR Buffer, 0.2 M KCl pH 8, 50% EtOH

[a] Deviating transfer coefficient.
[b] Half-wave potential for methyl acrylate.
[c] Reaction constants ρ^*_{zR} for these symmetrical molecules were calculated using σ^*_z instead of σ^*_{zz}.
[d] 0.005% Bromcresol Green and 0.005% Methyl Red.
[e] Extrapolated value.
[f] The half-wave potential for methylphenylstilbene, but not ethylphenylstilbene derivative, was included.
[g] Reduction of C—Cl bond.
[h] ρ^*_{zR}.
[j] Worse correlation than for other half-wave potentials.
[k] Radicals; change of transfer coefficient.
[l] Versus Hg pool.
[m] $Z^1 = Z^2 = CH_3$.
[n] For $\sigma^*_{COOH} = 1.10$, $\sigma^*_{CH_2CH_2COOH} = 0.14$.
[o] Probably steric effects.
[p] Wave is smaller than corresponds to two-electron transfer.
[r] In the aliphatic chain COOH or COO$^-$ is treated as CH_3 to the first approximation.
[s] $Z^1 = Z^2 = Z^3 = H$.
[t] Anodic waves.

Table V-2 (*continued*)

No.	Reaction series	Solution	Note	Equation	n	E_{CH_3}, (SCE)	$\rho^*_{\pi.r.}$, V	Fitting linear relation	Deviating	Ref. to source of $E_{1/2}$
20	$CH_2=C(CH_3)COOZ$	0.1 M NMe$_4$I, 50% EtOH		29	3	-1.82[b]	$+0.80$	CH_3[b]; C_2H_5; C_4H_9	—	24
21		0.1 M NMe$_4$Br, 50% EtOH		29	4	-1.95	$+0.60$	CH_3; C_2H_5; C_3H_7; C_4H_9	—	25
22		0.1 N Li$_2$SO$_4$, 10% MeOH		29	3	-1.88	$+0.76$	CH_3; C_2H_5; C_4H_9	—	26
23	C_6H_5COOZ	Unbuffered, pH 6.7–7	$f \neq$ pH	29	3	-2.12	$+0.30$	CH_3; C_2H_5; C_6H_5	C_4H_9	27
24	o-$C_6H_4(COOZ)_2$	0.5 M NMe$_4$Cl 20% EtOH	1st wave 0.001% gel.	29	4	-1.77	$+0.30$[c]	CH_3; C_2H_5; C_4H_9; C_8H_{17}	—	28
25		0.6 M NMe$_4$Cl 60% EtOH	1st wave 0.001% gel.	29	4	-1.73	$+0.42$[c]	CH_3; C_2H_5; C_4H_9; C_8H_{17}		28
26		Buffer, pH 10	$f \neq$ pH	29	4	-1.77	$+0.39$[c]	CH_3; C_2H_5; C_4H_9; C_6H_5	—	29
27	NO_2—Z	Buffers, pH 7.0		29	4	-0.88	$+0.08$	CH_3; C_2H_5; C_3H_7; i-C_3H_7	—	30
28		Buffers, pH 8.9		29	4	-0.92	$+0.22$	CH_3; C_2H_5; C_3H_7; i-C_3H_7	—	30
29		Glycine, pH 10.9	0.005% br.-g.[d]	29	4	-0.94	$+0.20$	CH_3; C_2H_5; C_3H_7; i-C_3H_7	—	30
30		Buffers, pH 11.9		29	4	-0.96	$+0.12$	CH_3; C_2H_5; C_3H_7; i-C_3H_7	—	30
31		Clark–Lubs, pH 5.0		29	4	-0.80	$+0.13$	CH_3; C_2H_5; C_3H_7; i-C_3H_7	—	31
32		Clark–Lubs, pH 6.0		29	4	-0.85	$+0.19$	CH_3; C_2H_5; C_3H_7; i-C_3H_7	—	31
33		Clark–Lubs, pH 7.0		29	4	-0.88	$+0.26$	CH_3; C_2H_5; C_3H_7; i-C_3H_7	—	31
34	$C_6H_5CH=C(C_6H_5)Z$ $\vert$ Z	0.1 M NEt$_4$I 75% dioxane		29	3	-2.3[e]	$+0.29$	C_2H_5; H; C_6H_5	—	32
35	p-OHC$_6$H$_4$CH=C(p-OHC$_6$H$_4$)Z $\vert$ Z	0.1 M NEt$_4$I 75% dioxane		29	4	-2.48[e]	$+0.30$	C_3H_7; C_5H_{11}; H; C_6H_5	i-C_3H_7	32

#	Z-compound	Conditions		Ref.			n	R groups		Ref.
36	$C_6H_5(C_2H_5)C=C(C_6H_5)$, Z (Z on double bond)	0.1 M NEt$_4$I 75% dioxane		29	+0.49	-2.56^d	3	C_2H_5; H; $C_6H_5^f$	—	32
37	Br—Z	0.05 M NEt$_4$Br		29	$+4.9_0$	-1.63	3	CH_3; C_2H_5; C_4H_9	—	33
38		0.01 M NEt$_4$Br N,N-dimethylformamide		29, 36	$+4.9_5$	-1.65^d	8	C_2H_5; C_3H_7; C_4H_9; i-C_4H_9; C_5H_{11}; (t-C_4H_9)CH_2; C_6H_{13}; $C_{10}H_{21}$	i-C_3H_7; t-C_4H_9	34
39		0.01 M NEt$_4$Br N,N-dimethylformamide		29	4.5	—	6	C_2H_5; C_3H_7; C_4H_9; i-C_4H_9; neo-C_5H_{11}; cyclo-C_6H_{11}	i-C_3H_7; t-C_4H_9; cyclo-C_5H_9	35
40	$C_6H_5SO_2$—Z	0.05 M NMe$_4$Cl 75% dioxane	$f \neq pH$	29	+0.21	-2.27	3	CH_3; $CH_2C_6H_5$; C_6H_5	CH_2Cl^g	36
41		0.05 M NMe$_4$Cl 75% dioxane	$f \neq pH$	25	$+0.90^h$	-2.27	4	CH_3; C_6H_5; OCH_3; OC_6H_5	—	36
42	Z—SS—Z	0.025 M N(Bu)$_4$OH 40% MeOH, 40% i-PrOH	pH 12.3	29	$+1.10^e$	-1.64^e	11	C_2H_5; C_3H_7; C_4H_9; i-C_4H_9; t-C_4H_9; C_5H_{11}; i-C_5H_{11}; C_6H_{13}; C_7H_{15}; C_8H_{17}; $CH_2C_6H_5$	C_6H_5	37
43		0.025 M N(Bu)$_4$OH 80% i-PrOH		29	$+1.5^c$	-1.67^e	3	C_4H_9; $CH_2C_6H_5$; C_6H_5	t-C_4H_9	38
44		0.05 M NBu$_4$OH 80% i-PrOH	$\delta_{\alpha,R} = 0.09_7^c$	29, 36	$+1.61^c$	-1.49^e	11	CH_3; C_2H_5; C_3H_7; i-C_3H_7; s-C_4H_9; C_5H_{11}; $(CH_3)_2C(C_2H_5)$; cyclo-C_5H_9; C_8H_{17}; $C_{10}H_{21}$; C_6H_5	t-C_4H_9; $CH_2C_6H_5$	39
45	Z—SSSS—Z	0.025 M NBu$_4$OH 80% i-PrOH		29	$+0.33^c$	-1.20^e	3	C_4H_9; t-C_4H_9; $CH_2C_6H_5$	—	38
46	HOO—Z	0.1 M H$_2$SO$_4$ 5-20% EtOH	$\delta_{\alpha,R} = -0.08$	29, 36	+1.5	—	8	C_4H_9; i-C_4H_9; s-C_4H_9; t-C_4H_9; C_5H_{11}; i-C_5H_{11}; $C_7H_7CHCH_3$; $(C_2H_5)_2CH$	C_6, C_7 and C_8^k	40

B. Reduction by Electrophilic Mechanism

#	Z-compound	Conditions		Ref.			n	R groups		Ref.
47	NO_2—Z	BR pH 1.8 3% MeOH		29	-0.52	-0.75	3	CH_3; C_2H_5; C_3H_7	—	41
48		pH 2.0 Clark–Lubs		29	-0.90	-0.71	3	CH_3; C_2H_5; C_3H_7	i-C_3H_7	31

Table V-2 (*continued*)

No.	Reaction series	Solution	Note	Equation	$\rho^*_{z,x}$, V	E_{CH_3} (SCE)	n	Substituents Fitting linear relation	Deviating	Ref. to source of $E_{1/2}$
49	NO_2-Z	Clark–Lubs pH 3.0		29	−0.19	−0.75	3	CH_3; C_2H_5; C_3H_7	$i\text{-}C_3H_7$	31
50		Clark–Lubs pH 4.0		29	−0.09	−0.78	3	CH_3; C_2H_5; C_3H_7	$i\text{-}C_3H_7$	31
51		0.05 M H_2SO_4		29	−0.90	−0.69	4	CH_3; C_2H_5; C_3H_7; C_4H_9	$i\text{-}C_3H_7$; $s\text{-}C_4H_9$	42
52		McIlvaine, pH 2.1	br.g.ᵈ	29	−0.15	−0.60	3	CH_3; C_2H_5; C_3H_7	$i\text{-}C_3H_7$	30
53		McIlvaine, pH 5.1	br.g.ᵈ	29	−0.10	−0.81	4	CH_3; C_2H_5; C_3H_7; $i\text{-}C_3H_7$	—	30
54		1 M NH_4OAc glacial HOAc		29	−0.20	−1.12ˡ	3	CH_3; C_2H_5; C_3H_7	$i\text{-}C_3H_7$	43
56	O_2NO-Z	0.5 M LiCl 11.2% EtOH	0.01% gel. $f \neq pH$	29	−4.8	—	3	C_2H_5; C_6H_{13}; cyclo-C_6H_{11}	—	44
57	$Z-C-N=N-C_6H_5$ $\Vert$ $NNH-C_6H_5$	Acetate pH 5 75% MeOH		29	−0.06₆	−0.18	3	CH_3; C_2H_5; C_6H_5	—	45
58	$CH_2ClCOOZ$		$f \neq pH$	29	−1.0	−1.54	3	CH_3; C_2H_5; C_4H_9	—	46
59	$OHC-CH(Br)-Z$	0.1 M LiCl		29	−0.70	−0.2ᶜ	3	C_2H_5; C_5H_{11}; H	—	47
60		0.1 M LiCl 50% dioxane		29	−0.52	−0.5ᶜ	3	C_2H_5; C_5H_{11}; H	—	47
61	$HOO-Z$	0.5 M H_2SO_4	$f \neq pH$	29	−0.76	(−0.64)	3	C_2H_5; $t\text{-}C_4H_9$; H	CH_3ᵃ	48
62		0.1 M Li_2SO_4 0.01 M LiOH		29	−0.88	(−0.95)	3	C_2H_5; $t\text{-}C_4H_9$; H	CH_3ᵃ	48
63	$CH_2=CH-C\equiv C-Z$	NBu_4I 75% dioxane		29	−0.46	−2.63	4	CH_3; C_2H_5; C_4H_9; C_6H_{13}	H	49

				E_{tang}	$f \neq pH$				
64	NZ_4^+	—	29	−0.92	—	3	CH_3;C_2H_5;C_4H_9	—	50
65	Z_2Tl^+		29	−3.6^e	—	3	C_2H_5;C_3H_7;i-C_3H_7	—	51
66	$Z-Hg^+$		29	−1.06	—	3	C_2H_5;C_3H_7;i-C_3H_7	—	51
67	Z_3Pb^+	pH 1	29	−0.18 (−0.53^g)	−0.75^e	5	C_2H_5;C_3H_7;C_4H_9;$CH_2C_6H_5$; C_6H_5	—	52

C. Extended Examples

					$f \neq pH$				
68	$Z^1-N(NO)-Z^2$	Buffer, pH 9.9	29	+0.61	−1.65,^{m}	6	$(CH_3)_2$;$(C_2H_5)_2$;$(C_3H_7)_2$; (CH_3,C_6H_5);(C_2H_5,C_6H_5);$(C_6H_5)_2$	—	53
69		0.1 N NaOH	29	+0.43	−1.68^m	3	$(CH_3)_2$;$(C_2H_5)_2$;$(C_6H_5)_2$	—	54
70	$Cl-CZ^1Z^2Z^3$	0.1 N H_2SO_4, 0.1 N K_2SO_4, 20% EtOH	38^f	+0.45^h	—	4	(H,CH_3,NO_2);(CH_3,NO_2,NO_2); (C_2H_5,NO_2,NO_2);$(NO_2)_3$	(H,H,NO_2)	55
71	$Br-CZ^1Z^2Z^3$	0.1 N H_2SO_4, 0.1 N K_2SO_4, 20% EtOH	38^f	+0.15^h	—	4	(H,CH_3,NO_2);(CH_3,NO_2,NO_2); (C_2H_5,NO_2,NO_2);$(NO_2)_3$	(H,H,NO_2)	55
72	$Z^1Z^2C=CZ^3Z^4$	0.17 M NBu_4I 75% dioxane	38"	+0.55	—	6	(H,H,C_6H_5);(H,C_6H_5,H,CH_3); (H,C_6H_5,C_6H_5,C_6H_5);(H,C_6H_5,C_6H_5,C_6H_5); $(C_6H_5,H,C_6H_5,C_6H_5,C_6H_5)$;$(C_6H_5)_4$	—	56
73	$Z-CH=CHCH_2Br$	Buffer, pH 4.9	25	+0.25^h	−1.10	4	CH_3;H;Cl;Br	—	57
74	$CH_2=C(Z)=CH_2Br$	Buffer, pH 4.9	25	+0.52^h	−1.08	3	CH_3;H;Cl	—	57
75	$I-CH_2-SiZ^1Z^2$ / CH_3	0.09 N KCl, 57% EtOH	38"	+0.19	−1.58^m	5	(CH_3,CH_3);(C_2H_5,C_2H_3); (C_3H_7,C_3H_7);(C_4H_9,C_4H_9); (CH_3,C_6H_5)	—	58
				E_{tang}					
76	$Z(CH_2)_2\overset{+}{N}H(C_2H_5)_2$	—	25	+0.80^h	−2.11	5	CH_3;H;OH;Cl;CN	Br	59

Table V-2 (*continued*)

No.	Reaction series	Solution	Note	Equation	$\rho^*_{z,R}$, V	E_{CH_3}, (SCE)	n	Substituents — Fitting linear relation	Substituents — Deviating	Ref. to source of $E_{1/2}$
77	Z^1-C-Z^2 $\underset{NOH}{\parallel}$	—		38^n	$+0.66$	—	9	$(CH_3,C_6H_5):(CH_3,COOH)^a$: $(CH_3,CH_2C_6H_5):(i\text{-}C_3H_7,CH_3CO)$: $(C_6H_5,H):(C_6H_5,COOH)^a$: $(CH_2C_6H_5,COOH)^a$: $(C_2C_6H_5,CH_3CO)$: $(COOH,CH_2COOH)^a$	—	60
78	Z^1-C-Z^2 $\underset{NNHCONH_2}{\parallel}$	Citrate–phosphate pH 6.0, 20% MeOH		38^n	$+0.09$	-1.42^m	7	$(CH_3,CH_3):(CH_3,C_2H_5):(CH_3,C_3H_7)$: $(CH_3,C_4H_9):(CH_3,C_5H_{11})$: $(CH_3,C_6H_{13}):(C_2H_5,C_2H_5)$	$(CH_3,i\text{-}C_3H_7)^p$: $(CH_3,i\text{-}C_4H_9)^{n,o}$	3
79	Z^1-C-Z^2 $\underset{NNHCSNH_2}{\parallel}$	Citrate–phosphate pH 6.0, 20% MeOH		38^n	$+0.06$	-1.42^m	6	$(CH_3,CH_3):(CH_3,C_2H_5)$: $(CH_3,C_4H_{11}):(CH_3,C_4H_9)$: $(CH_3,C_5H_{11}):(C_2H_5,C_2H_5)$	(CH_3,C_6H_{13}): $(CH_3,i\text{-}C_3H_7)^p$	3
80	Z^1-C-Z^2 $\underset{NNHCOCH_2N(CH_3)_3Cl}{\parallel}$	BR. 0.2 M KCl 50% EtOH	pH 8	29	$+0.25_5$	-1.50^m	17	$(CH_3)_2:(CH_3,C_2H_5):(CH_3,C_3H_7)$: $(CH_3,C_4H_9):(CH_3,i\text{-}C_3H_7)$: $(CH_3,t\text{-}C_4H_9):(CH_3,C_5H_{11})$: $(CH_3,\text{cyclo-}C_5H_{11}):(C_2H_5)_2$: $(C_2H_5,C_3H_7):(C_2H_5,i\text{-}C_3H_7)$: $(C_2H_5,t\text{-}C_4H_9):(C_2H_5,\text{cyclo-}C_6H_{11})$: $(C_3H_7)_2:(C_3H_7,i\text{-}C_3H_7)$: $(i\text{-}C_3H_7,\text{cyclo-}C_5H_{11})$	$(CH_3,i\text{-}C_4H_9)$	23
81	$NO_2\text{-}CH(C_2H_5)CH_2OZ$	Buffer, pH 5.0		29	-0.46	-0.71	3	$CH_3:C_2H_5:C_3H_7$	—	61
82		Buffer, pH 7.0		29	-0.50	-0.81	3	$CH_3:C_2H_5:C_3H_7$	—	61
83	$C_2H_5OOCCH=CHCONZ_2$	0.2 N LiCl		38^n	$+0.07^c$	-1.21^m	4	$(CH_3)_2:(C_2H_5)_2:(H)_2:(C_6H_5)_2$	$(C_4H_9)^p$	62
84		0.2 N LiOH		38^n	$+0.20^c$	-1.64^m	4	$(CH_3)_2:(C_4H_9)_2:(H)_2:(C_6H_5)_2$	$(C_2H_5)_2$	62
85	$C_6H_5SO_2N\text{-}Z^1Z^2$	0.05 M NMe$_4$Cl 75% dioxane	$f \neq$ pH	38^n	-0.16	-2.22^m	5	$(CH_3)_2:(H,CH_3):(H,CH_2C_6H_5)$: $(H,H);(H,C_6H_5)$	$(H:xyclo\text{-}C_6H_{11})$: (CH_3,C_6H_5): $(C_6H_5)_2$	36

No.	Structure	Medium		Temp.		Value 1	Value 2	Substituents		Ref.
86	HOOC(Br)C—Z¹Z²	Buffers, acid.	$f \neq pH$	38°	7	−0.39	−0.28^m	(CH₃)₂; (CH₃,C₂H₅); (C₂H₅)₂; (C₂H₅,C₄H₉); (H,H); (H,CH₃); (H,C₂H₅)	—	63
87		Buffer, pH 3	$f \neq pH$	38°	3	−0.49	−0.35^m	(H)₂; (H,CH₃); (CH₃)₂	—	64
88	⁻OOC(Br)C—Z¹Z²	Buffer, alkal.	$f \neq pH$	38°	5	−0.30	−1.06^m	(CH₃)₂; (CH₃,C₂H₅); (H,CH₃); (H,C₂H₅); (H,C₃H₇)	(C₂H₅)₂; (H)₂; (C₂H₅,C₄H₉)	63
89	C₂H₅OOC(Br)C—Z¹Z²		$f \neq pH$	38°	5	−0.25	−0.25^m	(CH₃)₂; (C₂H₅)₂; (C₂H₅,C₄H₉); (H; CH₃); (H,C₂H₅)	—	63
90	Br—Z Br...COO⁻	alkal.	$f \neq pH$	29	4	−1.1	—	CH₂COO⁻ ($\sigma^*_{C_2H_5}$); CH(CH₃)COO⁻ ($\sigma^*_{CH(CH_3)_2}$); CH(C₂H₅)COO⁻ ($\sigma^*_{CH(C_2H_5)CH_3}$); C(CH₃)₂COO⁻ ($\sigma^*_{C_4H_9}$)	—	63
91	Br...COOH	Buffer, acid	$f \neq pH$	29	4	−2.0	—	CH₂COOH ($\sigma^*_{C_2H_5}$); CH(CH₃)COOH ($\sigma^*_{i-C_3H_7}$); CH(C₂H₅)COOH ($\sigma^*_{i-C_4H_9}$); C(CH₃)₂COOH ($\sigma^*_{t-C_4H_9}$)	—	63
92	Br...COOC₂H₅		$f \neq pH$	29	3	−0.9	—	CH(CH₃)COOC₂H₅ ($\sigma^*_{i-C_3H_7}$); CH(C₂H₅)COOC₂H₅ ($\sigma^*_{t-C_4H_8}$); C(CH₃)₂COOC₂H₅ ($\sigma^*_{t-C_4H_8}$)	—	63
93	Z¹Z²C—CHZ³ Br Br	3 M NaOAc, 80% HOAc		38°	5	+1.0^b	−1.23^a	(H,H,H); (H,H,Cl); (H,Cl,Cl); (H,Br,Br); (Cl,Cl,Cl)	—	65, 66
94	Z¹—P(OZ²)₂ ‖ O	—		38°	5	+0.22	−1.25^m	(CH₃,CH₃,CH₃); (CH₃,C₄H₉,C₄H₉); (C₂H₅,C₂H₅,C₂H₅); (C₃H₇,C₃H₇,C₃H₇); (C₅H₁₁,C₅H₁₁,C₅H₁₁)	—	14
95	Z¹NHCSNHZ²	Borate, pH 9.3		38°	5	−0.85^{a,i}	—	(H,H); (H,C₆H₅); (CH₃,CH₃); (C₂H₅,C₂H₅); (C₆H₅,C₆H₅)	—	67
96		1 N NaOH		38°	4	−0.7^{a,i}	—	(H,H); (H,C₆H₅); (CH₃,CH₃); (C₂H₅,C₂H₅); (C₆H₅,C₆H₅)	—	67
97		0.2 N NaOH		38°	3	−1.1^{a,i}	−0.30^{a,i}	(H,H); (H,C₆H₅); (C₆H₅,C₆H₅)	—	68

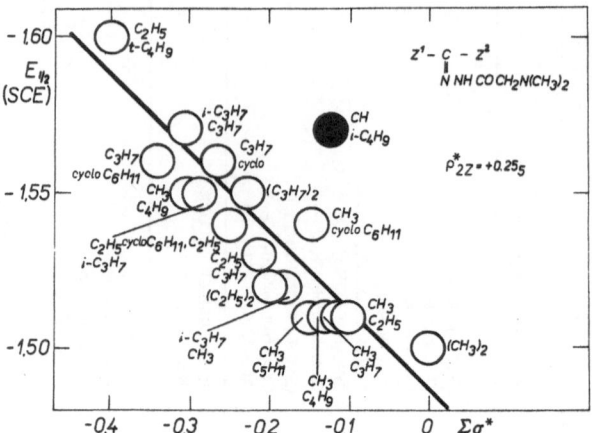

Fig. V-2. Relation of half-wave potentials for the reduction of dialkyl ketone betainyl-hydrazones to the sum of the Taft polar substituent constants $\Sigma\sigma_X^*$. Reaction Series No. 80, Table V-2. Full point deviates.

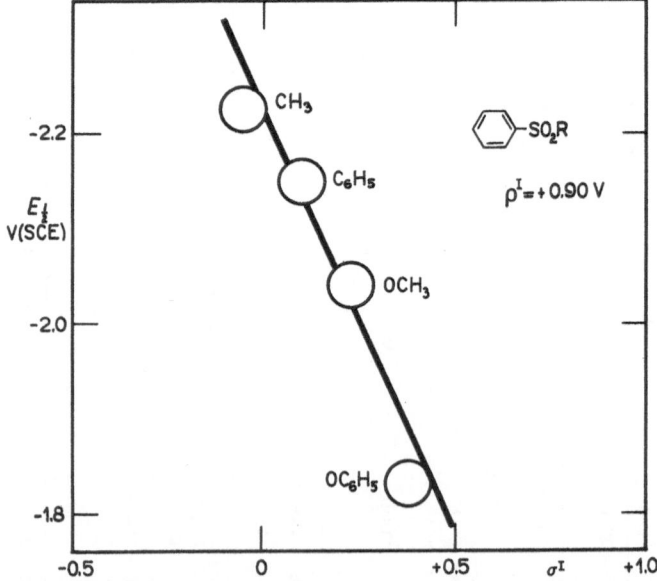

Fig. V-3. Relation of half-wave potentials for the reduction of substituted phenyl sulphones to inductive substituent constants σ_X^I. Reaction Series No. 41, Table V-2.

the electroactive group is not directly attached to the substituent, are summarized. Furthermore, some anodic waves and some other less frequent types of reaction series are given there.

In this chapter, the part of the molecule that is changed by substitution is denoted by Z. Symbol E_{CH_3} stands for the value of the half-wave potential of the parent substance of the reaction series for which $Z = CH_3$, n is the number of substances that fit the equation used, $f \neq pH$ is a symbol for pH-independent half-wave potentials (in the pH range studied). Some selected examples are given in Figs. V-1 to V-3.

3. Effect of Changes in the Mechanism of the Electrode Process Reflected in Deviations from Linear Free Energy Relations and Changes in the Sign of the Reaction Constant

Generally speaking, the validity of any discussion of reaction series for which the shifts of half-wave potentials are given by equation (29) is more restricted than in the case of series of benzene derivatives which fit equation (27). This is due to several factors: First, changes in the mechanism of the electrode process within a reaction series seem to be more frequent here than for benzene compounds. In some instances, not only the introduction of a bulky substituent, but even a change in the pH of the supporting electrolyte results in an inversion of the mechanism. Among reaction series of compounds in which an alkyl group is attached to the electroactive group, those for which $\rho < 0$ are relatively frequent. This type of behavior, which is associated with an electrode process with an electrophilic mechanism, is more frequent among reaction series of this type (and particularly among organometallic compounds) than among benzene compounds in which the reactive group is in the side chain (see Chapter III). The existence of a parallelism between the mechanism of the reduction of a reactive group attached to a benzene ring and that of the same group in an aliphatic compound is far from being general.

Another factor that limits the significance of the discussion of reaction series of aliphatic and related compounds is that these reaction series consist, on the average, of even smaller numbers of substituted compounds than the corresponding reaction series of benzene compounds. Furthermore, the most frequently studied substituents do not cover a sufficiently broad range of σ_X^* values. It frequently happens that a great number of half-wave potentials are measured for

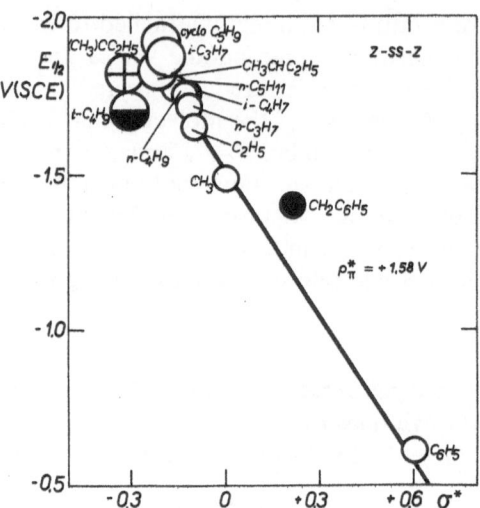

Fig. V-4. Relation of half-wave potentials for the reduction of dialkyl disulphides to Taft polar substituent constants σ_X^*. Reaction Series No. 44, Table V-2. Full point and halved point: deviating; crossed circle and halved point: steric effects operating.

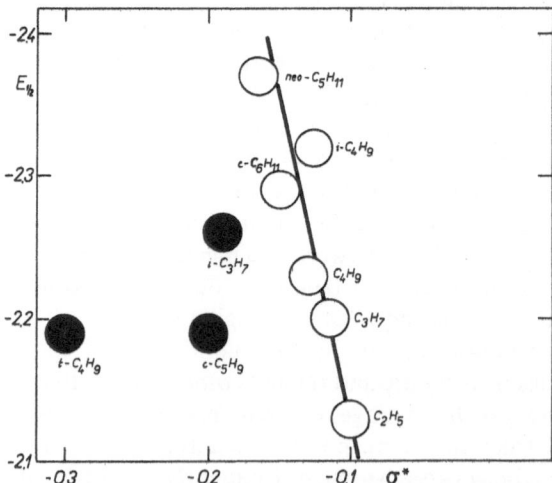

Fig. V-5. Relation of half-wave potentials for the reduction of alkyl bromides to Taft polar substituent constants σ_X^*. Reaction Series No. 39, Table V-2. Full points deviate.

compounds containing substituents that differ only a little in the value of σ_X^* (Fig. V-4).

We shall discuss first examples of reaction series which show deviations that can be explained by changes in reaction mechanism, and this will be followed by a discussion of series in which both the value and the sign of the reaction constant depend on the reaction conditions, in particular on the pH of the polarographed solution.

Marked deviations were observed for α-branched alkyl bromides (Fig. V-5). The deviations were still marked when the half-wave potentials were corrected for steric effects of the type considered by Taft.[1] The value for the steric reaction constant $(\delta_{\pi,R} = -0.04 \text{ V})$ was computed from a system of linear equations (36) using tabulated data on steric substituent constants $(E_S)_X$ (see Table V-1). When the corrected value $[\Delta E_{1/2} - \delta_{\pi,R}(E_S)_X = (\Delta E_{1/2})_{calc}]$ was plotted against $(E_{1/2})_{exp}$ obtained experimentally (Fig. V-6), deviations were observed which were similar to those of the uncorrected values. Since the half-wave

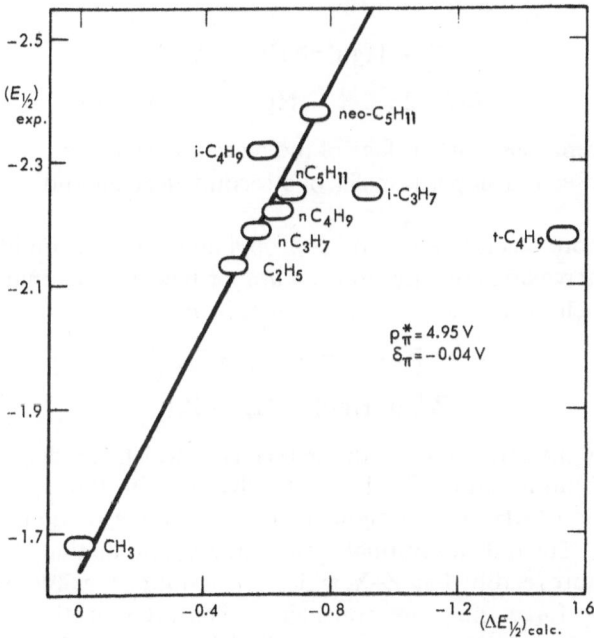

Fig. V-6. Relation of measured half-wave potentials $(E_{1/2})_{exp}$ for the reduction of alkyl bromides to the half-wave potentials calculated from the equation $(\Delta E_{1/2})_{calc} = \rho_{\pi,Br}^* \sigma_X^* + \delta_{\pi,Br}(E_S)_X$ for $\rho_{\pi,Br}^* = 4.9_5$ V and $\delta_{\pi,Br} = -0.04$ V. Reaction Series No. 39, Table V-2.

potentials observed were shifted toward more positive potentials than those predicted by equations (29) and (36), steric hindrance to the approach of the molecule to the electrode surface,[34] which would undoubtedly result in a more negative reduction potential, need not be considered.

For alkyl halides Elving[70] suggested the possibility of three different types of reduction mechanism: (*i*) analogous to $S_N 1$; (*ii*) analogous to $S_N 2$; (*iii*) a radical mechanism, analogous to $S_H 1$ or $S_H 2$. At that time, the available data did not allow[70] a decision to be made between these types of mechanisms.

The observed deviations can be explained on the basis of change in the relative contributions of $S_N 1$- and $S_N 2$-like mechanisms. For straight-chain alkyl halides and for alkyl halides branched at the β-carbon (Reaction Series No. 37–39, 73, 74) or at β-silicon (No. 75), the reaction constant ($\rho^*_{\pi, R}$) is found to be positive. On analogy to the numerous homogeneous nucleophilic reactions of such compounds, a predominating mechanism of the $S_N 2$ type is assumed. The potential is determined by the step:

$$Z - X + Hg(l) \rightleftharpoons Hg \ldots Z \ldots X$$

$$Hg \ldots Z \ldots X \rightarrow Hg - Z \cdot + X^-$$

where Hg represents the surface of the mercury electrode.

In the second step a transfer of a second electron and of a proton follows:

The more positive waves of isopropyl and *t*-butyl bromides result from the increasing contribution of another mechanism, analogous to $S_N 1$, in which the first steps in the reduction are

$$Z - X \rightleftharpoons Z^+ + X^-$$

$$Z^+ + Hg(l) \rightarrow Hg - Z.$$

followed by the same steps as in the former case. The tendency toward a more predominating $S_N 1$-like mechanism in the series $C_2H_5 < i\text{-}C_3H_7 < t\text{-}C_4H_9$ is analogous to that observed for homogeneous reactions.[71] The only additional assumption would be that Z^+ accepts electrons more readily than Z–X, which seems quite plausible even when electrostatic forces only are considered. The rates of dissociation of Z–X into Z^+ and X^- observed for alkyl bromides in homogeneous kinetics are too low to affect the electrode kinetics. Nevertheless, it is possible to assume an increase in the rate of this reaction in the electrical field of the dropping electrode. Similar conclusions were reached by Lambert and Kobayashi[35] on the basis of a qualitative comparison of half-wave potentials and rates of nucleophilic reactions.

The deviation for the cyclopentyl derivative and the fitting of the half-wave potential of cyclohexyl derivatives (Fig. V-5) also could have been explained on the view that the contribution of the mechanism of $S_N 1$ type is greater for the cyclopentyl derivative. Unfortunately, the support from homogeneous kinetics is not unambiguous, as effects due to the replacement of cyclopentyl by cyclohexyl are similar in $S_N 1$ and in $S_N 2$ reactions. This question will be treated in more detail in Chapter IX.

On the other hand, support can be found in the reaction series of haloalkanoic acids and their anions and esters and of halo aldehydes (Reaction Series No. 58, 59, 86–92). The more bulky the substituent, the more positive the potential. Hence, with increasing size of the alkyl a more substantial contribution of mechanisms analogous to $S_N 1$ can be expected. In homogeneous reaction kinetics a predominating $S_N 1$ mechanism was deduced for these substances.[71]

The resulting shift of the half-wave potential results in a change in the sign of the reaction constant $\rho_{\pi.R}$ (Table V-2).

The above deductions can be used as a basis for the discussion of the course of the reduction of the dibromo derivative (I) in Reaction Series No. 73:

$$CH{=}CH{-}CH_2{-}Br^1 \atop \underset{Br^2}{|} \tag{I}$$

In this compound either Br^1 or Br^2 can be reduced. For derivatives of compound (I) containing another substituent Z instead of Br^2, equation (25) satisfactorily expresses the observed shifts in half-wave potential corresponding to the reduction of the $C{-}Br^1$ bond. Since the half-wave potential of (I) satisfactorily fits the equation $\Delta E_{1/2} = \rho^I_{\pi,Br^1}\sigma^I_{Z^2}$ for $\sigma^I_{Z^2} = \sigma^I_{Br}$, it can be assumed that even for the dibromide (I) the bromine in position Br^1 is reduced. The observed positive value of the reaction constant ρ^I_{π,Br^1} (Table V-2) is that expected for the reduction of a bromo derivative in which the bromide is bound to a CH_2 group. Independent support for the suggested course of the reduction process is provided by the observation[72] that halogens bound to an olefinic carbon are usually reduced at more negative potentials than those bound to a saturated carbon. The reduction of the $CH_2{-}X$

bonds is, on the other hand, facilitated by an unsaturated bond in the β-position.

The reduction of aliphatic nitro compounds follows an essentially different course from the reduction of nitrobenzene, its derivatives, and analogous heterocyclic substances. If we leave the extraordinary behavior of some *ortho-* and *para*-substituted nitrobenzenes (e.g., nitrophenols and nitroanilines) out of consideration, then for nitrobenzene

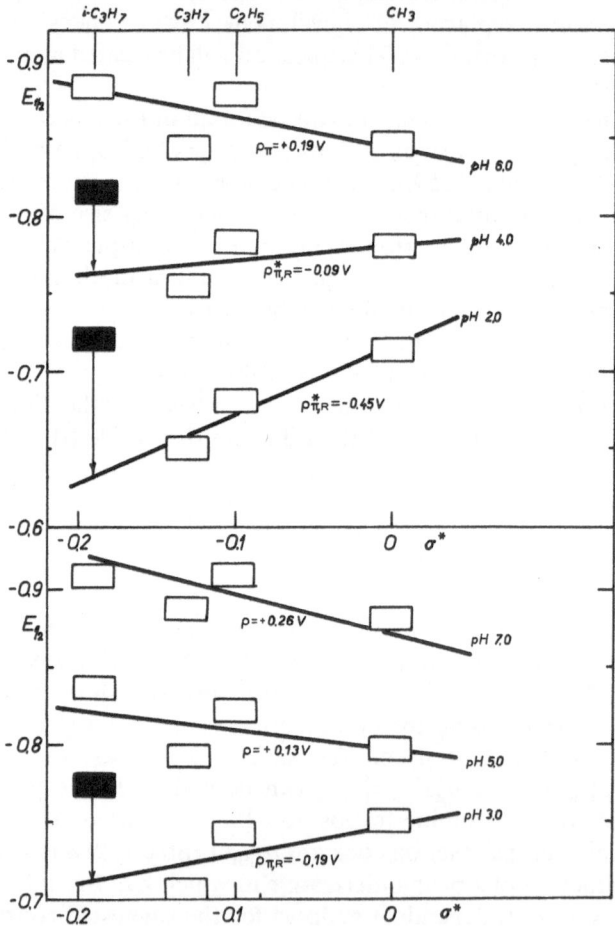

Fig. V-7. Relation of half-wave potentials for the reduction of aliphatic nitro compounds to Taft polar substituent constants σ_X^* at various pH values Reaction Series No. 31 to 33, 49, and 50, Table V-2. Full points deviate.

derivatives and nitroalkanes the same number of electrons is transferred in the electrode process. In both series, in the first step on the current–potential curve, reduction at the dropping mercury electrode occurs with the consumption of four electrons and the formation of hydroxylamine derivatives. But even though the number of the electrons transferred is the same, the mechanism is different. This is demonstrated by the difference in the shapes of the pH-dependence curves for the half-wave potentials of nitrobenzenes and nitroalkanes. Furthermore, the relation of $E_{1/2}$ to σ_X^* in the reaction series of nitroalkanes Z–NO$_2$ is not obeyed by the half-wave potential for Z $=$ C$_6$H$_5$ (i.e., of nitrobenzene). Nitroalkanes differ from nitrobenzene derivatives also in the change of the reaction constant ρ_{π,NO_2}^* with pH. For substituted nitrobenzenes in aqueous media (and similarly also for nitro derivatives of aromatic heterocycles) a positive value is always found for the reaction constant ρ_{π,NO_2}^*. For nitroalkanes both the value and the sign of the reaction constant ρ_{π,NO_2}^* changes with pH: in acid media (Reaction Series No. 47–54) ρ_{π,NO_2}^* is negative; in alkaline solutions (No. 27–31) it is positive (see Fig. V-7). Two mechanisms are thus operating in the reduction of nitroalkanes, one predominating in acid media, the other in alkaline. Little can be said about the nature of the electrophilic process taking place in acid media; the participation of proton transfer in the potential-determining step cannot be excluded—it would be analogous to the reduction of substituted nitrobenzenes in concentrated sulphuric acid media (see p. 98).

It is found that in several reaction series (No. 48–54) the half-wave potential of 2-nitropropane in acid media is shifted toward more negative values than those predicted by equation (29). A similar phenomenon is observed for 2-nitrobutane (No. 51). It can be assumed, on analogy to halogen derivatives, that for α-branched nitroalkanes in acid media the predominating mechanism differs from that applying to unbranched nitroalkanes. Moreover, since the half-wave potentials of 2-nitropropane satisfactorily fit the $E_{1/2}$–σ^* correlation at higher pH (Fig. V-7), it can be supposed that with 2-nitropropane the same mechanism operates both in acid and in alkaline media.

In the case of alkyl peroxides the situation is at present confused. A limited number of data covering the lower members of the reaction series (Reaction Series No. 61 and 62) give a negative value for the reaction constant, whereas, data for C$_4$ and C$_5$ derivatives in reaction series No. 46 give a positive value for the reaction constant $\rho_{\pi,OOH}^*$, determined by the use of equation (36). The comparability of the

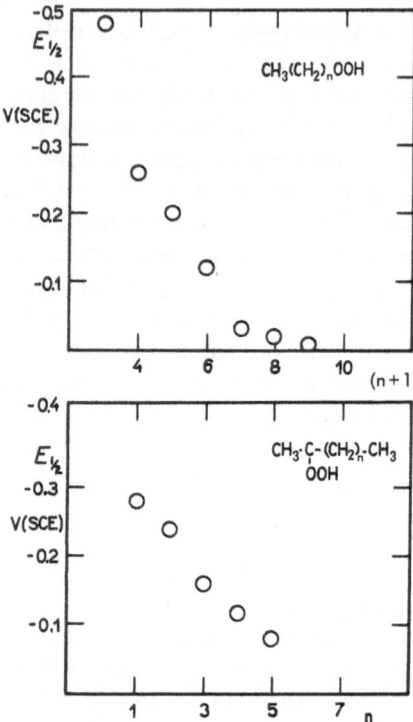

Fig. V-8. Relation of half-wave potentials for the reduction of alkyl hydroperoxides to chain length. Half-wave potentials from Ref. 40.

half-wave potentials in this reaction series seems doubtful, since there are indications that the transfer coefficient α does not remain constant throughout the series. In Refs. 40 and 48 there are no quantitative data on the wave shape, but the wave of methyl peroxide is described[48] as less steep, and the wave of t-butyl peroxide as more steep than the hydroperoxide wave. In Ref. 40 the waves of butyl and pentyl peroxides are described as drawn-out, and for the octyl and nonyl derivatives the values 0.026 and 0.024 are given for the logarithmic-analysis coefficients. The plot of the shifts in half-wave potential against chain length (Fig. V-8) possesses an unusual shape. Effects due to adsorption and the participation of a radical mechanism can possibly be involved, but these assumptions have not been adequately verified experimentally.

For unsaturated hydrocarbons in Reaction Series No. 63 $(Z-C{\equiv}C-CH{=}CH_2)$ there is an unexpected and marked shift for

$Z = H$ toward more positive potentials (Fig. V-9). A similar trend is observed for the difference between $Z = $ alkyl and $Z = H$ in the reaction series (all half-wave potentials from Ref. 49):

(a) $Z-C\equiv C-\underset{\underset{\displaystyle C(CH_3)_3}{|}}{C}=CH_2$

(b) $CH\equiv C-\underset{\underset{\displaystyle Z}{|}}{C}=CH_2$;

(c) $CH_3-C\equiv C-\underset{\underset{\displaystyle Z}{|}}{C}=CH_2$;

(d) $Z-C\equiv C-\underset{\underset{\displaystyle CH_3}{|}}{C}=CH_2$;

(e) $CH\equiv C-CH=CH-Z$;

(f) $CH\equiv C-\underset{\underset{\displaystyle CH_3}{|}}{C}=CH-Z$.

For all these reaction series, in which the number of compounds with different Z groups is too low to allow the computation of $\rho^*_{\pi,R}$, the half-wave potentials for $Z = $ alkyl are substantially more negative than for $Z = H$. For the difference $(E_{1/2})_{CH_3} - (E_{1/2})_H$ the following values have been found: Reaction Series No. 63: $+0.23$ V; (a) $+0.32$ V; (b) $+0.11$ V; (c) $+0.17$ V; (d) $+0.29$ V; (e) $+0.19$ V; (f) $+0.04_5$ V. A steric effect of the alkyl group can be involved, the effect being greater for substitution at a $C\equiv C$ bond than at a $C=C$ bond, and this would

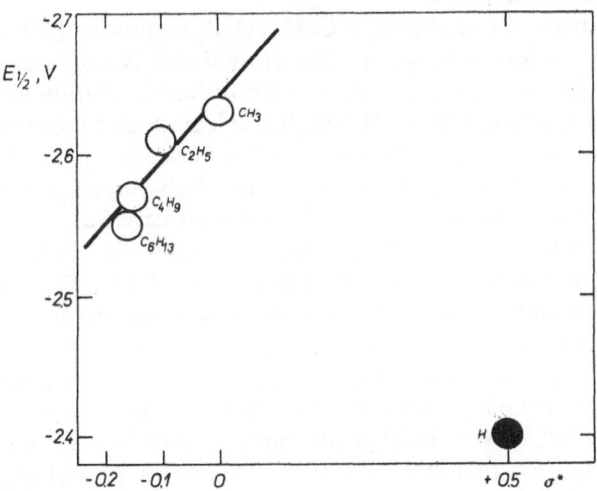

Fig. V-9. Relation of half-wave potentials for the reduction of hydrocarbons $Z-C\equiv C-CH=CH_2$ to Taft polar substituent constants σ^*_X. Half-wave potentials from Ref. 49. Full point deviates.

cause a shift in the half-wave potentials of alkyl derivatives toward more negative potentials. The linear course of the relation shown in Fig. V-9 for straight-chain alkyls, in which the whole observed line is shifted to more negative potentials when compared with the unsubstituted compound, is in agreement with the above assumption. This linear course, however, implies that the steric effects of all the straight-chain alkyls concerned are approximately the same when compared with that of hydrogen. Inspection of the E_S values in Table V-1 confirms this assumption.

A smaller shift toward more negative values was observed for the case of isopropyl substitution at a C=C bond in Reaction Series No. 35, but the authors mentioned[32] that this particular value of half-wave potential (-0.61 V) is not reliable.

In most of the examples discussed so far, equation (25) or (29) has been applied to molecules in which substituents are attached to carbon atoms. These equations also have been successfully applied to compounds containing substituents attached to oxygen (Reaction Series No. 20–26, 46, 56, 58, 61, 62, 81, and 94), to sulphur (No. 40–45), to silicon (No. 75), and to metals (No. 65–67). As well as for the nitro compounds discussed above, the validity of equation (29) has been proved for Reaction Series No. 64, 68, and 69. A few more complicated cases will now be discussed.

For substances of the type $C_6H_5SO_2Z$, equations (25) and (29) are valid for various substituents (Reaction Series No. 40 and 41) with predominantly polar effects. A completely different solution was found for substances of the type $C_6H_5SO_2NZ^1Z^2$. Equation (38b) describes the shifts in half-wave potential in Reaction Series No. 85 for benzene-sulfonamide and its N-methyl, N-benzyl, and N,N-dimethyl derivatives (Fig. V-10). The reduction of the N-phenyl derivative was not observed in the available potential range. This is in agreement with the shift toward negative potentials predicted by equation (38b). The N-methyl-N-phenyl sulfonamide, and also the N,N-diphenyl derivative (-2.03 V, not shown in Fig. V-10), is reduced at significantly more positive potentials than those predicted by equation (38b). The steric effect of the bulky phenyl group can be excluded as the cause, since the comparably bulky cyclohexyl group causes a shift in just the opposite direction. Hence, either the mesomeric effect of the phenyl group or a change in mechanism for N-methyl-N-phenyl and N,N-diphenyl derivatives can be assumed as the cause of the discrepancies. With the former explanation it would be necessary also to explain why the mesomeric effect of the phenyl group does not shift the half-wave potential of N-phenylbenzenesulfonamide into the useful potential range.

A similar situation was found for compounds of type $C_2H_5OOCCH{=}CHCONZ_2$. Whereas, for neutral (Reaction Series No. 83) and alkaline (No. 84) media equation (38b) can be used, the split in the values (Fig. V-11) suggests the possibility of two linear relations in which the $\rho^*_{\pi,R}$ values are of opposite sign, probably corresponding to two different mechanisms.

Similarly, the shifts in the half-wave potentials of nitro- and nitroso-guanidines[73] (not included in Table V-2) can be approximated by two linear $E_{1/2}{-}\sigma^*$ plots differing in the sign of $\rho^*_{\pi,R}$; this suggests different mechanisms for saturated and unsaturated substituents.

A change of reduction mechanism also can be predicted for (2-substituted ethyl) diethylammonium ions (Reaction Series No. 76), for which the "decomposition potentials," measured as tangent potentials, are shifted in accordance with equation (25). This holds true even for the chloro derivative, but for the bromo derivative the potential is about 0.5 V more positive than the value predicted by equation (25). Reductive substitution of bromine probably takes place in this compound.

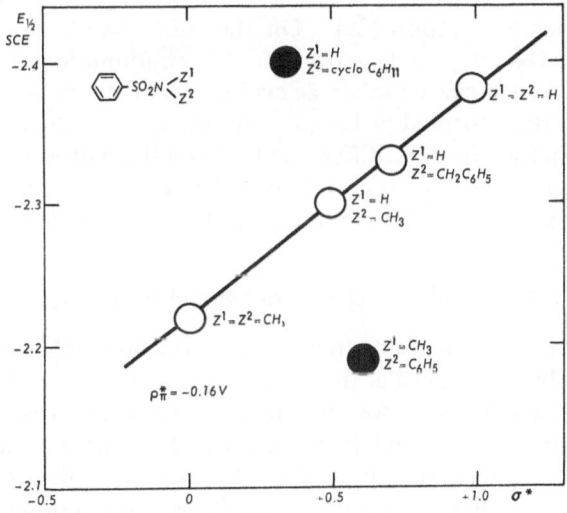

Fig. V-10. Relation of half-wave potentials for the reduction of N,N-dialkyl sulphonamides to the sum of Taft polar substituent constants $\Sigma\sigma^*_X$. Reaction Series No. 85, Table V-2.

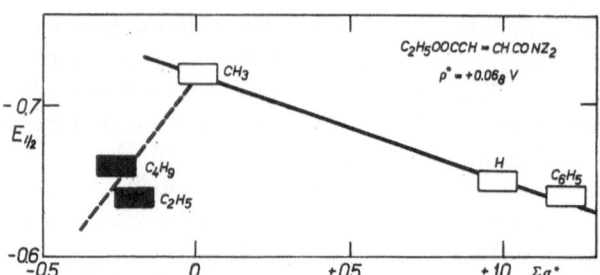

Fig. V-11. Relation of half-wave potentials of substituted amides of the type $C_2H_5OOCCH=CHCONZ_2$ to the sum of Taft polar substituent constants $\Sigma\sigma_X^*$. Reaction Series No. 83, Table V-2.

The role of mesomeric effects will be discussed in next paragraph. With bulky alkyl derivatives, shifts toward more negative potentials than those predicted by equation (29) were observed. These can be explained by the operation of steric hindrance in the reduction process, e.g., in the orientation of the reacting molecule at the surface of the electrode (because the steric hindrance to electron transfer seems improbable). Thus, for disulfides (Reaction Series No. 43 and 44) the value for the di-t-butyl derivative is 0.75 V more negative than the value calculated from equation (29).† On the other hand, in the case of tetrasulfides (No. 45) no deviations have been found for the di-t-butyl derivatives, which can be taken as evidence that the steric effect is no longer operating, probably because the bulky substituent is too far from the reactive site. The difference between the half-wave potentials of the di-n-butyl and di-t-butyl derivatives decreases in the following order: disulfide (-1.1 V), trisulfide (-1.2) $\gg$ tetrasulfide (-0.18 V).

4. Effect of Exchange of Methyl for Phenyl

Substances in which a phenyl group is attached to the electroactive center usually are reduced at the dropping mercury electrode at potentials more positive than those in which the same electroactive center is attached to an alkyl group. It has been usual to attribute the observed shift toward more positive potentials mainly to the effect of conjugation, i.e., to mesomeric interactions. In most of the reaction series given

---

† Similarly, in the case of dialkyl peroxides the nonreducibility of di-t-butyl peroxide has been claimed.[74]

in Table V-2, the value $\sigma^*_{C_6H_5} = 0.60$ for the polar substituent constant satisfactorily expresses the observed shifts in half-wave potential. If it is accepted that $\sigma^*_{C_6H_5} = 0.60$ describes the polar effect of the phenyl group quantitatively, it is necessary to correct our views on the effect of conjugation and the role of the phenyl group in polarography. In most instances, the shift caused by the introduction of a phenyl group should be ascribed predominantly to the polar effect of that group.

Among the few examples in which a different behavior is shown, the disulfides (Reaction Series No. 42) should be mentioned: the half-wave potential of diphenyl disulfide is shifted markedly from the predicted value toward more positive values (by about $+0.32$ V). The explanation of this effect as due solely to the mesomeric effect must be discounted in view of the statement by the authors[37] who studied these compounds: they state that the wave of diphenyl disulfide is steep, whereas, the waves of dialkyl disulfides are drawn-out. Hence, change in the value of transfer coefficient α prevents quantitative expression of the mesomeric effect. Moreover, in Reaction Series No. 43 and 44, which differ only slightly in reaction conditions, the values of the half-wave potentials of the diphenyl derivative do not show such marked deviations.

For some reaction series, which by reason of the fewness of their members are not included in Table V-2, the shifts due to the exchange of methyl for phenyl are given in Table V-3. In this table only those

Table V-3
Effect of the Exchange of Methyl for Phenyl (Z = the group exchanged)

No.	Reaction series	pH		Δ,[a] V	Ref. to source of $E_{1/2}$
1	Z—CO COO⁻	9.2		$+0.26$	75
2	C_6H_5CO CH=CH—Z	7.0		$+0.25$	76
3	Z—CO CH=CH—CO—Z	5.0		$+0.15^b$	77
4	p-OH—C_6H_4—COOZ	Unbuffered		$+0.11$	78
5	$C_6H_5C(=NOH)$—Z	0		$+0.22$	20
6	CH_3CO C(=NOH)—Z	0		$+0.18$	20
7	C_6H_5CO C(=NOH)—Z	0	α	$+0.09$	20
			β	$+0.02$	
8	$CH_3C(=NOH)C(=NOH)$—Z	1		$+0.19$	20
9	$C_6H_5C(=NOH)C(=NOH)$—Z	1	β	-0.08	20
10	Z—CO C(=NOH)CO—Z	1		$+0.05^b$	20

[a] $\Delta = (E_{1/2})_{Z=CH_3} - (E_{1/2})_{Z=C_6H_5}$.
[b] For one Z group.

series are listed for which no difference in the reduction mechanisms of the methyl and phenyl derivatives has yet been proved. It was, thus, necessary to exclude, e.g., data on the comparison of benzaldehyde and acetaldehyde and of nitrobenzene and nitromethane. The nonadditivity of the shifts due to the introduction of a phenyl group in Series No. 6 and 7, or 8 and 9 demonstrates the participation of another effect, in addition to the polar effect.

The effect of a phenyl group is sometimes—e.g., in ultraviolet spectroscopy—compared with the effect of one double bond. Even in polarography, the effects of these systems are comparable in some reaction series, for example,

Series	Z	$E_{1/2}$	Ref.
Z—COO CO—Z	$CH_3CH{=}CH$	−1.62	79
	C_6H_5	−1.62	79
Z—CHO	$CH{=}CH_2$	−1.32	76
	C_6H_5	−1.32	76
Z—C≡CH	$CH{=}CH_2$	−2.40	75
	C_6H_5	−2.41	75

(the complete coincidence is, of course, accidental). In these cases, however, we are actually dealing with a change in the electroactive system, and not merely with substitution as defined in Chapter I.

5. Effect of the Transfer Coefficient

In the deviation of equations (25), (29), and others used in this chapter, it was originally assumed that the values of the slope $dE_{1/2}/dpH$ and of the transfer coefficient α determined from the shape of the wave remain unchanged for all members of the reaction series under study. In fact, practically identical slopes of the pH-dependence curves were recorded by Elving[2] for substances of the type C_6H_5COZ:

Z	$dE_{1/2}/dpH$ (at pH 9)
C_6H_5	0.033
CH_3	0.037
C_2H_5	0.038
n-C_3H_7	0.036
i-C_3H_7	0.036
t-C_4H_9	0.015

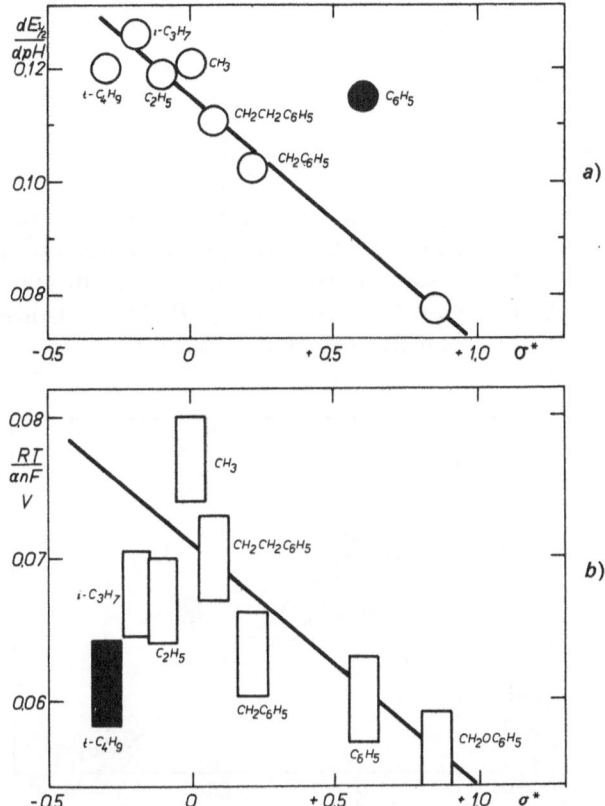

Fig. V-12. Reduction of ketone semicarbazones of the type $CH_3C(Z)$=$NNHCONH_2$. a) Plot of the slope of pH-dependence curve $dE_{1/2}/dpH$, and b) plot of the slope of logarithmic analysis curve $RT/\alpha nF$, against the Taft polar substituent constants σ_X^*. Reaction Series No. 15, Table V-2. Full point deviates.

(the value for the half-wave potential of the t-butyl derivative also shows deviation from the $E_{1/2}$–σ^* plot). Another example is provided by the identical slopes of pH-dependence curves of oximes.[20] On the other hand, the pH-dependence curves of nitro paraffins have different slopes. This is reflected in the change in the value of the reaction constant $\rho_{\pi,R}^*$ with pH and is discussed in Section 7 of this chapter.

Constant values for the transfer coefficient, determined from the wave shape, have been claimed[31,41] for nitro paraffins Z-NO_2 in acid media (Reaction Series No. 47 and 48):

Z	$RT/\alpha n\mathrm{F}$
CH_3	0.12
C_2H_5	0.12
C_3H_7	0.11
C_4H_9	0.11

On the contrary, it has been shown recently[80] that in some instances $dE_{1/2}/d\mathrm{pH}$ and $RT/\alpha n\mathrm{F}$ are approximately linear functions of the substituent constant σ_X^* (Fig. V-12). In some instances, plots of $(E_{1/2} \cdot dE_{1/2}/d\mathrm{pH})$ against σ_X^* or of $(E_{1/2} \cdot RT/\alpha n\mathrm{F})$ against σ_X^* show better correlations than $E_{1/2}$–σ_X^* plots. Thus, for

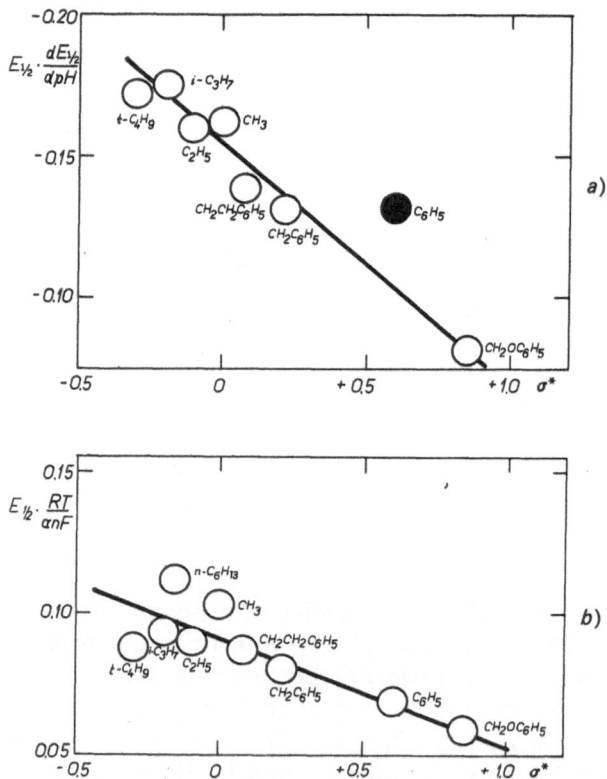

Fig. V-13. Reduction of ketone semicarbazones of the type $CH_3C(Z)=NNHCONH_2$ *a)* Relation of the product $E_{1/2} \times dE_{1/2}/d\mathrm{pH}$, and *b)* relation of the product $E_{1/2} \times RT/\alpha n\mathrm{F}$, to the Taft polar substituent constants σ_X^*.

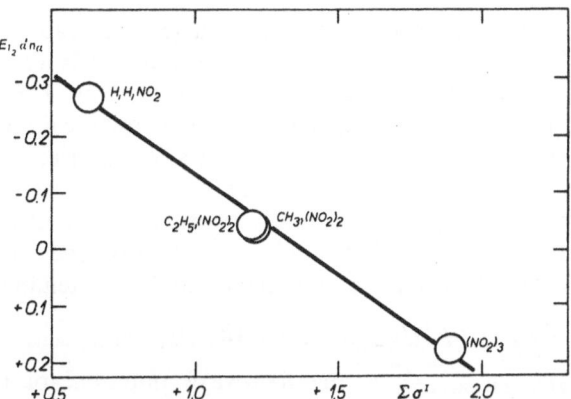

Fig. V-14. Reduction of the chloromethanes of the type $Z^1Z^2Z^3$ CCl. Relation of the product $E_{1/2} \times \alpha' n_a$ to the sum of inductive substituent constants $\Sigma\sigma_X^I$. Reaction Series No. 70, Table V-2.

ZC(CH$_3$)=NNHCONH$_2$ deviations observed[22] for n-C$_6$H$_{13}$ and (CH$_2$)$_2$C$_6$H$_5$ in an $E_{1/2}$–σ_X^* plot are not observed in Fig. V-13 and, similarly, for $Z^1Z^2Z^3$CCl the value for $Z^1 = Z^2 = $ H and $Z^3 = $ NO$_2$, while deviating from the relation of half-wave potentials to $\Sigma\sigma^I$, fitted the linear relation of $E_{1/2} \cdot \alpha n$ to $\Sigma\sigma^I$ satisfactorily (Fig. V-14).

On the other hand, more often the inaccuracy in the determination of the slope of the $E_{1/2}$–pH plot or of the polarographic curve is substantially greater than the inaccuracy in the measurement of half-wave potentials. In such cases, correlations using ($E_{1/2} \cdot dE_{1/2}/d$pH) or ($E_{1/2} \cdot RT/\alpha n$F) are worse than those using $E_{1/2}$. Such a situation was found, e.g., for data on the semicarbazones and thiosemicarbazones of aliphatic ketones[3] (Reaction Series No. 78 and 79). In some instances (e.g., in Reaction Series No. 75), linear correlation between $E_{1/2}$ and σ^* was observed even when the value of $RT/\alpha n$F varied between 0.13 and 0.23 and no simple correlation between the value of $RT/\alpha n$F and substituent constants was found.

6. Additivity of Substituent Effects

As a first approximation equations (38a) and (38b) can be used for the shifts in the half-wave potentials of compounds containing several substituents. This has been proved for Reaction Series No. 68–72, 75, 77–80, 83–89, and 93–97. Hence, the effects of substituents are in general additive.

The use of the approximate equation (38), instead of the more rigorous (39), assumes that the reaction constant ρ^* is approximately identical for all possible positions of the substituents. The validity of this assumption as a first approximation is shown, e.g., in Fig. V-2. When, however, a more detailed investigation is carried out, a more complicated situation is revealed. Instead of changing both substituents in $Z^1-C(=R)-Z^2$, let us keep one substituent constant and change only the other (Reaction Series No. 16–19). For the simultaneous changes of both substituents, equation (38) can be written in the form

$$\Delta E_{1/2} = \rho^*_{\pi,R}(\sigma^*_{Z_1} + \sigma^*_{Z_2}) + (E_{1/2})_{Z_1=CH_3,Z_2=CH_3} \qquad (55)$$

where $(E_{1/2})_{Z_1=CH_3,Z_2=CH_3}$ is the half-wave potential of the parent compound containing two methyl groups. This equation can be transformed into

$$\Delta E_{1/2} = \rho^*_{\pi,R}\sigma^*_{Z_1} + \rho^*_{\pi,R}\sigma^*_{Z_2} + (E_{1/2})_{Z_1=CH_3,Z_2=CH_3} \qquad (56)$$

When the substituent Z_2 remains unchanged in the now selected reaction series, the shift due to this substituent $\rho^*_{\pi,R}\sigma^*_{Z_2}$ remains constant. The shift is added to the value of half-wave potential of the dimethyl parent compound. Hence, a new additive factor $(E_{1/2})_{Z_1=CH_3,Z_2=Z_2}$ appears, and for the shift in the half-wave potentials in the reaction series with $Z_2 = $ constant the following equation should be used:

$$\Delta E_{1/2} = \rho^*_{\pi,R}\sigma^*_{Z_1} + (E_{1/2})_{Z_1=CH_3,Z_2=Z_2} \qquad (57)$$

According to equation (57) for reaction series of compounds containing the same electroactive group R, a substituent Z_1 which varies within a given series, and a substituent Z_2, which varies from one series to another (i.e., $Z_2=CH_3$, C_6H_5, C_3H_7, or i-C_3H_7) the slopes of $\Delta E_{1/2}$–$\sigma^*_{Z_1}$ plots for various Z_2 should be identical and, hence, also should be the values of the reaction constant $\rho^*_{\pi,R}$ found for each series. The linear $E_{1/2}$–$\sigma^*_{Z_1}$ plots should be parallel, merely being shifted along the potential axis by $\rho^*_{\pi,R}\sigma^*_{Z_2}$ relative to the plot of the dimethyl derivative.

Experimentally, it has been found, on the contrary, that the determined value of the reaction constant $\rho^*_{\pi,R}$ depends on the kind of the substituent Z_2. The value of $\rho^*_{\pi,R}$ is therefore affected to a certain extent by substitution in the part of the molecule not directly involving the reaction center. Moreover, the reaction constant is approximately a linear function of the half-wave potential of the parent compound of the particular series, which contains the substituents $Z_1=CH_3$ and $Z_2=Z_2$. Because $E_{1/2Z_1=CH_3,Z_2=Z_2}$ is a linear function of $\sigma^*_{Z_2}$, the

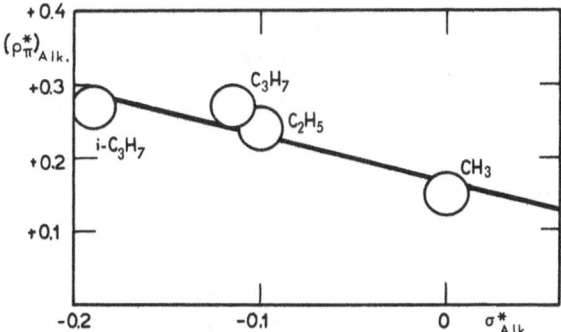

Fig. V-15. Reduction of ketone betainylhydrazones of the type Alk C(Z) = NNHCOCH$_2$ N(CH$_3$)$_3$Cl. Relation of the reaction constant $(\rho^*_\pi)_{Alk}$ for a given alkyl group to the Taft polar substituent constant σ^*_{Alk} for the same alkyl group. Reaction Series No. 16 to 19, Table V-2.

linear correlation between $\rho^*_{\pi,R}$ and $\sigma^*_{Z_2}$ found empirically (Fig. V-15) is understable. Such correlations make it possible to predict values of the changes in $\rho^*_{\pi,R}$. It can be concluded that a more detailed investigation of the limits of the additivity of substituent effects would be of importance.

7. Factors Affecting the Value of the Reaction Constant $\rho^*_{\pi,R}$

With the exception of the example mentioned in last paragraph, among the reaction series compared, it is impossible to correlate the values of the reaction constants with $(E_{CH_3})_R$ or σ^*_R in the way that this was carried out in the benzene series. The reason for such a limitation lies in the relatively small number of reaction series which fulfill the conditions for a comparison. Either the particular electroactive groups characterizing the reaction series are structurally too different, or some important values of the constant σ^*_R are not available, or finally the experimental conditions are not sufficiently comparable. Furthermore, various types of mechanisms can be involved. Moreover, the computed values of the reaction constants $\rho^*_{\pi,R}$ are less reliable than the values of the values of their counterparts $\rho_{\pi,R}$ in the benzene series: this arises from the smaller number of members in particular series and their less satisfactory selection.

The positive or negative sign of the reaction constant and its influence in the determination of the type of mechanism involved have been discussed earlier. Hence, in the present discussion we shall restrict

Table V-4

Effect of pH on the Values of Reaction Constants for the Reduction of Nitro Paraffins

$E_{1/2}$ from Ref. 30, McIlvaine buffer		$E_{1/2}$ from Ref. 31, Clark–Lubs buffer	
pH	$\rho^*_{\pi,R}$, V	pH	$\rho^*_{\pi,R}$, V
2.1	−0.42	2.0	−0.45
—	—	3.0	−0.19
—	—	4.0	−0.09
5.1	−0.10	5.0	+0.13
7.0	+0.08	7.0	+0.26
8.9	+0.22	—	—

ourselves in the first place to the problem of the sign of reaction constant $\rho^*_{\pi,R}$ for oxidations. For all three of the series No. 95–97, in which anodic waves corresponding to mercury salt formation and, thus, to oxidation of mercury are obtained, the reaction constant is negative, in agreement with the electrophilic properties of such systems. Even for the oxidation of alcohols of the type p-CH$_3$O-C$_6$H$_4$-CH(OH)-Z (not mentioned in Table V-2) for which unfortunately only three values of half-wave potentials[81] are available, a negative value ($\rho^*_{\pi,R} = -0.02_7$ V) was found.

In addition to the effects of the substituents Z_2 on the value of the reaction constant discussed in the preceding paragraph, in Reaction Series No. 34–36 a change in the value of the reaction constant with substitution in the molecule was again observed. Whereas, the substitution of stilbene by p-hydroxy groups has almost no effect on the value of the reaction constant (+0.29 V for stilbene and +0.30 V for 4,4′-stilbenediol), a marked change (+0.49 V) was found for substitution by alkyl on the olefinic carbon (No. 36). This indicates different values of $\rho^*_{\pi,R}$ for the first and second alkyl group attached at the double bond.

The role of the supporting electrolyte on the value of the reaction constant is so far little understood. For some reaction series the values of reaction constants obtained in different supporting electrolytes are almost identical, even when determined by various workers under varying conditions (*cf.* Reaction Series No. 2–5, 8–11, 20–22, 24–26, 37–39, 43, and 44). On the other hand, for other reaction series, e.g., for nitro paraffins (*cf.* Reaction Series No. 27–33 and 47–54), the reaction constant $\rho^*_{\pi,R}$ can undergo considerable changes, especially when the pH is changed. This is demonstrated in Table V-4 (some of the $E_{1/2}$ versus σ^* plots from which these values were obtained are given in Fig. V-7). Moreover, for values of $\rho_{\pi,R}^*$ obtained in this manner a correlation by equation (44) with the value of the half-wave potential of the parent compound E_{CH_3} has been found empirically (Fig. V-16).

Not only the pH, but even the composition of the buffer used can affect the values of the reaction constants $\rho_{\pi,R}^*$ to a certain extent. This is demonstrated in Table V-5, where the half-wave potentials involved were measured by a single group of authors.[31]

These values may have been affected by the lack of constancy of the ionic strength and the type and concentration of cations and anions. The considerable deviation observed for benzoate buffers may have been caused by the adsorption of the aromatic buffer components. Even though the values, with the exception of those obtained in the Clark–Lubs buffer, are only very approximate and are based on the half-wave potentials of nitromethane and nitroethane only, the effect of buffer composition can be clearly seen from Table V-5.

There have been no systematic studies on the role of the solvent. Comparison of Reaction Series No. 59 and 60 reveals that in the change from aqueous solution to 50% dioxane-water mixtures the value of the reaction constant $\rho_{\pi,Br}$ decreased from -0.70 V to -0.52 V. Similarly, in a comparison of Reaction Series No. 42 with Series No. 43 and 44 it can be observed that the value of $\rho_{\pi,SS}^*$ decreases from approximately 1.6 V in 80% isopropyl alcohol to 1.1 V in a mixture containing 40% of isopropyl alcohol and 40% of methyl alcohol.

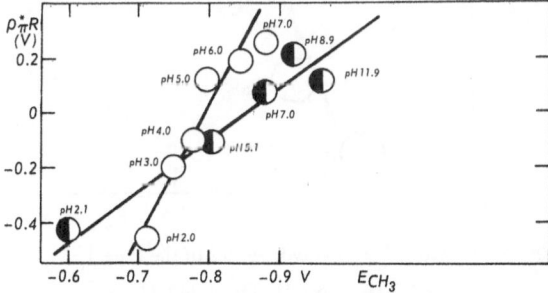

Fig. V-16. Reduction of nitroalkanes. Relation of the reaction constant $\rho_{\pi,R}^*$, determined at a given pH, to the half-wave potential of the unsubstituted parent compound E_{CH_3} at the same pH. Circles: Reaction Series No. 31–33 and 48–50; halved points: Reaction Series No. 27–30 and 52 and 53, Table V-2.

Table V-5
Effect of Buffer Composition on the Value of the Reaction
Constant for the Reduction of Nitro Paraffins
(pH 4.0 buffers of various composition)

Buffer	ρ^*_{π,NO_2}, V
Clark–Lubs	-0.09
McIlvaine	-0.16
Acetate	-0.08
Benzoate	-0.42

Finally, the effect of the skeleton on which the reactive group is bound (or of substituents other than the one that is changed) on the value of reaction constant $\rho^*_{\pi,R}$ can be discussed. The relation between the reaction constant $\rho_{\pi,R}$ and the substituent constant $\sigma^*_{Z_2}$ (Fig. V-15) was discussed in the preceding paragraph. Here, it remains for us to draw attention to the approximately linear relation between the constant $\rho^*_{\pi,R}$ and corresponding values of the half-wave potentials of the parent compounds of each series E_{CH_3}. Such a relation was found for the half-wave potentials of betainyl hydrazones (quaternized

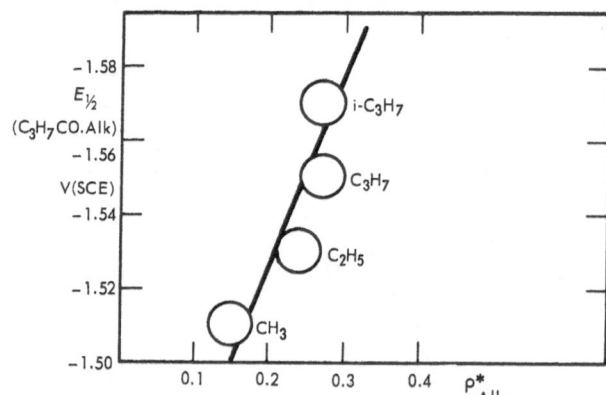

Fig. V-17. Reduction of ketone betainylhydrazones of the type

$$AlkC(C_3H_7)=NNHCOCH_2N(CH_3)_3Cl$$

Relation of half-wave potentials of betainylhydrazones containing the given alkyl in the place of Alk to the values of the reaction constants ρ^*_{Alk} computed for the given alkyl for betainylhydrazones of the type $Alk\ C(Z)=NNHCOCH_2N(CH_3)_3Cl$. Reaction Series No. 16–19, Table V-2.

glycyl-hydrazones) of dialkyl ketones (Fig. V-17) and is analogous to that predicted by equation (44). The validity of such a relation seems to be restricted to groups of structurally related substances studied under comparable conditions—which is true for Reaction Series No. 16–19, but is not generally true for aliphatic compounds.

8. Application of Equations (25) and (29) for the Determination of New Values of the Substituent Constants σ_X^l and σ_X^*

In some cases, it is possible, from measurements of half-wave potentials and by the use of an independently determined value of $\rho_{\pi,R}^*$, to compute or determine graphically the values of substituent constants which may not have been reported in the literature. These values of σ_X^l and σ_X^* are brought together in Table V-6. In most cases, these values have been based on one reaction series only and, moreover, because of the limited number of members in these series the values of $\rho_{\pi,R}^l$ and $\rho_{\pi,R}^*$, are not very reliable. Hence, the values given in Table V-6 should be considered only as rough approximations.

Wide scattering was observed for values of $\sigma_{CH=CH_2}^*$. The two positive values agree reasonably well with the tabulated[1] value for

Table V-6
Approximate Values of Substituent Constants Derived from Polarographic Measurements

Substituent	Value	Reaction series no.	Substituent	Value	Reaction series no.
σ_X^* Constants			σ_X^* Constants		
p-$CH_3C_6H_4$	$+0.55$	44	$NHCH_2C_6H_5$	-0.27	40
$2,4,6,$-$(CH_3)_3C_6H_2$	0	85	$N(CH_3)_2$	$+0.25$	40
p-OHC_6H_4	$+0.45$	34	$N(CH_3)C_6H_5$	$+0.40$	40
	$+0.39$	36	$N(C_6H_5)_2$	$+1.15$	40
$CH=CH_2$	$+0.74$	14	$NHOH$	$+0.06$	40
	$+0.53$	97	σ_X^l Constants		
	-0.10	22			
$COOH$	$+1.10$	77	$CH=CH_2$	$+0.03$	97
CH_2CH_2COOH	$+0.14$	77	$NHCH_3$	-0.08	40
$Br(CH_2)_4$	$+0.37^a$	—	$NHCH_2C_6H_5$	-0.11	40
$Br(CH_2)_5$	$+0.21_5{}^a$	—	$N(CH_3)C_6H_5$	$+0.05$	40
$Br(CH_2)_6$	$+0.1^a$	—	$N(C_6H_5)_2$	$+0.22$	40
NH_2	-0.50	40	$NHOH$	-0.03	40
$NHCH_3$	-0.13	40			

[a] Half-wave potentials from Ref. 66, and for $\rho_{\pi,Br}^* = 4.9$ V.

$\sigma^*_{C_6H_5CH=CH}$ ($+0.41$). The negative value -0.10, probably involves a change in the reduction mechanism or some other complicating factor. Values for $Br(CH_2)_n$ are restricted to higher members of this series: for the lower members a change in mechanism occurs. Values obtained are comparable with tabulated[1] data for $\sigma_{ClCH_2CH_2}$ ($+0.38_5$) and $\sigma_{CF_3(CH_2)_3}$ ($+0.12$).

An attempt[22] to determine the value of the substituent constant for cyclopropyl from the measurement of semicarbazones of alkyl methyl ketones was not successful, because the transfer coefficients of $C_6H_5CH\!\!-\!\!CHCOCH_3$ and $CH_2\!\!-\!\!CHCOCH_3$ derivatives, meas-
$\diagdown\!\!\diagup$ $\diagdown\!\!\diagup$
$\qquad CH_2 \qquad\qquad\qquad CH_2$
ured both from the wave shape and from the shifts in half-wave potential, differed from those of $Z\text{-}COCH_3$ semicarbazones. The fundamental condition for the application of equation (29) was not fulfilled and, therefore, it was impossible to use it for the determination of σ^*_X.

9. Other Empirical Quantitative Relations

For substituents for which substituent constants have not been tabulated, it is necessary to choose other reference magnitudes. If the half-wave potentials of compounds containing these substituents (but a different electroactive group) are available, the following equation can be used.

$$(E_{1/2})_{1,X} - (E_{1/2})_{1,H} = \rho'_\pi (E_{1/2})_{2,X} - (E_{1/2})_{2,H} \tag{40}$$

Index 1 refers to the reaction series studied, index 2 to the reference series, X to compounds containing the substituent X, and H to the parent unsubtituted compound. Correlations of this type have proved useful in the treatment of the half-wave potentials of nitrosamines with hydrogenated heterocyclic rings as substituents in alkaline media.[54] The reference series consisted of p-phenylenediamines carrying analogous substituents on one nitrogen, for which half-wave potentials were measured[82] at pH 11 (Fig. V-18).

It is useful, however, to correlate polarographic half-wave potentials with other physical quantities that are known to show linear free energy relations, such as equilibrium constants, rate constants, and certain spectral data, even in those cases in which substituent constants are available. Such treatment sometimes can reveal the nature of the deviation from the linear free energy relation. Such applications can be discussed for the case of dissociation constants: (i) If $E_{1/2}$–σ_X and

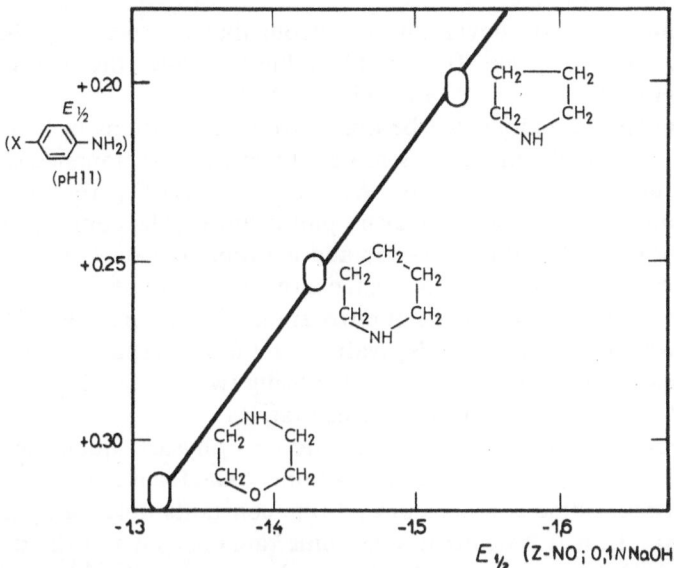

Fig. V-18. Relation of half-wave potentials (from Ref. 82 at pH 11) for the oxidation of cyclic p-phenylenediamines to half-wave potentials (from Ref. 54 in 0.1 M NaOH) for the reduction of nitrosamines derived from the same cyclic amines.

$E_{1/2}$–pK relations show deviations for the substituent X_i, but pK$_X$–σ_X is linear, the polarographic reduction of the substance $R-X_i$ must show an anomaly when compared with other substances $R-X_i$, e.g., a change in the mechanism or pronounced adsorption effects. (*ii*) If $E_{1/2}$–σ_X and pK$_X$–σ_X, relations shows deviations for the substituent X_i, but in the $E_{1/2}$–pK$_X$ plot values for X_i fit the linear, then the compound $R-X_i$ shows properties different from those predicted by the linear free energy relation. In this particular case, the substituent X_i exerts an effect that is not parallel to the effect of this substituent in other reaction series. Anomalies in the behavior of the particular compound $R-X_i$ also can be expected to arise in the measurement of other physical quantities. (*iii*) If $E_{1/2}$–σ_X is linear, but $E_{1/2}$–pK$_X$ and pK$_X$–σ_X show deviations for the substituent X_i, it is probable that the behavior of the substance $R-X_i$ in the dissociation process is anomalous.

An example of this type is the correlation of the half-wave potentials of halo nitro compounds[55] (*cf.* Series No. 70 and 71) with the acid dissociation constants of the corresponding halogen-free nitro compounds,[83] which act as C-acids. Both $E_{1/2}$ versus pK$_X$ and pK$_X$

versus σ_X plots show deviations for nitromethane derivatives, whereas, the $E_{1/2}$ versus σ_X plot (Fig. V-14) is linear. Hence the behavior of nitromethane as an acid seems to be anomalous.

Similar deductions can be drawn from correlations of half-wave potentials with chemical shifts in nuclear magnetic resonance spectra (δ).[84] Thus for nitroalkanes both $E_{1/2}$ versus σ^* and $E_{1/2}$ versus δ show deviations, whereas the δ versus σ^* plot is linear. The complication in this case lies in the polarography and is probably caused by changes in reduction mechanism, as indicated already on p. 186.†

The half-wave potentials of nitrosamines in alkaline media[54] have been correlated with the pK_a values of the corresponding amines. Experimental data are best correlated using two linear relations—one for cyclic and the other for acyclic derivatives.

Sometimes, even in the correlation of physical quantities it is possible to rely solely upon polarographic data, for it is possible then to determine some equilibrium and rate constants polarographically. For example, it is possible to determine rate constants of the decomposition of nitrosamines in acid media polarographically.[54] Logarithms of rate constants determined in this way have been shown to be a linear function of the half-wave potentials of these compounds. In an analogous way, it has been possible[30] to determine polarographically the rate constants of the transformation of the nitro form into the aci form for some nitroalkanes. The logarithms of rate constants found experimentally at pH 8.9 are a linear function of the half-wave potentials of these nitro paraffins measured in alkaline solutions. At other pH values the correlation was worse.

10. Two Functional Groupings in the Molecule

So far, we have restricted ourselves to compounds with one single electroactive grouping in the molecule. We now shall turn to those substances in which two or more groupings are electroactive in the available potential range.‡ The two groups can be either identical or different in nature. The latter case will be discussed first.

---

† Similarly, for iodobenzenes the cause of the deviations shown by amino and hydroxy derivatives from the linear relation established for other substituted iodobenzenes in both the $E_{1/2}$ versus σ and the $E_{1/2}$ versus δ plots, can be presumed to lie in the polarographic method. In this case,[84] the observed deviations are caused because the half-wave potentials of amino and hydroxy derivatives are still (in the pH range studied) affected by pH. It is in the participation of acid-base equilibrium in the electrode process that these derivatives differ from other iodobenzenes.

‡ Substances giving two or more cathodic, or two or more anodic waves, will be discussed here, not those giving one cathodic and one anodic wave.

Two (or more) waves usually can be observed on the polarographic curves of such substances. The degree of separation of these two waves depends on the structure of the compound electrolyzed and the composition of the supporting electrolyte.

For a molecule carrying two electroactive groups R^1 and R^2 on its skeleton [S] electrolysis can occur principally in two ways: either in two successive steps

$$R^1[S]R^2 + n_1 e \xrightarrow{E_1} P^1[S]R^2 \tag{A_1}$$

$$P^1[S]R^2 + n_2 e \xrightarrow{E_2} P^1[S]P^2 \tag{A_2}$$

(where P is the group formed by the reduction† of R in the electrolysis); or by two different paths

$$R^1[S]R^2 + n_1 e \xrightarrow{E_1} P^1[S]R^2 \tag{B_1}$$

$$R^1[S]R^2 + (n_1 + n_2)e \xrightarrow{E_2} P^1[S]P^2 \tag{B_2}$$

Most reductions of organic substances of the type $R^1[S]R^2$ occur in successive steps.‡ The interaction between the two groups R^1 and R^2 can be either small or large.

† For the sake of convenience further discussion will be restricted to reductions. All the deductions can be applied to oxidations as well, with appropriate changes in the signs and directions of potential shifts, etc.

‡ A special type of such reductions occurs for organic compounds for which in scheme (A) the potential of the step E_2 is more positive than that of the step E_1. In these cases, the reduction of $R^1[S]R^2$ occurs at potential E_1 in one single step with the consumption of $(n_1 + n_2)$ electrons. An example of this type is the reduction of p-benzoquinone monooxime (p-nitrosophenol),[85] in which in the first place the quinonimine is formed in a two-electron step. Because the oxidation-reduction potential of the two-electron reduction of the quinonimine is more positive than that of the reduction of the quinone monooxime, one single four-electron step, corresponding to the reduction of the quinone monooxime to p-aminophenol, is observed. An analogous explanation has been proposed for the reduction of p-nitrophenol and p-nitroaniline[86]:

Because the potential for the reduction of the quinonimine is more positive than that for the nitro group, one six-electron wave is observed, whereas, for other nitrobenzenes only a four-electron reduction occurs in the first step. The situation in this particular case is complicated because the dehydration of p-(hydroxyamino)phenol occurs at a finite rate and, hence, the increase of the current above that required for the four-electron reduction possesses[86] a kinetic character. Finally, the four-electron

When this interaction is small (i.e., in cases in which either the inductive or mesomeric effect of the group R^2 on the given skeleton [S] is small or when the susceptibility of R^1 toward the substituent effects of R^2 is small), the reduction of R^1 in $R^1[S]R^2$ occurs in the same potential region and follows the same mechanism as in other molecules of the type $R^1[S]X$, where X is an electroinactive substituent. An example of this type is the reduction of bromine atoms in N,N'-poly-methylenebis-C-bromosydnones (Fig. V-19) which occurs in the potential range in which other C—Br bonds in heterocyclic compounds are reduced.

The more unusual case of strong interactions between the electroactive groups will be demonstrated by reference to several examples. In most straight-chain amines the reduction of the C—N bond is at too negative a potential to produce a wave in the supporting electrolytes usually used. On the other hand, alkyl phenyl ketones are reducible over a broad pH range in the accessible potential range and form well developed waves (Curve 4, Fig. V-20). It would seem logical to predict that in compounds containing both phenacyl and amino groups the reduction of the carbonyl group will be affected by the presence of the amino group (usually an electroinactive substituent) and will occur before the reduction of the amino group. But the strong interaction between the carbonyl and amino groups in α-amino ketones[87,88] results in the reversal of this supposition: in these compounds the C—N$^+$ bond is reduced before the carbonyl group (Fig. V-20, Curves 1–3). Similarly, C—O and C—S bonds, whose reduction waves cannot be observed for most compounds before the electrolysis of the supporting electrolyte, are reduced before carbonyl as a result of strong interaction with carbonyl in compounds such as α-hydroxy ketones[89-91] and phenacylsulfonium salts.[92]

Another example of this type is that of α,β-unsaturated ketones. Even though the reduction of ethylenic bonds occurs at very negative potentials, the interaction of CO and C=C bonds in α,β-unsaturated ketones results in the reduction of the C=C bond[93,94] at a substantially more positive potential than the reduction of the C=O

---

reduction of the nitro group in the first step (the impossibility of detecting two separate two-electron waves and the absence of the corresponding nitroso compound as intermediate) is explained by the fact that the half-wave potentials of nitroso compounds are more positive than those of the corresponding nitro derivatives. Every nitroso group formed is immediately transformed further into the hydroxylamine.

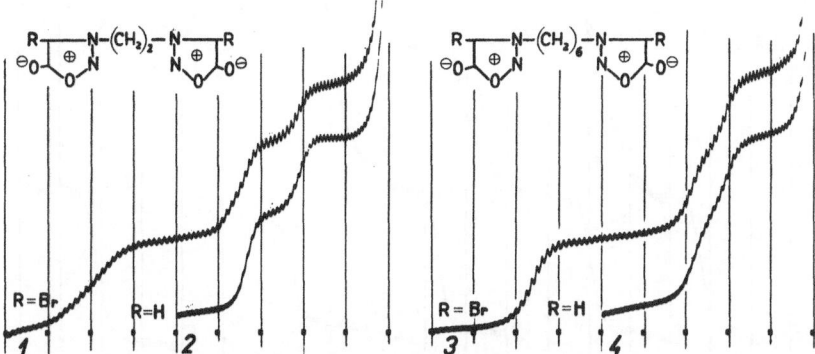

Fig. V-19. Comparison of polarographic curves for C-bromosydnones with curves for the corresponding unsubstituted sydnones. Borate buffer pH 9.3, 5×10^{-5} M sydnone. (1) N,N'-ethylenebis-C-bromosydnone; (2) N,N'-ethylenedisydnone; (3) N,N'-hexa-methylenebis-C-bromosydnone; (4) N,N'-hexamethylenedisydnone. Curves (1) and (3) start at -0.2 V, and (2) and (4) start at -1.0 V, S.C.E., 200 mV/absc., $h = 65$ cm, full scale sensitivity 1.2 μA.

group.† Similarly, the reduction of the ethylenic bond is claimed to be facilitated in α,β-unsaturated sulfones.[97]

We shall next turn our attention to the second, more negative reduction step, in which the substance $P^1[S]R^2$ is further reduced by the reaction (A_2). It should be stressed first that only in a small number of organic compounds is R^1 reduced (hydrogenolyzed) to P^1=H. In the majority of cases, only some of the bonds in the group R^1 undergo reductive cleavage, and a new group P^1 is formed. For example, R^1=NO$_2$ is reduced to P^1=NHOH and R^1=CO is reduced to P^1=CHOH. Hence, when the behavior of the second wave of the compound $R^1[S]R^2$ is examined it is usually necessary to compare it with the wave of $P^1[S]R^2$, and not of $H[S]R^2$.

† In the examples given above we discuss cases in which both of the interacting groups are situated in a straight chain. Some examples of this type have been discovered recently in which these two interacting groups were separated by a phenylene group. These include the reduction of benzonitriles containing a *para* aldehydic or ketonic group in acidic media. Unsubstituted benzonitrile does not give a wave in acid solutions, the only reported[95] reduction being in a tetraalkylammonium bromide solution at -2.3 V. On the other hand, both benzaldehyde and acetophenone give well developed waves in acid media. Thus, in *p*-cyanobenzaldehyde and *p*-cyanoacetophenone comparison with the parent substances would indicate that the carbonyl groups will be the more easily reducible but, in fact, the protonated cyano group is attacked first. A four-electron electrode process gives rise to *p*-aminomethyl derivatives, and this is followed by the reduction of the carbonyl group.[96]

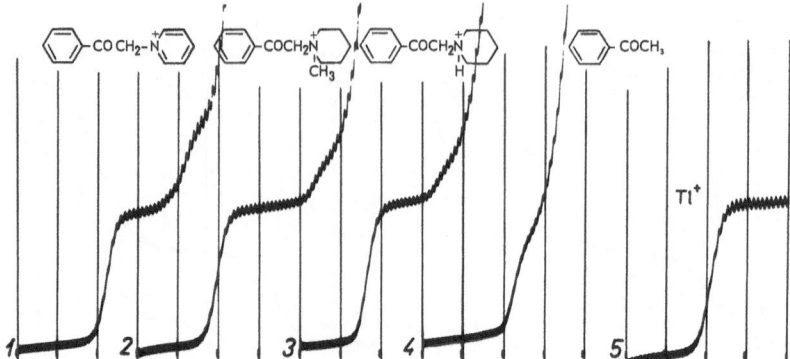

Fig. V-20. Comparison of polarographic curves of α-amino ketones. $2 \times 10^{-4} M$ α-amino ketone, acetate buffer pH 4.45, 0.015 % gelatin. (1) 1-Phenacylpyridinium chloride; (2) 1-methyl-1-phenacylpiperidinium iodide; (3) 2-piperidinoacetophenone hydrochloride; (4) acetophenone; (5) Tl^+. Curves (1) and (2) start at -0.4 V, (3) and (4) start at -0.6 V, and (5) at 0.0 V, S.C.E., 200 mV/absc., $h = 65$ cm, full scale sensitivity 2.1 μA.

Hence, when the second wave, corresponding to the reduction of electrolytically formed $P^1[S]R^2$, is compared with the wave of synthetic $P^1[S]R^2$, in most cases identical behavior is observed. For example, all three of the more negative waves of 2,3-dimethylquinoxaline 1-oxide and 1,4-dioxide are identical with those of the parent

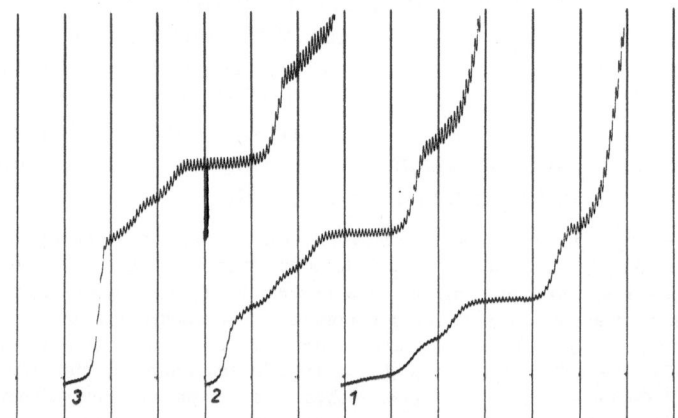

Fig. V-21. Comparison of the polarographic curves of quinoxaline and its oxides. $5 \times 10^{-4} M$ quinoxaline derivative, 0.05 M H_2SO_4. (1) 2,3-Dimethylquinoxaline; (2) 2,3-dimethylquinoxaline 1-oxide; (3) 2,3-dimethylquinoxaline 1,4-dioxide. Curves start at 0.0V, N.C.E., 200 mV/absc., full scale sensitivity 5.2 μA.

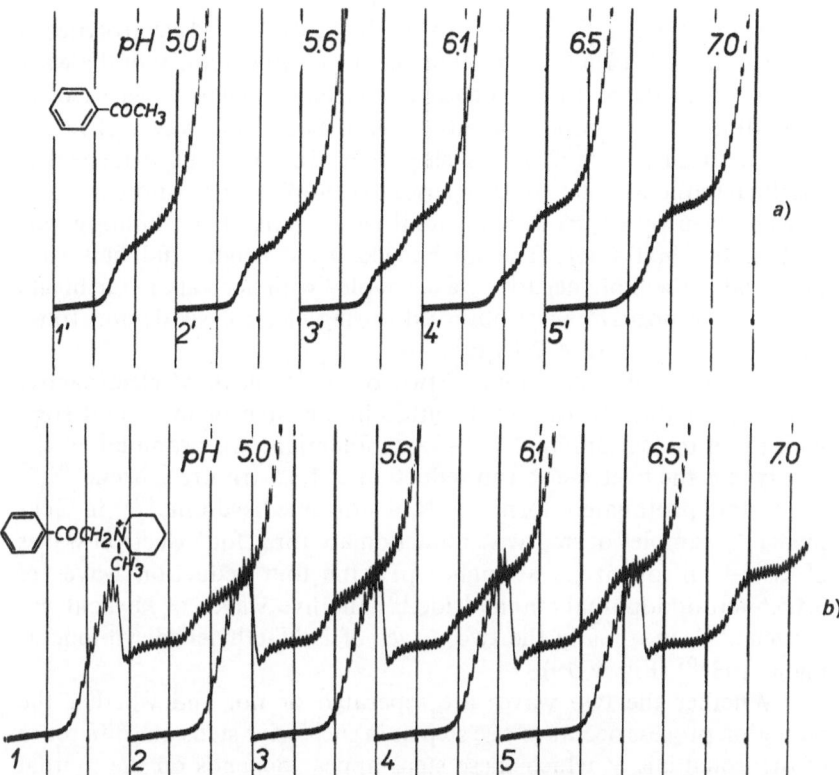

Fig. V-22. Comparison of the polarographic curves of 2-piperidinoacetophenone methiodide and of acetophenone. 2×10^{-4} M ketone, Britton–Robinson buffers, pH stated on the polarogram. (1)–(5) 2-Piperidinoacetophenone methiodide, curves start at -0.6 V; (1′)–(5′) acetophenone, curves start at -1.2 V. S.C.E., 200 mV/absc., full scale sensitivity 1.6 μA.

dimethylquinoxaline (Fig. V-21). Similarly, in N,N′-polymethylenebis-C-bromosydnones the two more negative waves are identical to the waves of the unsubstituted polymethylenedisydnones (Fig. V-19). In some examples,[98] a difference between the half-wave potentials of the second wave and the wave of synthesized $P^1[S]R^2$ has been claimed. One possible explanation would be that the orientation of the P^1SR^2 formed at the electrode can differ in the transition state of the electrolytic process from that of the $P^1[S]R^2$ transported toward the electrode by electrolysis. Another explanation can be based on difference in the solvation of these two species.

Additional proof of the identity of the entity formed electrolytically with synthetic $P^1[S]R^2$ can be obtained from measurement of the wave height. Since the pH-dependence of the more negative waves of 2-piperidinoacetophenone is identical with that of the waves of aceto-phenone itself (Fig. V-22), it is clear that acetophenone is formed in the first reduction step of 2-piperidinoacetophenone. Another proof can be given by controlled potential electrolysis at the limiting current of the first wave. The unchanged wave height and half-wave potential of the more negative wave coupled with decrease in the height of the more positive wave observed during electrolysis demonstrates the nature of substance $P^1[S]R^2$.

When a molecule contains two or more *identical* electroactive groupings, reduction can occur either in one step or in several steps with more or less different half-wave potentials. An example[†] of the first type is the twelve-electron reduction of 1,3,5-trinitrobenzene[99,100] or of the protonated form of N,N'-ethylenedisydnone[101] in acid media;[‡] examples of step-wise reduction are three four-electron waves of 1,3,5-trinitrobenzene at higher pH, the four reduction waves of 3,4,5,6-tetraiodophthalic anhydride,[102] the five waves of hexachloro-benzene,[103] and finally the two waves of N,N'-ethylenedisydnone at higher pH[101] (Fig. V-19).

Whether the two waves are separated or not and whether the reduction process occurs in one step or in successive steps, the difference of the potentials at which these steps appear depends on the mutual interaction of the electroactive groupings in the molecule, on the inter-action between the remaining electroactive groupings and groups that have been formed in an antecedent electrode process, and finally on the interaction between the electroactive groups and an electroinactive substituent. Furthermore, steric factors and the orientation of the molecule at the electrode surface can be of importance. The question of whether in an electrode process involving molecules with several electroactive groups the statistical factor is involved or not[§] remains open. The reduction of one of the $R-ONO_2$ groups occurs with the consumption of two electrons and the formation of NO_2^-. The reduc-

---

[†] Since examples of this type of behavior rarely occur in the aliphatic series, the examples cited here were selected from aromatic compounds.

[‡] An example of this type from the aliphatic series is the reduction of organic nitrates.[104]

[§] That is, whether the shift in the half-wave potential of a molecule containing several electroactive groups is affected by the probability of the occurrence of reaction at one end of the molecule.

tion of polynitrates occurs in one single step, with the slope decreasing with increasing number of nitrate groupings. The height of the total step increases with increase in the number of electroactive groupings. Moreover, the half-wave potential of this wave is shifted toward more positive values with increase in the number of nitrate groupings (by about 100 mV for each ONO_2 group). These shifts are probably affected by the interactions between the groupings, but the statistical factor may play a part. It seems that a factor of primary importance in deciding whether the reduction occurs in one single or in several successive steps is the orientation of the reacting molecule.

It can be assumed that for the simultaneous electroreduction of two or more electroactive groups the molecule must be oriented in such a way that the probability of attack by an electron is the same for each group. We can imagine that in this case the molecule being electrolyzed is placed in such a way that both (or all) of the electroactive groups are inside the double layer.

On the other hand, in the case of molecules for which the electrode process occurs as a separated step for each of the groupings involved we can assume that among the possible orientations of the molecule those predominate in which only one electroactive group is accessible. We can imagine that for this type of electrode process only one of the electroactive groups is placed inside the double layer at a time. Before the next electrode process can occur a reorientation is necessary.

A special factor affecting the orientation of a molecule containing several groups is the position and distance of these groups in the molecule undergoing electrolysis. As regards cyclic systems, the data available usually do not allow us to separate these effects in the case of aromatic compounds, and only a few alicyclic compounds of this type have been studied. For this reason the only general conclusions that can be reached are drawn from the effect of chain length in straight-chain compounds. From the experimental evidence available it can be deduced that shortness of chain, leading to increased interaction between the two groups, is associated with a more positive first wave and a greater difference in half-wave potential. The half-wave potential of the more negative wave approaches the value corresponding to a compound containing a single electroactive group attached to a larger aliphatic chain. Hence, the interaction between the groups is responsible for the shift of the reduction wave of the first reducible group toward more positive potentials. Some results demonstrating this effect of chain length are summarized in Table V-7.

Table V-7
Effect of Chain Length on the Difference in Half-Wave Potential between the First and Second Reduction Wave ($\Delta E_{1/2}$) of Compounds Containing Two Identical Electroactive Groups in a Straight Chain

Substance	n^a	$\Delta E_{1/2}$, V	Ref. to source of $E_{1/2}{}^c$
$Br(CH_2)_nBr$	4^b	0.14	66
	5	0.07	
	6	0.09	
$Sy(CH_2)_nSy^d$	2	0.24	101
	4	0.14_5	
	6	0.13_6	
$C_6H_5CO(CH_2)_nCOC_6H_5$	0	0.90^c	76
	1	0.50^c	
	2	0.08^c	

[a] n = number of methylene groups;
[b] for n = 3 an anomalous mechanism was observed;
[c] in the paper by Prevost[76] only values for the first wave were given, and therefore the value for acetophenone ($n = \infty$) was selected from the same paper as the reference half-wave potential for the more negative wave.
[d] Sy stands for sydnone ring.

It should not be forgotten when inspecting the values in Table V-7 that, even if one single wave were observed, $\Delta E_{1/2}$ would have a finite value, corresponding to $E_{1/4} - E_{3/4}$. This value is given by the expression $E_{1/4} - E_{3/4} = 0.056_5/n\alpha$ (in volts at 20°C). For the most common case of $n = 2$ and $\alpha \approx 0.5$, the value of $E_{1/4} - E_{3/4}$ would be 0.056 V. Hence, to express the mutual effect of ω-groupings it would be necessary to subtract approximately 0.06 V (depending on the values of n and α) from the $\Delta E_{1/2}$ value given in Table V-7. After such correction the validity of the above statement would be even more evident.

For the case of two separated waves, the electrode process consists of the steps (A$_1$) and (A$_2$). As in the case of the molecule $R^1[S]R^2$ containing different groups R^1 and R^2, it is important in the present case of $R^1 = R^2 = R$ that in the first step of the reduction of the molecule $R(CH_2)_nR$ the product $P(CH_2)_nR$ should be formed. In the second step, the group P in $P(CH_2)_nR$ affects the reduction of the second group R. Apart from some cases in which P = H (e.g., in the reduction of dihalogen compounds), the effect of P in $P(CH_2)_nR$ should not be neglected when identifying the nature of the second wave. On the other hand, for higher n in $P(CH_2)_nR$ the effect of the group P, acting through the long chain, is so small that a comparison of the second wave with that of $H(CH_2)_nR$ can be a good approximation.

The more negative wave is frequently identical with that of $P(CH_2)_nR$ prepared synthetically. Sometimes, however, the more negative wave and the wave of the synthetic substance are analogous, but not identical. For example, for 1,2,3-trichlorobenzene, the two-electron reduction of which can produce either o-dichlorobenzene or m-dichlorobenzene, the half-wave potential of the second reduction wave has been given[103] as $-2.53\,V$. The half-wave potentials of o- and m-dichlorobenzenes under the same conditions[103] are $-2.59\,V$ and $-2.55\,V$, respectively, so that the observed half-wave potential of the second wave of the trichloro compound is at least $20\,mV$ more positive than that of the synthetic compound. The reasons for this difference are probably the same as for the previously discussed compounds $R^1[S]R^2$, containing two different groups.

References

[1] R. W. Taft Jr., *Separation of Polar, Steric and Resonance Effects of Reactivity* in "Steric Effects in Organic Chemistry" (ed. M. S. Newman), John Wiley & Sons, New York, 1956.

[2] P. J. Elving and J. T. Leone, *J. Am. Chem. Soc.* **80**: 1021 (1958).

[3] Ju. P. Kitajev, G. K. Budnikov, and I. M. Skrebkova, *Izv. Akad. Nauk SSSR*; *Otdel. Khim. Nauk 1962*, 244.

[4] P. J. Elving and J. M. Markowitz, *J. Org. Chem.* **25**: 18 (1960).

[5] P. Zuman, D. Sc. Thesis, Czechoslovak Academy of Science, Prague, 1959.

[6] P. Zuman, *Collection Czech. Chem. Commun.* **25**: 3225 (1960).

[7] P. Zuman, *Ricerca Sci.* **30**, Contributi teor. spec. polarografia **5**: S 229 (1960).

[8] P. Zuman, "Advances in Polarography, Proc. Second Internat. Congr. Polarography," Cambridge 1959, Vol. 3, p. 812 (ed. I. S. Longmuir) Pergamon Press, Oxford, 1960.

[9] P. Zuman, "Proceedings of the International Symposium Microchemistry," Birmingham, 1958, p. 229, Pergamon Press, Oxford, 1960.

[10] P. Zuman, *Chem. Listy* **54**: 1244 (1960); *J. Polarog. Soc.* **7**: 66 (1961).

[11] P. Zuman, *Current Trends in the Study of the Influence of Structure on the Polarographic Behavior of Organic Substances*, in "Progress in Polarography" (ed. P. Zuman and I. M. Kolthoff) Vol. I, p. 319, Interscience Publ., John Wiley & Sons, New York, 1962.

[12] P. Zuman, *Rev. Polarog. (Kyoto)* **11**: 102 (1963).

[13] P. Zuman, "Correlations in Organic Chemistry," Vol. 2, 3, Tartu State University, Tartu, 1964.

[14] Ju. P. Kitajev and G. K. Budnikov, *Usp. chim.* **31**: 670 (1962).

[15] A. B. Hoefelmeyer and C. K. Hancock, *J. Am. Chem. Soc.* **77**: 4746 (1955).

[16] H. Adkins and F. W. Cox, *J. Am. Chem. Soc.* **60**: 1151 (1938).

[17] C. Calzolari and C. Furlani, *Ann. Chim.* **44**: 356 (1954); *Bull. Sci. Fac. Chim. Ind.* (Bologna) **12**: 14 (1954).

[18] C. Calzolari, *Ann. Triestini* **1954**: No. 1, 16 pp.

[19] P. J. Elving and C. M. Callahan, *J. Am. Chem. Soc.* **77**: 2077 (1955).

[20] P. Souchay and S. Ser, *J. Chim. Phys.* **49**: C172 (1952).

[21] D. M. Coulson, *Anal. Chim. Acta* **19**: 284 (1958).

[22] J. Krupička, unpublished results.

[23] J. R. Young, *J. Chem. Soc.* **1955**: 1516.

[24] M. I. Bobrova and A. N. Matvejeva, *Zh. Obshch. Khim.* **24**: 1741 (1954).

[25] B. Matyska and K. Klier, *Collection Czech. Chem. Commun.* **21**: 1592 (1956).

[26] V. D. Bezuglyj and V. N. Dmitrijeva, *Zh. Fiz. Khim.* **30**: 744 (1957).

[27] J. Nakaya, H. Inosito, and S. Ono, *J. Chem. Soc. Japan Pure Chem. Sect.* **78**: 935, 940 (1957); Referat. Zh. Khim., *1958*, 70257.

[28] G. C. Whitnack, J. Reinhart, and E. S. Gantz, *Anal. Chem.* **27**: 359 (1955).

[29] A. Ryvolová, private communication.

[30] E. W. Miller, A. P. Arnold, and M. J. Astle, *J. Am. Chem. Soc.* **70**: 3971 (1949).

[31] P. E. Stewart and W. A. Bonner, *Anal. Chem.* **22**: 793 (1950).

[32] F. Goulden and F. L. Warren, *Biochem. J.* **42**: 420 (1948).

[33] M. von Stackelberg and W. Stracke, *Z. Elektrochem.* **53**: 118 (1949).

[34] F. L. Lambert and K. Kobayashi, *Chem. Ind.* **1958**: 949.

[35] F. L. Lambert and K. Kobayashi, *J. Am. Chem. Soc.* **82**: 5324 (1960).

[36] L. Horner and H. Nickel, *Chem. Ber.* **89**: 1681 (1956).

[37] M. E. Hall, *Anal. Chem.* **25**: 557 (1953).

[38] J. H. Karchmer and M. T. Walker, *Anal. Chem.* **26**: 271 (1954).

[39] R. L. Hubbard, W. E. Haines, and J. S. Ball, *Anal. Chem.* **30**: 91 (1958).

[40] D. A. Skoog and B. H. Lauwzecha, *Anal. Chem.* **28**: 825 (1956).

[41] F. Petrů, *Collection Czech. Chem. Commun.* **12**: 620 (1947).

[42] T. De Vries and R. W. Iwett, *Ind. Eng. Chem. Anal. Ed.* **13**: 339 (1941).

[43] I. Bergman and J. C. James, *Trans. Faraday Soc.* **48**: 956 (1952).

[44] F. Kaufman, H. J. Cook, and S. M. Davis, *J. Am. Chem. Soc.* **74**: 4997 (1952).

[45] G. Giacometti, *Ricerca Sci.* **26**: 2167 (1956).

[46] I. Rosenthal, Ch. S. Tang, and P. J. Elving, *J. Am. Chem. Soc.* **74**: 6112 (1952).

[47] A. Kirrman and P. Federlin, *Bull. Soc. Chim. France* **1958**: 944.

[48] H. Brüschweiler and G. J. Minkoff, *Anal. Chim. Acta* **12**: 186 (1955).

[49] A. A. Petrov and V. P. Petrov, *Zh. Obshch. Khim.* **29**: 3987 (1959).

[50] V. Gutmann, G. Schöber, and K. Utvary, *Monatsh. Chem.* **88**: 887 (1957).

[51] G. Costa, *Ann. Triestini* **20**: Ser. 2 (1951).

[52] M. K. Saikina, *Uč. Zapiski Kanansk. Inst.* **116**: 129 (1956); *R. Ž. Ch.* **1957**: Nr. 14927.

[53] L. Holleck and R. Schindler, *Z. Elektrochem.* **62**: 942 (1958).

[54] R. Zahradník, E. Svátek, and M. Chvapil, *Chem. Listy* **51**: 2232 (1957).

[55] S. G. Majranovskij, A. A. Fajnzilberg, S. S. Novikov, and V. A. Klimova, *Dokl. Akad. Nauk SSSR* **125**: 351 (1959).

[56] H. A. Laitinen and S. Wawzonek, *J. Ann. Chem. Soc.* **64**: 1765 (1942).

[57] M. Kleine-Peter, *Compt Rend.* **240**: 517 (1955).

[58] S. G. Majranovskij, V. A. Ponomarenko, N. V. Baraškova, and A. D. Snegova, *Dokl. Akad. Nauk SSSR* **134**: 387 (1960).

[59] H. Maloney, B. Sc. Thesis, Univ. Sydney, 1950; quoted according to H. J. Gardner and L. E. Lyons, *Rev. Pure Appl. Chem.* **3**: 134 (1953).

[60] J. Nakaya, K. Mori, H. Kinoshita, and S. Ono, *Nippon Kagaku Zasshi* **80**: 1212 (1959).

[61] W. J. Seagers and P. J. Elving, *J. Am. Chem. Soc.* **72**: 3241 (1950).

[62] I. A. Koršunov, Jm. V. Vodzinskij, N. S. Vjazankin, and A. I. Kalinin, *Zh. Obshch. Khim.* **29**: 1364 (1959).

[63] P. J. Elving, J. M. Markowitz, and I. Rosenthal, *J. Electrochem. Soc.* **101**: 195 (1954).

[64] E. Saito, *Bull. Soc. Chim. France* **1951**: 960.

[65] V. Medonos, *Collection Czech. Chem. Commun.* **23**: 1465 (1958).

[66] J. Závada, J. Krupička, and J. Sicher, *Collection Czech. Chem. Commun.* **28**: 1664 (1963).

[67] M. Fedoroňko, O. Manoušek, and P. Zuman, *Collection Czech. Chem. Commun.* **21**: 672 (1956); *Chem. Listy* **49**: 1494 (1955).

[68] L. Jenšovský, *Collection Czech. Chem. Commun.* **21**: 459 (1956).

[69] L. P. Hammett, "Physical Organic Chemistry," p. 184 McGraw-Hill, New York, 1940.

[70] P. J. Elving, *Record Chem. Progr. Kresge-Hooker Sci. Lib.* **14**: 99 (1953).

[71] C. K. Ingold, "Structure and Mechanism in Organic Chemistry," Cornell University Press, New York, 1953.

[72] P. Zuman, *Chem. Listy* **48**: 94 (1954).

[73] P. Lanza, A. Delmarco, A. F. McKay, and G. Semerano, *Ricerca Sci.* **26**: 116, 129, 148 (1956).

[74] C. O. Willits, C. Riccuiti, H. B. Knight, and D. Swern, *Anal. Chem.* **24**: 785 (1952).

[75] P. Zuman, *Collection Czech. Chem. Commun.* **15**: 1107 (1950).

[76] C. Prévost and P. Souchay, *Chim. Anal.* **37**: 3 (1955).

[77] A. Ryvolová, *Chem. Listy* **50**: 1918 (1956).

[78] V. N. Dmitrijeva and V. D. Bezuglyj, *Aptečn. Delo* **2**: 17 (1959).

[79] C. Ricciuti, C. O. Willits, H. B. Knight, and D. Swern, *Anal. Chem.* **25**: 933 (1953).

[80] P. Zuman, unpublished results.

[81] H. Lund, *Acta Chem. Scand.* **11**: 491 (1957).

[82] R. L. Bent, et al., *J. Am. Chem. Soc.* **73**: 3100 (1951).

[83] V. I. Sloveckij, A. A. Fajnzilberg, and S. S. Novikov, *Izvest. Akad. Nauk., Otdel. Chim. Nauk 1962*, 989.

[84] C. A. Bennett and P. J. Elving, *Collection Czech. Chem. Commun.* **25**: 3213 (1960).

[85] R. M. Elofson, J. G. Atkinson, *Can. J. Chem.* **34**: 4 (1956).

[86] D. Stočesová, *Collection Czech. Chem. Commun.* **14**: 615 (1950).

[87] P. Zuman and V. Horák, "Advances in Polarography. Proceedings of the Second International Congress Polarography," Cambridge, 1959, p. 804, Pergamon Press, Oxford, 1960.

[88] P. Zuman and V. Horák, *Collection Czech. Chem. Commun.* **26**: 176 (1961).

[89] M. Fedoroňko, *Chem. Zvěsti* **12**: 17 (1958).

[90] P. Kabasakalian, J. McGlotten, *Anal. Chem.* **31**: 1091 (1959).

[91] H. Lund, *Acta Chem. Scand.* **14**: 1927 (1960).

[92] P. Zuman and S. Tang, *Collection Czech. Chem. Commun.* **28**: 829 (1963).

[93] B. Pasternak, *Helv. Chim. Acta* **31**: 753 (1948).

[94] P. Zuman and J. Michl, *Nature* **192**: 655 (1961).

[95] M. von Stackelberg and W. Stracke, *Z. Elektrochem.* **53**: 118 (1949).

[96] P. Zuman and O. Manoušek, *Collection Czech. Chem. Commun.* in press.

[97] C. W. Johnson, C. G. Overberger, and W. J. Seagers, *J. Am. Chem. Soc.* **75**: 1495 (1953).

[98] J. Volke, R. Kubíček, and E. Šantavý, *Collection Czech. Chem. Commun.* **25**: 871 (1960).

[99] J. Pearson, *Trans. Faraday Soc.* **44** : 683 (1948).

[100] L. Holleck and G. Perret, *Elektrochem.* **59** : 114 (1955).

[101] P. Zuman, *Collection Czech. Chem. Commun.* **25** : 3245 (1960).

[102] P. J. Elving and C. L. Hilton, *J. Am. Chem. Soc.* **74** : 3368 (1952).

[103] E. S. Levin and Z. I. Fodiman, *Zh. Fiz. Khim.* **28** : 601 (1954).

[104] G. C. Whitnack, J. M. Nielsen, and E. St. Clair Gantz, *J. Am. Chem. Soc.* **76** : 4711 (1954).

VI: Effects of Substituents in Condensed Polycyclic Hydrocarbons

1. General

In passing from benzene derivatives to derivatives of polycyclic aromatic hydrocarbons, we meet the additional possibility of substitution in a ring other than that carrying the electroactive group. Furthermore, the reduction or oxidation can take place in the aromatic nucleus. And finally, it is possible to compare the polarographic behaviors of compounds containing the same polarographically active group in various positions in the ring system and even of compounds containing the same group in various ring systems, e.g., benzene, naphthalene, and anthracene. The limited experimental data allow only an approximate treatment of such comparisons.

2. Classification of Reaction Series

Reaction series of condensed polycyclic aromatic hydrocarbons can be classified, in the first place, according to the nature of the reaction center. The electrolytic attack can occur either (A) on the aromatic ring system or (B) in the side chain.

(A) when the aromatic ring system is attacked during the electrode process, the whole ring system can be considered as one single electroactive group. Since most of the bonds between carbon and a hetero atom are reduced at more positive potentials than multiple carbon-carbon bonds in the ring system, the verification of equations derived for this group of substances is practically restricted to various alkyl derivatives.

The reaction series of this group can be further divided into two subgroups according to whether the electroactive ring system is substituted in one (Aa) or more (Ab) positions.

(Aa) This subgroup comprises reaction series of compounds containing a substituent X in a given position in the electroactive ring system. Since resonance effects can be neglected for the alkyl substituents concerned, equation (29p) can be considered to apply as a first approximation when it is possible to neglect steric interactions between the ring system and the substituent.

$$\Delta E_{1/2} = \rho^*_{\pi,P,n}\sigma^*_X \qquad\qquad (29p)$$

In this equation, the substituent constant σ^*_X (for $X = CH_3$ as reference substance) possesses the same meaning as in Chapter V, $\rho^*_{\pi,P,n}$ is a reaction constant expressing the susceptibility of the reduction of the aromatic ring system to the effect of the substituent X in the position n. Its value depends on the kind of hydrocarbon ring system, on the conditions used for polarographic electrolysis, and particularly on the position n. For electroinactive substituents it is independent of the kind of substituent.

(Ab) Substances having substituents in various positions of the aromatic ring system belong to this subgroup. When one single substituent is placed in various positions in the aromatic ring system, the treatment is equivalent to the comparison of various electroactive groups attached to this substituent. For this case, equation (29p) and similar equations cannot be expected to be valid. The value of the constant $\rho^*_{\pi,P,n}$ cannot be expected to be the same for different positions n, and, therefore, this case will not be considered here in any detail.

When values of half-wave potential are available for compounds substituted in two positions (e.g., positions $n = 1$ and $n = 2$ of the aromatic system) and sufficient values are available for substances with substituents X_1 in position $n = 1$, for substances with substituents X_2 in position $n = 2$, and for polysubstituted compounds with substituents in both position $n = 1$ and position $n = 2$, the two reaction constants $\rho^*_{\pi,P,1}$ and $\rho^*_{\pi,P,2}$ are determined first.† Two possibilities can now occur:

---

† The treatment can of course be extended to a higher number of nonequivalent positions $n = 3, 4, \ldots$

First, the values of $\rho^*_{\pi,P,1}$ and $\rho^*_{\pi,P,2}$ may be approximately equal. The simplified equation (38p) can be used:

$$\Delta E_{1/2} = \rho^*_{\pi,P,n}\Sigma\sigma^*_{X_n} \qquad (38p)$$

Second, $\rho^*_{\pi,P,1}$ may differ from $\rho^*_{\pi,P,2}$, and the following equation should be used:

$$\Delta E_{1/2} = \rho^*_{\pi,P,1}\Sigma\sigma^*_{X_1} + \rho^*_{\pi,P,2}\Sigma\sigma^*_{X_2} \qquad (39p)$$

Some examples are cited below.

(B) Somewhat more frequently, half-wave potentials reported in the literature can be grouped into reaction series in which the polycyclic aromatic ring carries an electroactive grouping. We shall restrict ourselves to the discussion of three types of reaction series of this kind.

(Ba) The first type includes derivatives of an aromatic polycyclic hydrocarbon, itself electroinactive in the potential range studied, whose ring system carries an electroactive group R which is constant in position for the whole reaction series and also a second group as a substituent, which is not reduced at more positive (or oxidized at more negative) potentials than the electroactive group R.

We shall restrict ourselves to substances in which the substituent X is not in the *ortho* or *peri* position relative to the electroactive group R. Steric interaction and direct field effects between the groups X and R then are excluded. The conditions for the application of equation (27) are now fulfilled, and it can be used in the form

$$\Delta E_{1/2} = \rho_{\pi,P,R}\sigma_{P,X} \qquad (27p)$$

In this equation $\sigma_{P,X}$ is the substituent constant, dependent on the nature of the substituent. Its value depends on the position of the substituent X relative to the electroactive group R and, in principle, it can change according to the type of aromatic system. Its value is expressed relative to $X = H$.

$\rho_{\pi,P,R}$ is the reaction constant, expressing the susceptibility of reduction in the side chain to the effect of the substituent X. The value of this constant depends on the composition of the electroactive group R, on its position, on the type of aromatic system, and on the reaction conditions in the polarographic electrolysis, but it is independent of the kind and position of the substituent X.

Values of substituent constants $\sigma_{P,X}$ can be most logically derived from the dissociation constants of substituted carboxylic acids, $\rho_{P,COOH}$ being defined as equal to unity for this reaction series. Some data are available for substituted naphthoic acids,[1,2] but the number of

Table VI-1

Values of Substituent Constants $\sigma_{1\text{-naphth},X}$ for Naphthalene Derivatives with the Reactive Group in Position 1 (Mean $\sigma_{P,X}$ values from various reactions[a] computed from the relation
$$\Delta\sigma_{P,X} = \Delta pK/\rho_R)$$

Substituent	n^b	$\sigma_{1\text{-naphth},X}$	σ_X	$\Delta(\sigma_{1\text{-naphth},X} - \sigma_X)$
3-Cl	1	0.50	0.37 m	+0.13
4-Cl	5	0.25[c]	0.23 p	+0.02
5-Cl	1	0.30	—	—
7-Cl	2	0.12	—	—
4-Br	3	0.28	0.23 p	+0.05
5-Br	1	0.30	—	—
4-F	1	−0.06[c]	0.06 p	−0.12
3-SO$_3^-$	1	0.26	0.15 m	+0.11
4-SO$_3^-$	1	0.42	0.38 p	+0.04
5-SO$_3^-$	1	0.10	—	—
6-SO$_3^-$	1	0.04	—	—
7-SO$_3^-$	1	0.10	—	—
3-CH$_3$	1	−0.08[c]	−0.07 m	−0.01
4-CH$_3$	1	−0.12[c]	−0.17 p	+0.05
4-OCH$_3$	1	−0.67[c]	−0.27 p	−0.40
5-OCH$_3$	1	−0.23	—	—
3-NO$_2$	7	0.67[c]	0.71 m	−0.04
4-NO$_2$	5	0.80[c]	0.78 p	+0.02
4-NO$_2$	5	1.25–1.73[c,d]	1.04–1.27(σ^-)	...
5-NO$_2$	8	0.50	—	—
6-NO$_2$	5	0.40	—	—
7-NO$_2$	3	0.39	—	—
4-COC$_6$H$_5$	1	0.97[c,d]	—	—
4-CN	1	1.10[c,d]	1.00(σ^-)	+0.10
4-CHO	1	1.35[c,d]	1.13(σ^-)	+0.23
4-NO	1	0.58[c,d]	—	—

[a] If not otherwise stated, $\sigma_{1\text{-naphth},X}$ values were computed from data given in Ref. 3.
[b] Number of reaction series compared.
[c] Data given in Ref. 2 are included.
[d] Data are obtained from pK$_a$ values of substituted 1-naphthols for substituents for which mesomeric interaction may occur with the hydroxy group via the aromatic ring; they may be denoted by $\sigma_{P,X}^-$ (cf. σ_X^- values computed for phenols of benzene series). The indices m and p refer to $\sigma_{m\text{-}X}$ and $\sigma_{p\text{-}X}$, respectively.

σ_{P,X_n} values accessible in this way is too limited to be of practical importance.

The other possibility, suggested by Wells and Ward,[3] is based on two assumptions: First, that in the polycyclic series, the linear free energy relation is valid for substances substituted in an analogous way; and second that $\rho_{P,R}$ values for the naphthalene series are equal to ρ_R

values for the benzene series and, hence, the differences between these two series lie in the value of E_H and in the values of the special substituent constants $\sigma_{P,X}$. This assumption makes it possible to calculate values of $\sigma_{P,X}$ from equilibrium and rate constants in the naphthalene series using data for ρ_R.

The validity of the first assumption has been proved from the linear relation between $\log k/k_H$ for the hydrolysis of substituted ethyl 1-naphthoates and $\log K/K_H$ for the ionization of the corresponding 1-naphthoic acids. The validity of the second assumption can be shown by plotting the pK_a values of 3- and 4-substituted 1-naphthols (excluding those substituents for which σ^- values have been shown to be appropriate, see p. 72) against σ_X values from benzene series. These plots are linear and show the usual spread of experimental data. The values of $\rho_{P,COOH} = 1.05$ and $\rho_{P,OH} = 2.03$ are in good agreement with the values obtained for aqueous solutions of benzene derivatives, namely, $\rho_{COOH} = 1.00$ and $\rho_{OH} = 2.11$.

Available values of substituent constants $\sigma_{P,X}$ computed under assumption that $\rho_{P,R} = \rho_R$ are summarized in Tables V-1 and V-2.

Table VI-2

Values of Substituent Constants $\sigma_{2\text{-naphth},X}$ for Naphthalene Derivatives with the Reactive Group in Position 2 (mean $\sigma_{P,X}$-values from various reactions cited in Ref. 3 computed from the relation $\Delta\sigma_{P,X} = \Delta pK/\rho_R$)

Substituent	n^a	$\sigma_{2\text{-naphth},X}$	$\sigma_{m\text{-}X}$	$\Delta(\sigma_{2\text{-naphth},X} - \sigma_X)$
4-Cl	1	0.46	0.37	+0.09
6-Cl	1	0.30	—	—
4-SO$_3^-$	1	0.18	0.15	+0.03
5-SO$_3^-$	1	0.07	—	—
6-SO$_3^-$	1	0.12	—	—
7-SO$_3^-$	1	0.08	—	—
8-SO$_3^-$	1	0.10	—	—
6-OCH$_3$	1	−0.16	—	—
7-OCH$_3$	1	−0.04	—	—
5-NH$_2$	1	−0.06	—	—
8-NH$_2$	1	−0.20	—	—
8-N(CH$_3$)$_2$	1	−0.07	—	—
4-NO$_2$	3	0.62	0.71	−0.09
5-NO$_2$	5	0.36	—	—
6-NO$_2$	4	0.54	—	—
7-NO$_2$	4	0.36	—	—
8-NO$_2$	5	0.40	—	—

[a] Number of reaction series compared.

With one single exception (namely, the value for $\sigma_{1\text{-naphth},4\text{-OCH}_3}$) values of $\sigma_{P,X}$ for derivatives with the substituent in the same ring as the reactive group agree within the limits of the accuracy of the method with values derived for benzene derivatives, σ_X. Hence, for compounds of this type equation (27'p) can be used (in those instances in which values $\sigma_{P,X}$ are not available) as a good approximation:

$$\Delta E_{1/2} = \rho_{\pi,P,R}\sigma_X \qquad (27'p)$$

Instead of $\sigma_{1\text{-naphth},3\text{-X}}$ and $\sigma_{2\text{-naphth},4\text{-X}}$ values of σ_{m-X}, can be used and instead of $\sigma_{1\text{-naphth},4\text{-X}}$ values of $\sigma_{p\text{-X}}$, can be used. An inspection of Table V-1 shows that for groups in position 4 in mesomeric interaction with the group R in position 1 the application of σ_X^- values (Table III-1) is possible, even though with lower degree of accuracy.

Effects of substituents in the annelled ring, as expressed by their $\sigma_{P,X}$-values, have the same sign in all cases studied and are generally smaller than the effects of such substituents when in the ring carrying the reactive group. For the reactive group in position 2 $\sigma_{2\text{-naphth},X}$ has similar values for 5-X and 7-X, and also for 6-X and 8-X.

Application of the substituent constant $\sigma_{m\text{-X}}$ or $\sigma_{p\text{-X}}$ for the substituent X in position 5, 6, 7, or 8 using the trial-and-error method cannot be expected to be successful, as will be seen from an inspection of Tables V-1 and V-2. In some cases, $(\sigma_{m\text{-X}} + \sigma_{p\text{-X}})/2$ can be used for $\sigma_{1\text{-naphth},5\text{-X}}$ (good agreement for 5-Cl and 5-Br, acceptable for 5-OCH$_3$ and 5-SO$_3^-$, poor for 5-NO$_2$) and for $\sigma_{2\text{-naphth},6\text{-X}}$ (good agreement for 6-Cl and 6-OCH$_3$, acceptable for 6-SO$_3^-$, poor for 6-NO$_2$). Another treatment has been derived by Jaffé.[4] Only a few reaction series of simple derivatives of naphthalene and anthracene with a polarographically active group in the side chain have been studied, and some examples are cited below.

(Bb) To this group belong reaction series of compounds containing only a polarographically active group R in various positions in an electroinactive (in the given potential range) polycyclic aromatic ring. To express the effect of the structure of the hydrocarbon on the side-chain reduction by the use of linear free energy relations three possibilites exist: (*i*) the extension of the aromatic system is considered as a perturbation similar to those caused by substituents; (*ii*) the whole aromatic system is treated as a substituent; (*iii*) the assumption is made that the interaction between the aromatic ring system and the polarographically active group (which generally need not be constant) is not very different in various reaction series for reductions that follow a nucleophilic mechanism.

(*i*) If the extension of the aromatic ring system is treated as a substitution, then for the rigorous application of linear free energy relations this change in the molecule should not affect the electron density on the reducible group directly. Since, for the condensed aromatic hydrocarbons treated in this chapter the π-electrons are to a substantial degree shared by the whole condensed system carrying the electroactive group, this supposition is not correct. It is possible to make a further supposition, namely, that the mesomeric interaction of the polarographically active group and the particular aromatic ring system does not vary substantially for different reaction series (i.e., for different polarographically active groups). It is then possible formally to introduce a substituent constant σ_B^- which includes also the contribution of the mesomeric interaction between the "substituent" (i.e., the aromatic ring) and the group reduced by a nucleophilic mechanism. Formally, the shifts in half-wave potential resulting in the exchange of the aromatic ring can be expressed by

$$\Delta E_{1/2} = \rho_{\pi,P,R}^- \sigma_B^- \tag{30p}$$

$\rho_{\pi,P,R}^-$ is a reaction constant expressing the susceptibility of the given electroactive group R to the effect of the exchange of the aromatic rings. It depends on the kind of the polarographically active group R and on the reaction conditions, and it is independent of the kind of hydrocarbon and, in principle, even of the position in which this group R is bound to the aromatic system.

The value of substituent constant σ_B^- should express the effect of the kind of the aromatic ring system on the reducibility of the group in the side chain. Its value is expressed relative to the effect of the phenyl group and can be dependent on the type of the ring system and even on the position in which the electroactive group is bound to the aromatic system. If the above conditions are fulfilled, the value of σ_B^- should be practically independent of the kind of the polarographically active group. As in the derivation of substituent constants for the Hammett equation, it would be logical to define values of σ_B^- by

$$\sigma_B^- = \frac{\log K/K_H}{\rho_{P,R}^-}$$

By definition for the dissociation of carboxylic acids it is possible to write $\rho_{P,R}^- = \rho_R$. Values of σ_B^- obtained in this way, together with

Table VI-3
Values of Substituent Constants σ_B^- Expressing the Effect
of Aromatic Rings on Side-Chain Reactivity

Ring / Group Condition ρ_R	COOH[5]	COOH[4,6]	COOH[1]	OH[7]	NH$_3^{+}$[8]
Condition	H$_2$O	average	78 % EtOH	—	—
ρ_R	1.0	—	2.07	2.0	2.73
Phenyl	0	0	0	0	0
1-Naphthyl	0.51	—	0.12	0.05	0.22
2-Naphthyl	0.03	0.17	—	0.01	0.17
1-Anthryl	0.51	—	—	0.03	—
2-Anthryl	0.02	—	—	0.01	—
9-Anthryl	0.55	—	—	—	—

values obtained for some other reactions for which the same assumption
was made (for comparison), are given in Table VI-3.

The variation of the values for σ_B^- (Table VI-3) is considerably
greater than is usual for other types of substituent constants. This led
Otsuji, Kubo, and Imoto[9] to the observation that these variations
are especially marked for α-derivatives, and to the supposition that
they are caused by the steric effect of the *peri* hydrogen and are, therefore, analogous to *ortho* effects.

To cope with this situation, the Japanese authors introduced
another set of substituent constants denoted as σ_a. The values of these
constants for β-carboxylic acids were computed from

$$\sigma_a = \frac{\log K/K_H}{\rho}$$

but those for α-carboxylic acids from†

$$\sigma_a = \frac{\log K/K_H - 0.41}{\rho}$$

The factor -0.41 should account for the *peri* effect. The values of
σ_a derived are collected in Table VI-4. When σ_a constants were applied
to various reactions[9] the resulting log k–σ_a relations were split into
two (for 1 and 2-derivatives) or more linear plots. It can be assumed

---

† As well as from dissociation constants, σ_a values were determined from kinetic data
and by quantum chemical calculations.[9]

that the contribution of the *peri* effect will be different for various reaction series. This limits the applicability of σ_a values.

Instead of comparing the $E_{1/2}$ and σ_B^- or σ_a values, when the validity of the discussion of deviations would be hampered by the limited number of members in these reaction series, we used a different approach. By the use of equation (30p) $\sigma_{\pi,B}^-$, values were computed from polarographic data using the approximation $\rho_{\pi,P,R} = \rho_{\pi,R}$, and the results are discussed in Section 6 of this chapter. It will be shown that only in reaction series for which the reducible group is of a similar type (e.g., for carbonyl derivatives) comparable values of $\sigma_{\pi,B}^-$ are obtained. Marked deviations were observed, particularly for nitro compounds. Values of $\sigma_{\pi,B}^-$ can be used also in the informative discussion of reactions for which the value of the reaction constant $\rho_{\pi,R}$ for the corresponding benzene derivatives containing the reducible group R is not available.

(*ii*) If the whole aromatic ring system is regarded as a substituent exerting both polar and mesomeric effects, then

$$\Delta E_{1/2} = \rho_{\pi,P,R}^* \sigma_P^* + M_\pi \tag{33p}$$

or

$$\Delta E_{1/2} = \rho_{\pi,P,R}^I \sigma_P^I + \rho_{\pi,P,R}^R \sigma_P^R$$

can be used. In these equations σ_P^* is a substituent constant characterizing the polar effect of the aromatic ring system on the reducibility of an electroactive group in the side chain. Its value depends on the structure of the aromatic ring and on the position in which the electroactive group is bound to the ring system, but is independent of the kind

Table VI-4
Values of Substituent Constants σ_a (Aryl Values)

Ring	σ_a	Ring	σ_a
Phenyl	0.0	2-Phenanthryl	0.04
1-Naphthyl	0.17	3-Phenanthryl	0.08
2-Naphthyl	0.05	9-Phenanthryl	0.17
1-Anthryl	0.28		
2-Anthryl	0.14		
9-Anthryl	0.44		

of the electroactive group. As the reference substance, that for which $P = CH_3$ or $P = C_6H_5$ can be chosen.

As before σ_P^I and σ_P^R are similarly defined inductive and mesomeric substituent constants expressing contributions of the inductive or mesomeric effect to the total effect of the aromatic nucleus.

Constants $\rho_{\pi,P,R}^*$, $\rho_{\pi,P,R}^I$, and $\rho_{\pi,P,R}^R$ are reaction constants expressing the susceptibility of the reducible group to the polar, inductive, and mesomeric effects of the aromatic ring system. Their values depend on the kind of the electroactive group R and on reaction conditions, and they are independent both of the kind of aromatic ring and of the position of the electroactive group in the ring system.

Symbol M_π expresses the mesomeric contribution of the substituent effect of the aromatic ring system.

Values of substituent constants, determined from a comparison of acid- and base-catalyzed ester hydrolysis or by other methods, are not available. Neither can these values, which are required for such a rigorous treatment, be obtained at present from polarographic measurements because of difference in the mechanism of electrode process.

(iii) The determination of substituent constants from polarographic measurements directly appears to be of some importance. Such a treatment is based on the assumption that for reductions following a nucleophilic mechanism the relative contribution of the resonance interaction between the electroactive group and the aromatic ring remains constant and independent of the nature of the electroactive group. The validity of this assumption was tested for various carbonyl derivatives. Deviations were observed for nitro compounds, which can be explained by varying mesomeric and steric interactions between the nitro group and the aromatic system.

As the reference series aromatic aldehydes, for which the greatest amount of experimental data is available, were chosen.

To avoid assumptions about the magnitude of the reaction constant $\rho_{\pi,CHO}$ in the reaction series of aldehydes, half-wave potentials were compared directly using

$$(E_{1/2})_{1,P} - (E_{1/2})_{1,B} = \rho_{\pi,2}'[(E_{1/2})_{2,P} - (E_{1/2})_{2,B}] \qquad (49p)$$

The half-wave potentials with the index 2 are values measured for the reference series, i.e., for aldehydes, and those with the index 1 are for the reaction series studied. The index B refers to the parent phenyl compound, and the index P to the polycyclic compound.

Table VI-5
Effect of Substituents on the Half-Wave Potentials of
Anthrone and Bianthrone

($\Delta E_{1/2}$ in volts vs. X = H)

Substituent X	Anthrone	Bianthrone[a]
1-OH	−0.01	—
2-OH	−0.10[b]	—
3-OH	−0.05	+0.03[b]
4-OH	+0.04	+0.01
1-OCH$_3$	—	−0.17[b]
2-OCH$_3$	+0.01[b]	−0.05[b]
1-Cl	0.0[b]	+0.15
4-Cl	+0.02	+0.07[b]

[a] Disubstitution in positions 1, 1'; 2, 2'; etc; $\Delta E_{1/2}$ expresses the effect of two groups.
[b] Computed from values for polysubstituted derivatives on the assumption of additivity.[6]

In a similar way, Nakaya, Kinoshita, and Ono[10]† derived so-called "nucleous constants" σ_n from polarographic measurements of the half-wave potentials of carbonyl compounds, using values for $CH_3COCH = CH$ and $OHCCH = CH$ as reference compounds $(\sigma_n = 0): \sigma_{n,C_6H_5} = 1.5, \sigma_{n,\alpha\text{-furyl}} = 2.27$ and $\sigma_{n,\alpha\text{-thienyl}} = 2.55$.

(Bc) To this type belong reaction series of compounds containing both a polarographically active group R and a substituent Z in a side chain of an electroinactive polycyclic aromatic ring. For substituents exerting predominantly polar effects equation (29) can be used in the form

$$\Delta E_{1/2} = \rho_{\pi,R}^* \sigma_Z^* \tag{29}$$

3. Electroactive Aromatic System with One Substituent

The verification of equation (29p) for reaction series of this type can be demonstrated for alkyl-naphthalenes,[12] -anthracenes,[13] and -azulenes.[14]

For 1-substituted naphthalenes[12] equation (29p) with $\rho_{\pi,\text{naphth},1}^* = 0.054$ V is fitted by values for X = H, CH_3, C_2H_5, and C_3H_7, and the value for X = t-C_4H_9 deviates. For 2-substituted

† Because the original paper was available to the present author in Japanese only, he had to rely upon a review article.[11]

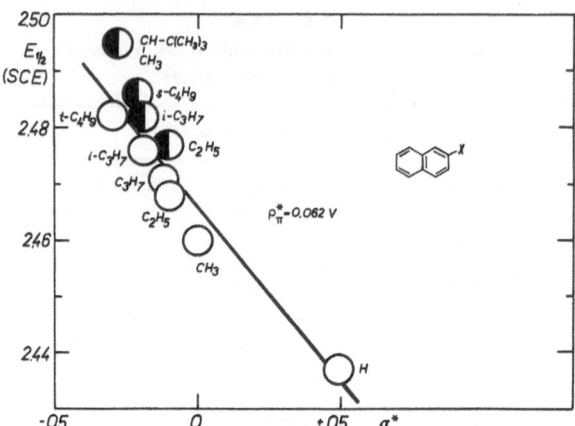

Fig. VI-1. Relation of half-wave potentials for the reduction of 2-alkylnaphthalenes to Taft polar substituent constants σ_X^*. Half-wave potentials in 75% dioxane, $0.1\,M$ $N(C_4H_9)_4I$, from Ref. 12, 15. Circles: alkylnaphthalenes; halved points: second waves of $Ar-C(X)=CH_2$.

naphthalenes[12] equation (29p) ($\rho_{\pi,\text{naphth},2}^* = 0.062$ V) is fitted (Fig. VI-1) not only by the half-wave potentials of the compounds in which $X = H$, CH_3, C_2H_5, $n\text{-}C_3H_7$, $i\text{-}C_3H_7$ and $t\text{-}C_4H_9$, but also by the half-wave potentials of the second reduction waves of 1-alkyl-1-(2-naphthyl) ethylenes[15] in which $X = C_2H_5$, $i\text{-}C_3H_7$, and $CH(CH_3)C(CH_3)_3$.

   For 1-substituted anthracenes,[13] equation (29p) was found valid for $X = H$, CH_3, and C_2H_5 with $\rho_{\pi,\text{anthr},1}^* = 0.035$ V and for 9-substituted anthracenes for $X = CH_3$, C_2H_5, C_3H_7, C_4H_9 with $\rho_{\pi,\text{anthr},9}^* = 0.038$ V. For 2-substituted anthracenes[13] for $X = H$, CH_3, C_2H_5, $t\text{-}C_4H_9$, the value of $\rho_{\pi,\text{anthr},2}^* = 0.066$ V was found.

   For 1-substituted azulenes† equation (29p) is valid for the half-wave potentials[14] of the compounds in which $X = H$, CH_3, C_2H_5, and $i\text{-}C_3H_7$ with $\rho_{\pi,\text{azul},1}^* = 0.17$ V; for 5-substituted azulenes it is valid for $X = H$, CH_3, C_2H_5, and $i\text{-}C_3H_7$ with $\rho_{\pi,\text{azul},5}^* = 0.03_6$ V; for 6-substituted azulenes it is valid for $X = H$, CH_3, and $i\text{-}C_3H_7$ with $\rho_{\pi,\text{azul},6}^* = 0.13$ V.

   Differences in $\rho_{\pi,P,n}^*$ for various values of n express quantitatively the susceptibility of the reduction of the carbocyclic ring to the effect of the position of the substituent.

---

† Azulenes, which usually are treated among alicyclic compounds, are discussed here because of the similarity of the treatment with that for other polycondensed hydrocarbons.

The limited data available do not allow such a quantitative treatment of the half-wave potentials of the anodic waves of the oxidation of hydrocarbons.[16] A comparison of the half-wave potentials of 1- and 2-methylnaphthalenes suggests that, as in the case of reduction, there is little difference between the effects of substituents in positions 1 and 2 of the naphthalene ring. Furthermore, the introduction of a methyl group seems to cause a much greater shift in the oxidation half-wave potential than in the reduction potential. Finally, the direction of the shifts indicates that the reaction constant corresponding to the oxidation process is positive, as in the case of nucleophilic reduction. For this last observation, no explanation can be given at present.

4. Polysubstituted Electroactive Aromatic Systems

Both types of treatments mentioned for (Ab) in Section 2 of this chapter can be demonstrated.

For naphthalenes the values for $\rho^*_{\pi,\text{naphth},1} = 0.054$ V and $\rho^*_{\pi,\text{naphth},2} = 0.062$ V (see Section 3) differ by no more than the accuracy of the treatment would allow. Hence, it is possible to correlate both substituents in position 1 and in position 2 using the same reaction constant $\rho^*_{\pi,R,n}$. A plot of data[12] using equation (38p) in the form

$$\Delta E_{1/2} = 0.054 \, \Sigma \, \sigma^*_{X_n}$$

is shown in Fig. VI-2.

$E_{1/2}$ for naphthalene was put equal to $(0.054 \cdot 3\sigma^*_H)$. Although several values show deviations, most of them fit the above equation within a tolerance of ± 0.01 V. Deviations were observed for $1\text{-}t\text{-}C_4H_9$ for steric reasons and for di- and tri-methylnaphthalenes with methyl groups simultaneously present in positions 1 and 4, 1 and 7, or 1 and 8.

On the other hand, for polysubstituted anthracenes the values of the reaction constant differed substantially according to the position of the substituent. For 1- and 9-alkyl-substituted anthracenes, it was possible to use a common value $\rho^*_{\pi,\text{anthr},1,9} = 0.037$ V, but for 2-substituted derivatives the application of $\rho^*_{\pi,\text{anthr},2} = 0.066$ V was necessary (see Section 3). The shifts in half-wave potential for anthracenes[13] substituted in all three classes of positions were then calculated using equation (39p) in the form

$$(\Delta E_{1/2})_{\text{calc}} = \rho^*_{\pi,\text{anthr},1,9} \Sigma\sigma^*_{X_{1,9}} + \rho^*_{\pi,\text{anthr},2} \Sigma\sigma^*_{X_2}$$

that is,

$$(\Delta E_{1/2})_{\text{calc}} = 0.037\Sigma\sigma^*_{X_{1,9}} + 0.066\Sigma\sigma^*_{X_2}$$

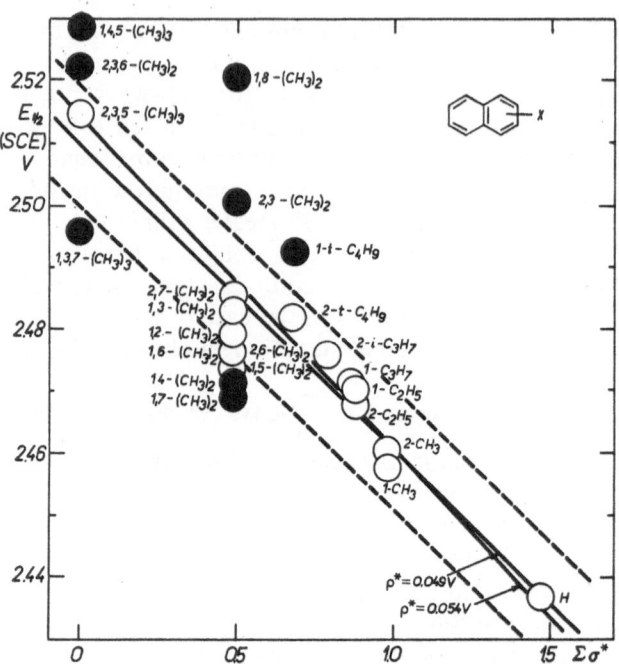

Fig. VI-2. Relation of half-wave potentials for the reduction of mono- and poly-alkyl-naphthalenes to the sum of Taft polar substituent constants σ_X^*. Half-wave potentials in 75% dioxane, 0.1 M N(C$_4$H$_9$)$_4$I, from Ref. 12, 15. Two possible regression lines with $\rho_{\pi,R}^* = 0.049$ V and 0.054 V were drawn. The dotted line delimits the region of approximate validity of the relation (± 10 mV). Full points show greater deviation.

The half-wave potential of anthracene was put equal to $(0.037 \cdot 2\sigma_H^* + 0.066 \cdot 4\sigma_H^*)$ to take into account the highest number of substituted hydrogen atoms possible.† Values of shifts $(\Delta E_{1/2})_{calc}$ computed using this equation show very good correlation‡ with measured $E_{1/2}$ values (Fig. VI-3).

Data for the half-wave potentials of alkylazulenes[14] have been treated similarly. Data for $\rho_{\pi,azul,n}^*$ for $n = 1$, 5, and 6, mentioned in Section 3 of this chapter, vary substantially. Since only the half-wave potentials for methyl derivatives have been measured[14] for azulenes substituted in positions 4 and 8, an exact determination of the reaction

---

† This treatment, which in fact represents only a change of the coordinate system, is necessary because of the use of E_{CH_3} as a standard.

‡ The treatments given here are somewhat different from those advanced by the authors of Ref. 12, 13.

constant $\rho_{\pi,\text{azul},4}^*$ is impossible. Comparison of shifts due to a methyl group in position 4 or 8 with shifts resulting from methyl substitution in other positions indicated that the value of $\rho_{\pi,\text{azul},4}^*$ will be similar to that of $\rho_{\pi,\text{azul},6}^*$. Hence, shifts in the half-wave potentials of alkylazulenes were treated using the equations:

$$(\Delta E_{1/2})_{\text{calc}} = \rho_{\pi,\text{azul},1,3}^* \Sigma\sigma_{X_{1,3}}^* + \rho_{\pi,\text{azul},2}^* \sigma_{X_2}^*$$

$$+ \rho_{\pi,\text{azul},4,6,8}^* \Sigma\sigma_{X4,6,8}^* + \rho_{\pi,\text{azul},5,7}^* \Sigma\sigma_{X5,7}^*$$

that is,

$$(\Delta E_{1/2})_{\text{calc}} = 0.1\Sigma\sigma_{X1,3}^* + 0.17\sigma_{X_2}^* + 0.13\Sigma\sigma_{X4,6,8}^* + 0.03_6\Sigma\sigma_{X5,7}^*$$

The half-wave potential of unsubstituted azulene was put equal to $(0.1 \cdot 2\sigma_H^* + 0.17 \cdot \sigma_H^* + 0.13 \cdot 3\sigma_H^* + 0.03_6 \cdot 2\sigma_H^*)$. Values of shifts in

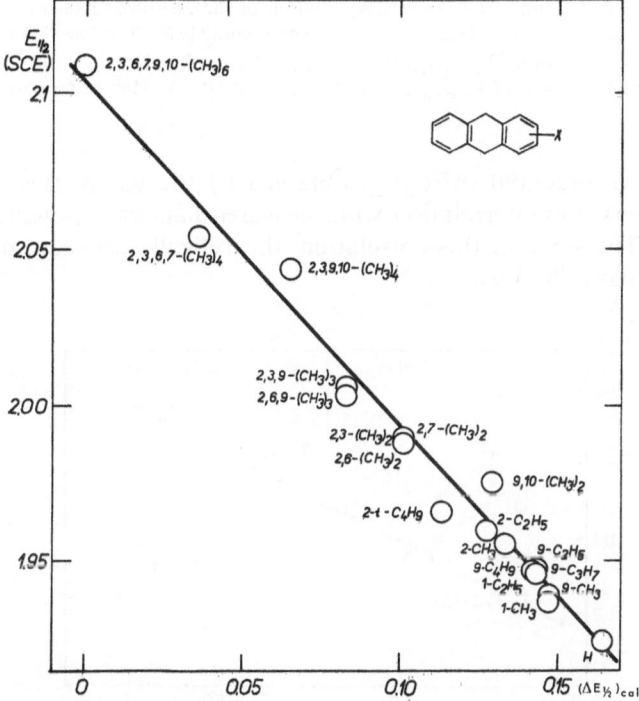

Fig. VI-3. Reduction of mono- and poly-alkylanthracenes. Relation of the measured half-wave potentials $E_{1/2}$ to the shifts in half-wave potential $(\Delta E_{1/2})_{\text{calc}}$ calculated using the equation $(\Delta E_{1/2})_{\text{calc}} = \rho_{1,9}^* \Sigma\sigma_{1,9-x}^* + \rho_2\Sigma\sigma_{2-x}^*$ for $\rho_{1,9}^* = 0.037$ V and $\rho_2^* = 0.066$ V. The value for unsubstituted anthracene was calculated using the equation $(\Delta E_{1/2})_{\text{anthracene}} = \rho_{1,9}^*2\sigma_4^* + \rho_2^*4\sigma_4^*$. Half-wave potentials from Ref. 13,

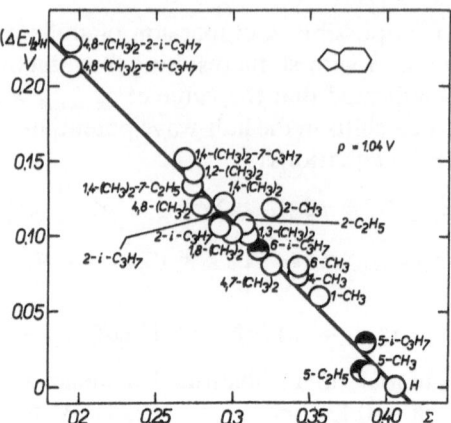

Fig. IV-4. Reduction of alkylazulenes. Relation of the measured half-wave potentials, expressed relative to that of the unsubstituted compound $(\Delta E_{1/2})_H$ to the shift Σ calculated using the equation $\Sigma = \rho^*_{1,3}\Sigma\sigma^*_{1,3\text{-}x} + \rho^*_2\Sigma\sigma^*_{2\text{-}x} + \rho^*_{4,6,8}\Sigma\sigma^*_{4,6,8} + \rho^*_{5,7}\Sigma\sigma_{5,7\text{-}x}$ for $\rho^*_{1,3} = 0.1$ V; $\rho^*_2 = 0.17$ V; $\rho^*_{4,6,8} = 0.13$ V; $\rho^*_{5,7} = 0.03_6$ V. Half-wave potentials from Ref. 14.

half-wave potential $(\Delta E_{1/2})_{calc}$ obtained by the use of this equation show very good correlation with measured half-wave potentials (Fig. VI-4). The slope of this correlation, theoretically expected to be 1.0, was found to be 1.04.

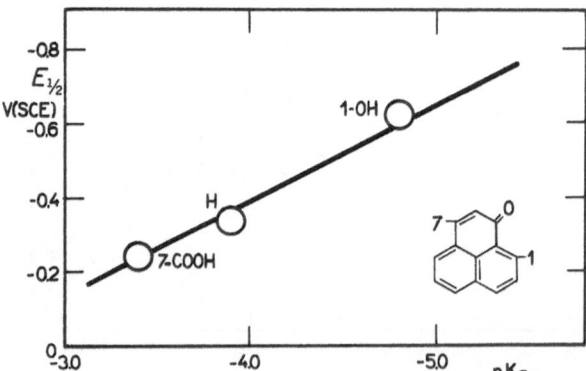

Fig. IV-5. Relation of half-wave potentials for the reduction of substituted perinaphthenone to logarithms of dissociation constants of conjugated acids. Experimental data form Ref. 17.

The above examples, especially those for alkyl-anthracenes and -azulenes, show a good additivity of the effects on half-wave potentials. Similarly, even for the oxidation waves of methylazulenes, it seems that shifts in half-wave potential due to substitution by methyl groups in various positions are additive.

5. Effects of a Substituent in a Condensed Aromatic Hydrocarbon on the Reduction of a Functional Grouping in Another Position in the Ring System

The only relevant reaction series that has been found so far [for which a quantitative treatment of half-wave potential shifts using equation (27p) is possible] is that of the perinaphthenone (1H-phenalen-1-one) (I) substituted in positions 3 and 9. The half-wave potentials[17]

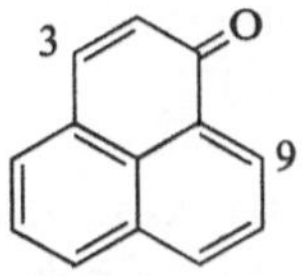

3 O 9

(I)

were compared with the values of dissociation constants of conjugated acids (Fig. VI-5).

A qualitative comparison is possible for substituent effects in anthrone (II) and bianthrone† (III), where there are no data for the standard series. Shifts in half-wave potentials[18] observed in a 10% solution of sulfuric acid in glacial acetic acid are summarized in Table VI-5. It is surprising that substitution by hydroxyl in position 3 or by methoxyl in position 2 in anthrone and in bianthrone results in shifts in half-wave potential which are opposite in sign. The effect of a chlorine atom in position 1 or 4 is only a little different from the effect of a hydroxy group in the same position.

In the 1,8-dihydroxyl and 1,8-dichloro derivatives, probably steric hindrance to coplanarity is involved. The shifts are more negative by 0.07 V for the 1,8-dichloro derivative and by 0.17 V for the 1,8-dihydroxy derivative than the values predicted on the basis of the simple additivity of substituent effects.

† The inclusion of these compounds into this subgroup is debatable.

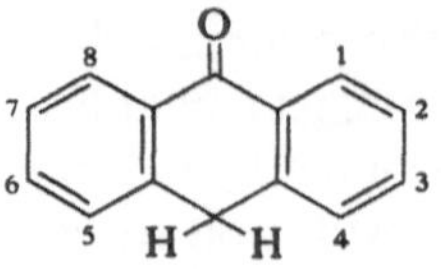

(II)

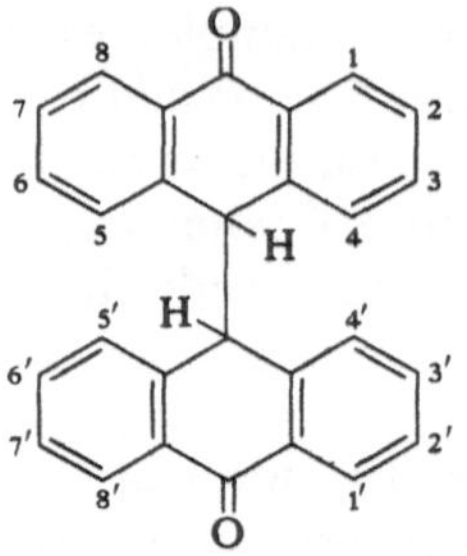

(III)

Differences in the effects of alkyl groups can be observed also for azulene ketones.[19] Table VI-6 shows the dependence on the position and the medium used.

The existence of an effect due to steric hindrance by alkyl groups in the azulene ring of azulene ketones to coplanarity is doubtful, since such an effect has not been observed for acetophenones substituted in vicinal positions.

6. Effect of the Nature of the Aromatic Ring on Reduction in the Side Chain

(i) Computation of substituent constants $\sigma_{\pi,B}^-$ from polarographic data and application of the aryl values σ_a. For the calculation of values of substituent constants $\sigma_{\pi,B}^-$ from polarographic measurements, an assumption was made similar to that made by Hammett[36] in the derivation of the σ-value for the 2-naphthyl group. This assumption is

Table VI-6

Effect of Substituents on Half-Wave Potentials of Azulene Ketones
(Britton–Robinson buffers, 50 % ethanol; $\Delta E_{1/2}$ in volts vs. X = H)

	$(\Delta E_{1/2})_{pH=0}^a$	$(\Delta E_{1/2})_{alkal}^b$
1-acetylazulene		
2-CH$_3$	−0.045	−0.090
3-CH$_3$	+0.062	−0.027
3-CH$_3$, 5-i-C$_3$H$_7$, 8-CH$_3$	−0.065	−0.127
1,3-diacetylazulene		
2-CH$_3$	−0.036	−0.097

[a] Extrapolated value for pH = 0.
[b] pH-independent waves at higher pH.

equivalent to the supposition that the value of the reaction constant remains unchanged during the transition from substituted benzene derivatives ($\rho_{\pi,R}$) to compounds containing the same electroactive group bound to various aromatic polycyclic systems ($\rho_{\pi,P,R}^-$). On the assumption that $\rho_{\pi,R} \approx \rho_{\pi,P,R}^-$ (in volts), the values of the substituent constants $\sigma_B^- = \Delta E_{1/2}/\rho_{\pi,R}$ given in Table VI-7 were computed.

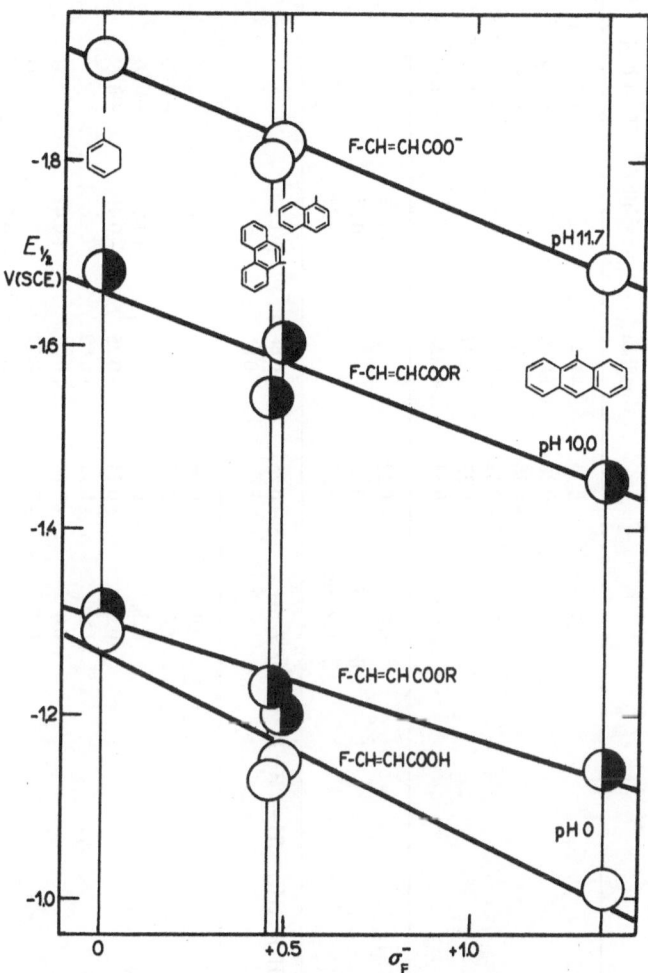

Fig. VI-6. Relation of half-wave potentials of arylacrylic acids (F—CH=CH—COOH) and their esters at pH 0, of their esters at pH 10.0, and of their anions at pH 11.7 to polar aromatic constants σ_F^-. Half-wave potentials from Ref. 38.

Table VI-7

Values of Constants $\sigma^-_{\pi,B}$ Calculated under Assumption that in equation (30p) $\rho_{\pi,P,R} = \rho_{\pi,R}$; $\therefore$ $\sigma^-_{\pi,B} = \Delta E_{1/2}/\rho_{\pi,R}$

Reaction series	pH	$\rho_{\pi,R}$, V	$\sigma^-_{\pi,B}$				Ref. to source of $E_{1/2}$
			1-naphthyl	2-naphthyl	9-anthryl	9-phenanthryl	
B-CHO	2.9	+0.32	+0.47	+0.35	+1.4		20
	5.0	+0.26	+0.46	+0.31	+1.8₅		20
	7.5₅	+0.26	+0.65	+0.46	+1.4₅		20
	10.3	+0.33	+0.48	+0.35	+1.1		20
	0.0	+0.32	+0.47		+1.2₅	+0.47	21
	13.0	+0.33	+0.35		+1.2	+0.42₅	21
B-COCH₃	0.0	+0.22		+0.77			22
	Alk. $f \neq$ pH	+0.20		+0.37ᵃ			22
	Alk. $f \neq$ pH	+0.22		+0.23			23
4-N(CH₃)₂C₆H₄—CH=CH—B	$f \neq$ pH	+0.25	+0.68	+1.9			24
			+0.8ᵃ				
B-NO₂	2.9	+0.16	+0.75	+0.88	+0.25		20
	5.0	+0.14	+1.07	+1.07	+0.50		20
	7.5₅	+0.14	+0.79	+1.07	+0.36		20
	10.3	+0.24	+0.25	+0.41₅	+0.29		20
	HAc	+0.15	+0.60				25
	0.0	+0.16	+0.47	+0.47			26
	0.0	+0.14	+0.77		+0.03₅		27
	2.0	+0.16	+0.74	+0.55	+0.68		27
C₆H₅—N=N—B	4.7	+0.16	+0.09				28
			+0.11ᵇ				28
			+0.14ᶜ				28

Compound	Conditions					Ref.
$4\text{-}N(CH_3)_2C_6H_4N{=}N\text{-}B$	5.0	+0.16	−0.31	−0.37		29
	7.0	+0.16	−0.25	−0.25		29
B-I	$f \neq pH$	+0.30	+0.37	+0.17		30
		+0.31	+0.28[a]	+0.12[a]		30
		+0.30	+0.14			31
		+0.30	+0.17[a]			31
		+0.30		+0.20		32
B-Cl	$f \neq pH$	+0.20		> +1.1	+0.31[e]	32
B-NCS	2.2	+0.29[d]	+0.28	+0.31		33
		+0.29[f]	+0.31	+0.38		33
	9.1	+0.33	+0.18	+0.33		33
$N^{+}\text{-B}$ (pyridinium)	$f \neq pH$	+0.30	+0.26	+0.42		34
B-CH$_2$OH[g]	Unbuffered acetonitrile	+0.91	+0.12	+0.12		35

[a] Values of half-wave potentials from a graph.
[b] Phenyl and naphthyl carry OCH$_3$ in position 4.
[c] Phenyl and naphthyl carry OCOCH$_3$ in position 4.
[d] First wave.
[e] 2-Anthryl.
[f] Second wave.
[g] Oxidation waves.

Table VI-8

Mean Values of $\sigma_{\pi,B}^{-}$ Substituent Constants (from reactions of carbonyl compounds)

B	$\sigma_{\pi,B}^{-}$	$\dfrac{\sigma_{\pi,B}^{-}}{\sigma_{a}}$
Phenyl	0.0	—
1-Naphthyl	+0.48	2.8
2-Naphthyl	+0.39	7.8
9-Anthryl	+1.3$_5$	3.1
9-Phenanthryl	+0.45	2.6

The values collected in Table VI-7 show that the assumption of constant relative mesomeric interactions of different reducible groups with given mono-, di-, and tri-cyclic aromatic systems is not generally fulfilled. Similar scattering of σ-values was observed by Jaffé[37] for values for 2-naphthyl. The wide scattering may arise in part from the fact that for some reaction series the value of $\rho_{\pi,R}$ was determined under conditions which were not identical, but only similar; but deviations caused by this cannot provide an adequate explanation for the degree of variability of the $\sigma_{\pi,B}^{-}$ values.

When the reactions used for determination of values of $\sigma_{\pi,B}^{-}$ constants are restricted to carbonyl compounds, the mean values given in Table VI-8 can be derived.

Comparison with values of σ_a (see Table VI-4) shows that the ratio $\sigma_{\pi,B}^{-}/\sigma_a$ is fairly constant for 1-naphthyl, 9-anthryl, and 9-phenanthryl, whereas, for 2-naphthyl a significantly different ratio is obtained. The practical application of the constants $\sigma_{\pi,B}^{-}$ is demonstrated in Figs. VI-6 and VI-7. The experimental data treated in these figures are for cases in which the value of $\rho_{\pi,R}$ has not been determined from half-wave potentials of m- and p-substituted benzene derivatives and this prevents the application of these half-wave potentials for the computation of the $\sigma_{\pi,B}^{-}$ values summarized in Table VI-7. The relatively good correlation shown in Figures VI-6 and 7 for these data demonstrates that values of $\sigma_{\pi,B}^{-}$ (Table VI-8) can be used as a first approximation for the treatment of polarographic data in these reaction series.

For the half-wave potentials of unsaturated acids and esters[38] the value of the reaction constant $\rho_{\pi,P,R}^{-}$ is only slightly dependent on the esterification or dissociation of the carboxyl (Fig. VI-6). The dependence of the reaction constant on the composition of the supporting electrolyte can be observed from a comparison of Figs. VI-7a and 7b, which differ only in the replacement[32] of 0.05 M $N(CH_3)_4Br$ in Fig. VI-7a by 0.05 M $N(C_2H_5)_4Br$ in Fig. VI-7b. Finally, the reversed slope of the curve in Fig. VI-7c is in agreement with the fact

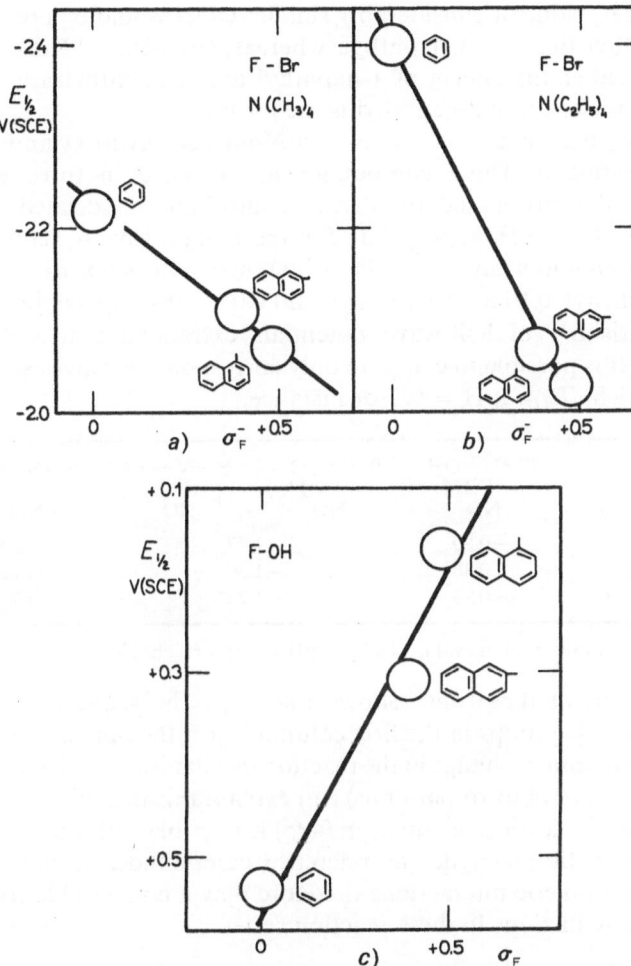

Fig. VI-7. Relation of half-wave potentials to polar aromatic constants σ_F^-. a) and b) Reduction of aryl bromides (half-wave potentials from Ref. 32); a) in a supporting electrolyte containing $N(CH_3)_4^+$; b) in a supporting electrolyte containing $N(C_2H_5)_4^+$; c) oxidation of phenols (half-wave potentials from Ref. 39).

that in this series half-wave potentials[39] for the oxidation of phenols were compared and, hence, the reaction constant $\rho_{\pi,P,R}^{-}$ has a negative value.

It can be understood from the above discussion that the correlation of half-wave potentials with σ_a values is, in general, worse than with $\sigma_{\pi,B}^{-}$ values. This can be deduced merely by an inspection of Table VI-7. From σ_a values a considerably smaller effect would be predicted for 2-naphthyl than for 1-naphthyl, whereas, the data in Table VI-7 show that either the effects of 1-naphthyl and 2-naphthyl are comparable, or the effect of 2-naphthyl is the greater.

Finally, the reduction of N-aryl-N-nitroso hydroxylamines[40] can be mentioned. These compounds are reduced in three steps, probably in the protonated, dipolar, or neutral and dissociated form. The choice of the pH appropriate for the comparison of half-wave potentials is of importance here. The difference in the slope $dE_{1/2}/d$pH (0.132 for phenyl, 0.10 for 1-naphthyl, and 0.081 for 4-biphenyl) makes the comparison[41] of half-wave potentials extrapolated to pH = 0 unsuitable. It is possible to compare only corresponding waves at a pH value at which $dE_{1/2}/d$pH = 0. For instance,

	B—N—OH $\mid$ }H$^+$ NO	B—N—O$^-$ $\mid$ }H$^+$ or NO	B—N—OH $\mid$ NO	B—N—O$^-$ $\mid$ NO
phenyl	−0.80	−1.47		−1.85
1-naphthyl	−0.88	−1.50		−1.82
4-biphenylyl	−0.93	−1.24[a]		−1.75

[a] Under the conditions used the value of $dE_{1/2}/d$pH had not reached zero.

Inspection of these data shows that, e.g., the sequence of the effects of the aryl groups in the first column is just the opposite of that in the third column. A change in the reaction mechanism with increasing pH (as in the case of nitro paraffins) can explain this behavior.

(iii) The application of equation (49p) is possible with the use of a series of aromatic aldehydes as reference compounds, as shown in Fig. VI-8. Mesomeric interactions therefore play a comparable role in nucleophilic reductions in these reaction series.

7. Substituent in a Reducible Side Chain

The half-wave potentials of 1-alkyl-1-(2-naphthyl)ethylenes[15] can be correlated using equation (29p) for Z = H, CH$_3$, C$_2$H$_5$, C$_3$H$_7$,

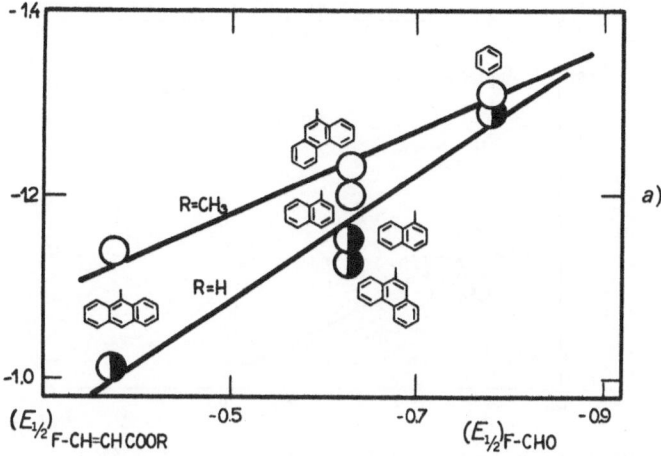

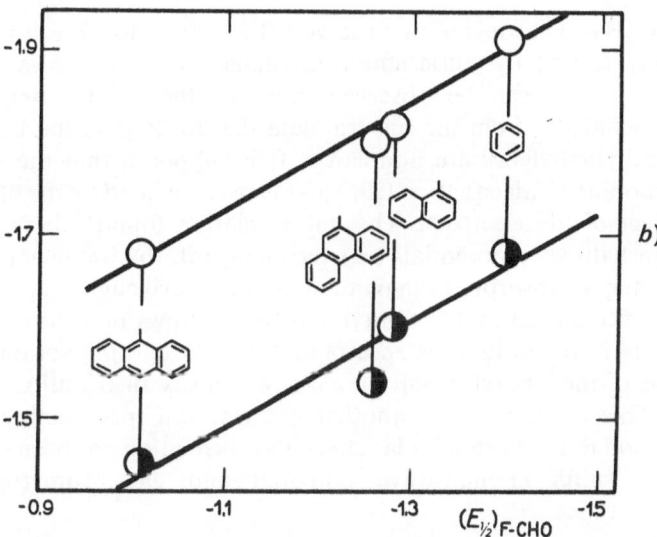

Fig. VI-8. Relation of half-wave potentials for the reduction of arylacrylic acids and their esters to half-wave potentials of the corresponding aldehydes. *a*) Half-wave potentials of acids (halved points) and esters (circles) as well as of aldehydes at pH 0; *b*) half-wave potentials of anions (halved points) and esters (circles) of arylacrylic acids and of aldehydes at pH > 10. Half-wave potentials from Ref. 38.

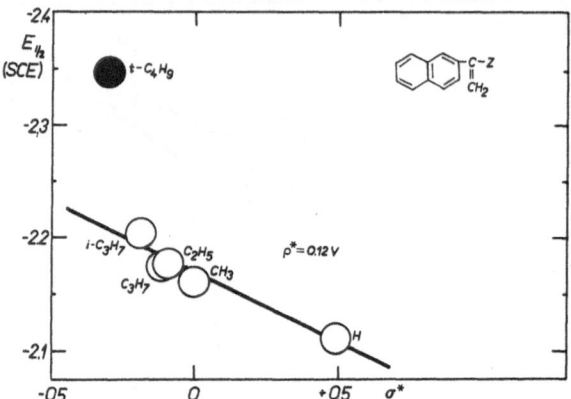

Fig. VI-9. Relation of half-wave potentials of substituted vinylnaphthalenes to Taft polar substituent constants σ_X^*. Half-wave potentials from Ref. 15. Full point deviates.

and $i\text{-}C_3H_7$. The deviation observed (Fig. VI-9) for $Z = t\text{-}C_4H_9$ is probably caused by steric effects (a smaller deviation toward more negative values can be observed even for the $i\text{-}C_3H_7$ derivative). It is concluded[15] from the spectral data that for $Z \neq H$ the 1-alkyl-1-(2-naphthyl)ethylenes are nonplanar. It is supposed that the validity of the modified Taft equation (29p) in this case is caused by the planarizing effect of the electrode. The linear relation found[15] between the shifts in half-wave potentials and the change in the frequency $\Delta\gamma\beta$ of the principal absorption maximum in the wavelength region 210–258 mμ, fitted even by the t-butyl derivative, shows that the effects, as expressed in the ultraviolet spectra and in polarography, of the introduction of the t-butyl group on electron-density distribution are the same. This effect includes another contribution (presumably steric) in addition to the normal polar effect. No correlation has been found[15] between shifts in half-wave potential and gas-chromatographic retention times.

References

[1] C. C. Price, E. C. Mertz, and J. Wilson, *J. Am. Chem. Soc.* **76**: 5131 (1954).

[2] L. K. Cremer, A. Fisher, B. R. Mann, J. Packer, R. B. Richards, and J. Vaughan, *J. Org. Chem.* **26**: 3148 (1961).

[3] P. R. Wells and E. R. Ward, *Chem. Ind.* **1958**: 528.

[4] H. H. Jaffé, *J. Am. Chem. Soc.* **76**: 4261 (1954).

[5] K. Lauer, *Ber.* **70B**: 1288 (1937).

[6] L. P. Hammett, "Physical Organic Chemistry," p. 184, McGraw-Hill, New York, 1940.

[7] H. Schenkel, *Experientia* **4**: 383 (1948).

[8] N. F. Hall and R. M. Sprinkle, *J. Am. Chem. Soc.* **54**: 3469 (1932).

[9] Y. Otsuji, M. Kubo, and E. Imoto, *Bull. Univ. Osaka Prefecture* **7A**: 61 (1959).

[10] J. Nakaya, H. Kinoshita, and S. Ono, *J. Chem. Soc. Japan* **78**: 940 (1957).

[11] E. Imoto, *Rev. Polarog.* **9**: 185 (1961).

[12] L. H. Klemm and A. J. Kohlik, *J. Org. Chem.* **28**: 2044 (1963).

[13] L. H. Klemm, A. J. Kohlik, and K. B. Desai, *J. Org. Chem.* **28**: 625 (1963).

[14] L. H. Chopard-dit-Jean and E. Heilbronner, *Helv. Chim. Acta* **36**: 144 (1953).

[15] L. H. Klemm, W. C. Solomon, and A. J. Kohlik, *J. Org. Chem.* **27**: 2777 (1962).

[16] J. W. Loveland and G. R. Dimeler, *Anal. Chem.* **33**: 1196 (1961).

[17] P. Beckmann and H. Silberman, *Chem. Ind.* **1955**: 1635.

[18] L. Stárka and A. Vystrčil, *Chem. Listy* **51**: 1449 (1957).

[19] R. Gerdil and E. Heilbronner, *Helv. Chim. Acta* **40**: 141 (1957).

[20] E. Imoto, R. Motoyma and H. Kakiuchi, *Bull. Naniwa Univ.* **3A**: 203 (1955).

[21] R. N. Schmid and E. Heilbronner, *Helv. Chim. Acta* **37**: 1453 (1954).

[22] C. Calzolari and C. Furlani, *Ann. Chim.* **44**: 356 (1954), *Bull. Sci. Fac. chim. ind.* (Bologna) **12**: 14 (1954).

[23] C. Calzolari, *Ann. Triestini* **1954**: No. 1 16 pp.

[24] F. Goulden and F. L. Warren, *Biochem. J.* **42**: 420 (1948).

[25] I. Bergman and J. C. James, *Trans. Faraday Soc.* **48**: 956 (1952).

[26] L. Holleck and H. J. Exner, *Naturwissenschaften* **39**: 159 (1952).

[27] R. Zahradnik and K. Boček, *Collection Czech. Chem. Commun.* **26**: 1733 (1961).

[28] I. F. Vladimircev and I. Ja. Postovskij, *Dokl. Akad. Nauk SSSR* **83**: 855 (1952).

[29] L. Moelants and R. Janssen, *Bull. Soc. Chim. Belg.* **66**: 209 (1957).

[30] E. Gergely and T. Iredale, *J. Chem. Soc. 1951*, 3502.

[31] E. L. Colichman and S. K. Liu, *J. Am. Chem. Soc.* **76**: 913 (1954).

[32] E. S. Levin and Z. I. Fodiman, *Zh. Fiz. Khim.* **28**: 601 (1954).

[33] D. Vlachová, R. Zahradník, K. Antoš, B. Kristián, and A. Hulka, *Collection Czech. Chem. Commun.* **27**: 2826 (1962).

[34] K. Schwabe, *Chem. Tech.* (Berlin) **9**: 126 (1957).

[35] H. Lund. *Acta Chem. Scand.* **11**: 491 (1957).

[36] L. P. Hammett, "Physical Organic Chemistry," p. 188, McGraw-Hill, New York, 1940.

[37] H. H. Jaffé, *Chem. Rev.* **53**: 1919 (1953).

[38] S. Ono and M. Uehara, *Bull. Univ. Osaka Prefect* **5A**: 139 (1957).

[39] R. A. Nash, D. M. Skauen, and W. C. Purdy, *J. Am. Pharm. Assoc. Sci. Ed.* **47**: 433 (1958).

[40] L. Holleck and R. Schindler, *Z. Elektrochem.* **60**: 1138, 1142 (1956).

[41] P. J. Elving and E. C. Olson, *J. Am. Chem. Soc.* **79**: 2697 (1957).

VII: Effects of Substituents in Polycyclic Heterocyclic Compounds

1. General

For polycyclic heterocyclic compounds containing one or more heterocyclic nuclei of aromatic or nonaromatic character, together with one or more benzene or hydroaromatic rings, the systematic treatment of constitutive effects is rather involved. The types of systems are more numerous than in the case of monocyclic heterocyclic compounds: the electroactive group may be either in a heterocyclic or in a benzene ring; similarly, the substituent also may be either in a heterocyclic or in a benzene ring and, moreover, it may be either in the same ring as the electroactive group or in another ring. Finally, the number of types of reaction series is increased by the possibility that in the system of condensed rings the electro-reduction or -oxidation of one of the rings may be considerably facilitated.

On the other hand, because of the presence of the hetero atom, the number of ways in which polycyclic heterocyclic compounds can be classified is greater than for condensed polycyclic hydrocarbons: the rings become nonequivalent, and it is necessary to express the positions of the electroactive group and the substituent relative to the hetero atom.

2. Classification of Reaction Series

As a first criterion in the classification of the reaction series of polycyclic heterocyclic compounds the position of the electroactive

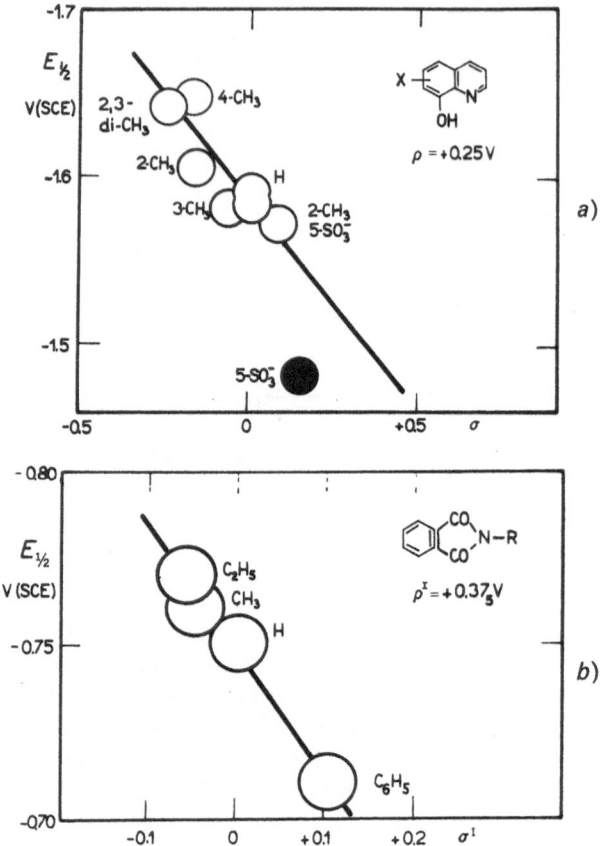

Fig. VII-1. *a*) Relation of half-wave potentials for the reduction of substituted 8-hydroxy-quinolines to Hammett substituent constants σ_X^*. Reaction Series No. 9, Table VII-2. *b*) Relation of half-wave potentials for the reduction of N-alkylphthalimides to inductive substituent constants σ_X^I. Reaction Series No. 28, Table VII-2.

group is used. Reaction series are divided into (A) those in which the reduction occurs in the heterocyclic ring, and (B), those in which it occurs in the side chain.

(A) For compounds for which the reduction or oxidation occurs in the heterocyclic ring the substitution can be carried out either (Aa) in the same heterocyclic ring (symbol Benzo-*Het*-X), (Ab) in the benzene ring (X-Benzo-*Het*), or (Ac) in a phenyl group attached either to the heterocyclic or to the benzene ring (symbols Benzo-*Het*-C_6H_4-X and X-C_6H_4-Benzo-*Het*). In the latter group systems in which the phenyl group is exchanged for other aromatic rings (Benzo-*Het*-B) are included.

(B) The reduction of the side chain can occur when it is attached to either (B₁) the heterocyclic or (B₂) the benzene ring.

(B_1) With compounds containing a reducible side chain R on the heterocyclic ring the substitution can be carried out either (B_1a) in the

same heterocyclic ring $\left(\text{Benzo-Het} \underset{R}{\overset{X}{\diagup}} \right)$, or (B_1b) in the benzene ring

(X-Benzo-Het-R). Finally, in this group it is also possible to include

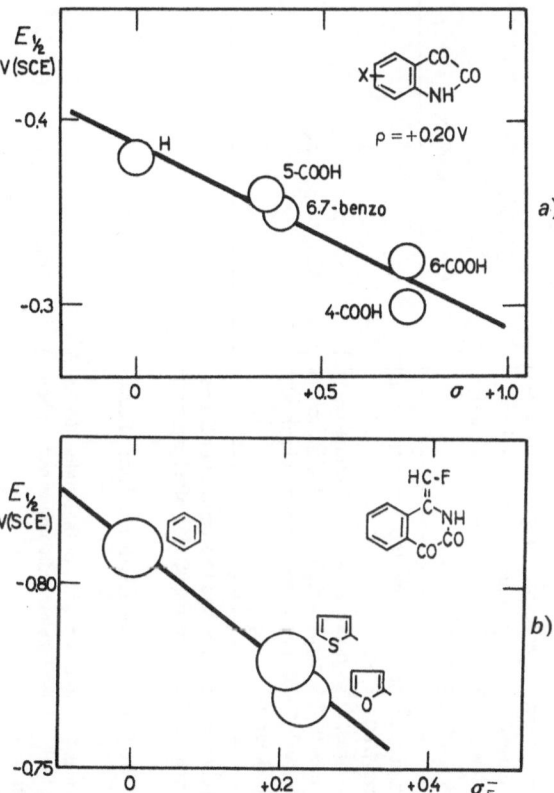

Fig. VII-2. *a*) Relation of half-wave potentials for the reduction of substituted isatins to Hammett substituent constants σ_X. Reaction Series No. 45, Table VII-2. *b*) Relation of half-wave potentials for the reduction of the substituted compounds given in the figure to aromatic substituent constants σ_F^-. Reaction Series No. 27, Table VII-2.

Table VII-1

Separation Reaction Series for Polycyclic Heterocyclic Compounds

Examples	Symbol	Equation	Applied constants
A Reduction in Heterocyclic Ring			
Aa Substitution in heterocyclic ring (Benzo-*Het-X*)			
Coumarins,[1,2] isoquinolines,[3] quinoxalines,[4,5] pteridines,[6] N-alkylphenazinones,[7] phenothia-zines,[a,8] 9-alkylacridines[9,10]	29 hp 25 hp 27 hp	$\Delta E_{1/2} = \rho^*_{\pi,het}\sigma^*_X$ $\Delta E_{1/2} = \rho^I_{\pi,het}\sigma^I_X$ $\Delta E_{1/2} = \rho_{\pi,het}\sigma_{p-X}$	σ^*_X σ^I_X σ_{p-X}
Ab Substitution in benzene ring (and in both rings) (X-Benzo-*Het*)			
Quinolines,[11,12] quinoxalines,[5,13] coumarins,[1,2] acridines,[10,14] acridinizinium salts,[15] phen-azines,[5,7,9,16] phenothiazines,[7] phenothiazinones,[8,9] furocoumarins[1]	27 hp	$\Delta E_{1/2} = \rho_{\pi,het}\sigma_X$	$\sigma_m, \sigma_p, \frac{1}{2}(\sigma_m + \sigma_p),$ $(\sigma_o), \sigma^-_p, (\sigma^+)$
Ac Substitution in phenyl bound to a benzene or heterocyclic ring; exchange of phenyl for other aromatic rings (Benzo-*Het*-C$_6$H$_4$-X; X-C$_6$H$_4$-Benzo-Het; Benzo-*Het*-B)			
Flavones,[17] isoflavones.[18] phthalimide derivatives,[19,20] 3,3'-benzylidenedicoumarins,[2] phen-azines[16]	27	$\Delta E_{1/2} = \rho_{\pi,R}\sigma_X$	$\sigma_m, \sigma_p, (\sigma_o), (\sigma^-_p),$ σ^-_B
B Reduction in the Side Chain			
B$_1$ Side chain on the heterocyclic ring			
B$_1$a Substitution in the heterocyclic ring $\left(\text{Benzo-}Het\diagdown_{X}^{R}\right)$			

Description	Relation	Ref.	σ
Phthalimides,[20,22] phthalonimides,[20] naphthalimides,[20] quinoxaline N-oxides,[23] 9-alkylacridine N-oxides,[10] phenazine N-oxide[24]	$\Delta E_{1/2} = \rho_{\pi,het}^I \sigma_X^I$ $\Delta E_{1/2} = \rho_{\pi,het}^I \sigma_{het}^I$	25 27 h	σ_X^I $\sigma_m, \sigma_p, (\sigma_o)$
B₁b Substitution in the benzene ring (X-Benzo-Het-R) Phthalimides,[20,21] N-methylphthalimides,[20,22] phthalonimides,[20] isatins,[25] xanthones,[25] chloroacridines,[24] acridine N-oxides,[10,24] phenazine N-oxides[9,27]	$\Delta E_{1/2} \doteq \rho_{\pi,Benzo} \sigma_{Benzo}$ $\Delta E_{1/2} \doteq \rho_{\pi,Benzo} \sigma_X$	27 bh 27 b	$\sigma_m, \sigma_p, (\sigma_m + \sigma_p)/2$
B₁c Position and nature of the heterocyclic ring (R-Benzo-Het) Benzothiophene 1,1-dioxides,[28] (arylmethylene)homophthalimides[20,22]	$\Delta E_{1/2} = \rho_{\pi,het}^* \sigma_{het}^*$ $\Delta E_{1/2} = \rho_{\pi,het}^* \sigma_B^-$ $\Delta E_{1/2} = \rho_{\pi,R} \sigma_B$	29 ph — 51	σ_B^-
B₂ Side chain on the benzene ring			
B₂a Substituent in the heterocyclic ring (R-Benzo-Het-X)	$\Delta E_{1/2} = \rho_{Benzo,het}^* \sigma_{Benzo,het}^*$ $\Delta E_{1/2} = \rho_{Benzo} \sigma_{het,X}$ $\Delta E_{1/2} \doteq \rho_{Benzo} \sigma_X$	29 hb — 29 b	
B₂b Substituent in the benzene ring (R-Benzo-Het) $\left(\begin{smallmatrix} R \\ X \end{smallmatrix}\!\!\!\searrow\!\!\text{Benzo-Het}\right)$	$\Delta E_{1/2} = \rho_{\pi,R} \sigma_X$	29	
B₂c Position and nature of the heterocyclic ring (R-Benzo-Het) Nitrophthalides,[29] nitroquinolines,[30] nitroimidazole,[30] acridine isothiocyanates[31]	$\Delta E_{1/2} = \rho_{\pi,R} \sigma_B^-$	51	σ_B^-

ª Oxidation, anodic waves.

Table VII-2
Reaction Series of Polycyclic Heterocyclic Compounds with Half-Wave Potentials Correlated by Modified Forms of Equation (20)

No.	Reaction series	Solution	Equation	$\rho_{r.r.}$, V	E_H (SCE)	n^a	Substituents Fitting	Substituents Nonfitting	Ref. to source of $E_{1/2}$
(Aa) Substitution in Reducible Heterocyclic Ring									
1		0.1 M LiCl, 25% EtOH, $f \neq$ pH	22 hp	+0.28	-1.68^b	3	CH_3; H; $C_6H_5(\sigma^*)$	—	1
2		0.1 M LiCl, 25% EtOH, $f \neq$ pH	22 hp	+0.20	-1.64^b	3	CH_3; H; $C_6H_5(\sigma^*)$	—	1
3		Acetate buffer, EtOH,		+0.69	-0.62	4	H: OC_6H_5; $NHNH_2$; NH_2	—	9
4		NH_3, NH_4Cl, pH 9.25		+0.36	-0.86	4	H: OC_6H_5; Cl: $NHCH_2CHOHCH_2N(C_2H_5)_2(\sigma_{p\text{-}NHCH_3})$	—	10
5		NH_3, NH_4Cl, pH 9.25		+0.19	-0.83^c	3	Cl: OC_6H_5; $NHCH_2CHOHCH_2N(C_2H_5)_2(\sigma_{p\text{-}NHCH_3})$	—	10
6		NH_3, NH_4Cl, pH 9.25		+0.15	-0.92^c	3	Cl: OC_6H_5; $NHCH_2CHOHCH_2N(C_2H_5)_2(\sigma_{p\text{-}NHCH_3})$	—	10
7		Acetate, pH 4.5		+0.37	-0.65^c	3	Cl: OC_6H_5; $NHCH_2CHOHCH_2(C_2H_5)_2(\sigma_{p\text{-}NHCH_3})$	—	10
8		NH_3, NH_4Cl, pH 9.25		+0.26	-0.89	3	Cl: OC_6H_5; $NHCH_2CHOHCH_2N(C_2H_5)_2(\sigma_{p\text{-}NHCH_3})$	—	10

9	Buffer, pH 12	22 hp[d]	+0.25	−1.59	6	H; 2-CH$_3$(σ_a); 3-CH$_3$(σ_m); 4-CH$_3$(σ_p); (2,3-diCH$_3$); (2-CH$_3$,5-SO$_3^-$)(σ_m)	5-SO$_3$; 2-C$_6$H$_5$	11, 12
10	Buffer pH 2	20 hp	+0.10	−0.34	5	H; OCH$_3$; OC$_2$H$_5$; Cl; Br(σ_{p-x})	—	13
11	Buffer pH 10	20 hp	+0.09$_4$	−0.86	6	H; OCH$_3$; OC$_2$H$_5$; NH$_2$; Cl; Br(σ_{p-x})	—	13
12	Buffer, lower pH, f ≠ pH	20 hp	+0.33	−1.53	9	H; 6-OH; 7-OH; 7-OCH$_3$; 6,7-OCH$_2$O; 8-OCH$_3$; (6-OCH$_3$,7-OH); (6-OCH$_3$,7,8-diOH);(6,7-diOCH$_3$,8-OH) (Position 6 and 8: σ_m; position 7: σ_p)	(6,7-diOH); (7,8-diOH); 7-COOCH$_3$	1
13	NH$_4$Cl, pH 6		+0.27	−1.58	4	H; 7-OH; 6,7-diOH; (6-OCH$_3$,7,8-di-OH)	—	2
14	Buffer, f ≠ pH	20 hp	+0.20	−1.7^c	3	OH; O$^-$; OCH$_3$(σ_m)	—	1

[a] Number of substances fitting the quoted form of equation (20) cited.
[b] Half-wave potential of the methyl derivative E_{CH_3}.
[c] Extrapolated value.
[d] Better correlation found for the $E_{1/2}$–pK relation.
[e] Half-wave potential of the phenyl derivative $E_{C_6H_5}$.
[f] $\rho_{\pi,R}^l$.
[g] $\rho_{\pi,R}^\ddagger$.

[h] σ_m was used, but correlations for $\dfrac{\sigma_m - \sigma_p}{4}$ and even for σ_p are not bad : only value of $\rho_{\pi,R}$ changes.

[j] According to Tirouflet.[22]
[k] Purity of these substances is doubtful.[15]

Table VII-2 (*continued*)

No.	Reaction series	Solution	Equation	$\rho_{n,B}$, V	E_B(SCE)	n^a	Substituents Fitting	Nonfitting	Ref. to source of $E_{1/2}$
15a		$N(C_4H_9)_4I$, 50% tetrahydrofuran 1st wave		+0.84	−0.81	5	$H;CH_3;Cl;Br;I(\sigma_p)$	F	15
15b		2nd wave		+0.25	−1.02	4	$H;CH_3;Cl;Br;(\sigma_p)$	F;I	15
16a		$N(C_4H_9)_4I$, 50% tetrahydrofuran 1st wave		+0.20	−0.81	4	$H;CH_3;OCH_3^k;OH;(\sigma_p)$	—	15
16b		2nd wave		+0.07	−1.02	4	$H;CH_3;OCH_3^k;OH;(\sigma_p)$	—	15
17a		$N(C_4H_9)_4I$, 50% tetrahydrofuran 1st wave		+0.60	−0.81	7	$H;CH_3^k;OCH_3;F;Cl;Br:I(\sigma_p)$		15
17b		2nd wave		+0.20	−1.02	4	$H;CH_3^k;OCH_3;Cl(\sigma_p)$	F;Br;I	15
18		Acetate buffer, EtOH		+0.17	−0.36	5	$H;OH;NH_2;Cl;NO_2(\sigma_p)$	—	9
19		Buffers pH 2.9		+0.18	−0.18	4	$H;OCH_3;F;Cl(\sigma_p)$	—	16
20		0.1 N H_2SO_4, EtOH, Au electrode, oxidation		+0.14	+0.31	4	$H;Cl;CH_3CO;CF_3\left(\dfrac{\sigma_m+\sigma_p}{2}\right)$	—	8

No.	Conditions		value	value		Substituents		Ref
21	0.1 N H_2SO_4, EtOH, Au electrode, oxidation		+0.36	+0.54	4	H; Cl; CH_3CO; CF_3 $\left(\dfrac{\sigma_m + \sigma_p}{2}\right)$	—	8

(Ac) Substitution in a Phenyl Ring Attached to a Reducible Heterocyclic Ring; Exchange of Rings Attached to the Heterocyclic Ring

No.	Conditions	n	value	value		Substituents		Ref
22	Buffer pH 6.2	20	+0.15	−1.47	3	H; OH; $OCH_3(\sigma_p)$	—	17
23	Buffer pH 6.2	20	+0.11	−1.62^c	3	H; OH; $OCH_3(\sigma_p)$	—	17
24	Buffer pH 6.0	20	+0.75	−1.57^c	3	4'-OH(σ_p); (3',4'-diOH); (3',4',5'-triOH) (σ_m and σ_p)	—	18
25	NH_4Cl, pH 6.0		+0.28	−1.61	3	H; 2'-OH(σ_p); 3'-OH(σ_m)	4'-$CH_3(\sigma_p)$	2
26	Buffers pH 2.9		+0.06$_2$	−0.24	4	H; OCH_3; F; $Cl(\sigma_p)$	—	16

Table VII-2 (*continued*)

No.	Reaction series	Solution	Equation	$\rho_{x,R}$, V	E_H(SCE)	n^a	Substituents — Fitting	Substituents — Nonfitting	Ref. to source of $E_{1/2}$
27		Buffer pH 2, 50% EtOH	23 f	+0.15	-0.81^e		C_6H_5; 2-thienyl($\sigma_F^- = +0.23$); 2-furyl($\sigma_F^- = +0.21$)	—	19, 20
28		Buffer pH 1, 10% EtOH	19 h	$+0.37_5^f$	-0.75	4	$H; CH_3; C_2H_5; C_6H_5(\sigma^I)$	Br	21, 22
29		Buffer pH 4.2, 50% EtOH		$+0.27^f$	-1.10	4	$H; CH_3; C_2H_5; C_6H_5(\sigma^I)$	$CH_2C_6H_5$	20
30		Buffers pH > 9, 50% EtOH, $f \neq$ pH		$+0.04^g$	-1.20^b	3	$H; CH_3; C_6H_5(\sigma^*)$	—	20
31		Buffers pH 4.2, 10% EtOH		$+0.05^g$	-0.29^b	3	$H; CH_3; C_2H_5(\sigma^*)$	—	20
32		NH_3, NH_4Cl, pH 9.25		+0.13	-1.02	4	$H; OC_6H_5; Cl(\sigma_p)$; $NHCH_2CHOHCH_2N(C_2H_5)_2(\sigma_{p\text{-NHCH}_3})$	—	10

No.	Substituents	n			Conditions		R	Ref.
33	OC_6H_5; $Cl(\sigma_p)$; $NHCH_2CHOHCH_2N(C_2H_5)_2(\sigma_{p\text{-}NHCH_2})$	3	+0.13	-0.91^c	NH_3, NH_4Cl, pH 9.25		—	10
34	OC_6H_5; Cl; (σ_p) $NHCH_2CHOHCH_2N(C_2H_5)_2(\sigma_{p\text{-}NHCH_2})$	3	+0.07	-0.96^c	NH_3, NH_4Cl, pH 9.25		—	10
35	OC_6H_5; Cl; (σ_p) $NHCH_2CHOHCH_2N(C_2H_5)_2(\sigma_{p\text{-}NHCH_2})$	3	+0.07	-0.87^c	NH_3, NH_4Cl, pH 9.25		—	10

(B, b) Effect of Substitution in the Benzene Ring on Reduction in the Side Chain of the Heterocyclic Ring

No.	Substituents	n			Conditions		R	Ref.
36	H; NH_3^+; $NHCOCH_3$; Cl; $Br(\sigma_m)^g$	5	+0.06	-0.67	Buffer pH 0, 10% EtOH	20 bh	OH	21
37	H; OH; NH_3^+; $NHCOCH_3$; Cl; $Br(\sigma)^g$	6	+0.09	-0.75	Buffer pH 1.0, 10% EtOH	20 bh	—	21
38	H; $NHCOCH_3$; Cl; $Br(\sigma_m)^g$	4	+0.14	-0.84	Buffer pH 2.0, 10% EtOH	20 bh	OH	21
39	H; $NHCOCH_3$; Cl; $Br(\sigma_m)^g$	4	+0.14	-0.98	Buffer pH 4.2, 10% EtOH	20 bh	OH	21
40	H; NH_2; $NHCOCH_3$; Cl; $Br(\sigma_m)^g$	5	+0.16	-1.07	Buffer pH 5.7, 10% EtOH	20 bh	OH	21
41a	H; Cl; Br; $(\sigma_m)^g$	3	+0.17	-1.12	Buffer pH 9.2, 10% EtOH 1st wave	20 bh	NH_2	21
41b	H; NH_2; Cl; $Br(\sigma_m)^g$	4	+0.17	-1.30	2nd wave	20 bh	—	21
42	H; O^-; NH_2; $NHCOCH_3$; Cl; Br; $(\sigma_m)^g$	6	+0.17	-1.35	Buffer pH 11.3, 10% EtOH	20 bh	—	21

Table VII-2 (*continued*)

No.	Reaction series	Solution	Equation	$\rho_{\pi,R}$, V	E_H(SCE)	n^a	Substituents Fitting	Nonfitting	Ref. to source of $E_{1/2}$
43		Buffer pH 2–7	20 bh	$+0.20^j$	—	9	$H:OH:NH_3^+:NHCOCH_3:NHCOC_6H_5:$ $Cl:Br:I:NO_2\left(\dfrac{\sigma_m+\sigma_p}{2}\right)$	—	22
44		pH 9	20 bh	$+0.26^j$	—	7	$H:O^-:NH_2:NHCOCH_3:Cl:Br:I$ $\left(\dfrac{\sigma_m+\sigma_p}{2}\right)$	—	22
45		Acetate pH 4.7, 50% EtOH	20 bh	$+0.10$	-0.38	4	$4\text{-COOH}(\sigma_p):5\text{-COOH}(\sigma_m):6\text{-COOH}(\sigma_p^-):$ $6,7\text{-benzo}(\sigma_{2\text{-naphthyl}}=-0.39)$	$5\text{-}C_6H_5:6\text{-}C_6H_5:$ $4,5\text{-benzo}(\sigma_{1\text{-naphthyl}})$	25
46		Buffer pH 3, 1st wave	20 bh	$+0.23$	-0.91	3	$H:2\text{-OCH}_3(\sigma_m):3\text{-OCH}_3(\sigma_p)$	—	26
47a		Acetate, EtOH, 1st wave	20 bh	$+0.17$	-0.75	3	$H:2\text{-OCH}_3(\sigma_p):3\text{-Cl}(\sigma_m)$	—	24
47b		2nd wave (C–Cl)	20 bh	$+0.69$	-1.51	3	$H:2\text{-OCH}_3(\sigma_m):[2\text{-OCH}_3,7\text{-Cl}(\sigma_p)]$	—	24

those cases (B_1c) in which one given electroactive group is attached to the heterocyclic ring and the effect of the kind of heterocyclic ring is studied (Benzo-Het-R).

(B_2) When the reducible side chain is attached to the benzene ring, either (B_2a) the substituent may be in the heterocyclic part of the molecule (symbol R-Benzo-Het-X) or (B_2b) both groups, electroactive

and substituent, may be in the benzene ring $\left(\begin{array}{c} X \\ \diagdown \\ \diagup \;\text{Benzo-Het} \\ R \end{array}\right)$, or

(B_2c) the nature of the heterocyclic ring and its bonding can be varied and the effect on the reducibility of a side chain on the benzene ring can be followed, in this case the heterocyclic ring can be considered as a substituent (R-Benzo-**Het**).

The treatment of reaction series of type (Aa) is identical with that for the effect of substitution in a monocyclic compound containing an electroactive heterocyclic ring. For type (B_2b) the treatment does not differ in principle from the treatment of substituent effects in benzene derivatives with the electroactive group in the side chain. For all the other types, the treatment required is characteristic for this group of compounds. Of the large number of possible forms of equation (20) for these complex systems the simplest are summarized in Table VII-1. In these equations, apart from those mesomeric effects which are taken into account in the substituent constants σ_X^- or σ_B^-, only polar substituent effects are considered. The conditions for the application of particular forms of equation (20) are similar to those discussed in previous chapters. Some examples of such applications are summarized in Table VII-2.

In Tables VII-1 and VII-2, R stands for the electroactive group, X for the substituent, Benzo for a benzene ring, Het for an electroinactive heterocyclic ring, *Het* for an electroactive heterocyclic ring, and **Het** for an exchangeable heterocyclic ring.

The substituent constants used in the evaluation of the particular form of equation (20) in Table VII-2 are the same as in previous chapters, and so are the conditions for the selection of the values of half-wave potential, for the detection of deviations from linear relations, and for the determination of the values of reaction constants. The treatment is illustrated by the examples in Figs. VII-1 and VII-2.

3. Effect of a Substituent in an Electroactive Heterocyclic Ring

Linear relations have been demonstrated for polar effects in coumarin derivatives,[1] but for 9-substituted acridines the application of $\sigma_{p\text{-}X}$ constants has proved to be superior, which indicates that there is a mesomeric contribution to the substituent effect. In Reaction Series No. 4–6 it was impossible to correlate the data obtained[10] at pH 4.5 in an acetate buffer, and also the half-wave potentials of the second, more negative acridine wave,[10] did not show linear correlation either at pH 4.5 or at pH 9.25. The differences in the $\rho_{\pi,\text{acridine}}$ values for 3- and 7-substituted compounds show that the substituent effects are not simply additive. No explanation for such behavior is known, and it would be of interest to verify the polarographic behavior of substituted acridines.

In addition to examples given in Table VII-2, there are some other classes of compounds mentioned in Table VII-1 for which either the available data are too sparse, or values of the substituent constants are unknown. Thus, for quinoxalines[4,5] the shifts in half-wave potential caused by substituents in positions 2 and 3 are in the same direction and in the same sequence as the $\sigma_{p\text{-}X}$ values of benzene derivatives. For 2-substituted 3,4-dihydro-1(2H)-isoquinolines (I) a linear correlation between the half-wave potentials of the first one-electron wave and pK_a values was found.[3] For phenothiazines substituted on nitrogen by an aliphatic chain containing various amino groups, a qualitative relation was found between the half-wave potential and the number of carbon atoms separating the chain nitrogen from the ring nitrogen.[8] The σ_X^+ values for complicated side chains are at present not available

$$R^1 = R^2 = R^3 = H, \, R^4 = CH_3$$

$$R^1 + R^2 = OCH_2O, \, R^3 = CH_3O \; R^4 = CH_3$$

$$R^1 + R^2 = OCH_2O, \, R^3 = CH_3O \; R^4 = C_6H_5$$

$$R^1 + R^2 = OCH_2O, \, R^3 = CH_3O \; R^4 = \text{〈} \rangle\text{—}OCH_3$$

and half-wave potentials of the oxidation waves of simpler derivatives were not measured, and hence it is impossible to give a quantitative treatment.

By a combination of polarographic data with the dissociation constants of the 4-hydroxy group it was possible[6] to determine values of the substituent constants σ_{Pt-X} for the effects of substituents in various positions of the pteridine ring on the reactivity of that ring. The results obtained are summarized in Table VII-3, which also gives half-wave potential increments ($\Delta E_{1/2}$) expressing the shifts in half-wave potential in the reference state (the undissociated molecule at pH 5). For those compounds (6-pteridineethanol, 6-pteridinecarboxaldehyde, and 6-pteridinecarboxylic acid), for which, owing to recombination a reduction of the cation occurred in this pH range, an extrapolation was carried out.

The sequence of σ_{Pt-X} is similar to that of σ_{p-X}. The values given in Table VII-3 show a good additivity and $E_{1/2}$ versus $\Sigma\sigma_{Pt}$ is linear (Fig. 2, Ref. 6). The half-wave potentials were also correlated with ultraviolet spectra. Since for pteridines the $n \rightarrow \pi$ band is overlapped by those corresponding to the first $\pi \rightarrow \pi$ transition, it was found useful to correlate values of $E_{1/2}$ with the wave number v of the second $\pi \rightarrow \pi$ band. Figure 3 in Ref. 6 shows a good correlation.

Table VII-3
(According to Ref. 6)
Polar Substituent Constants (σ_{Pt-X}) for Substituents in the Pteridine Ring, and Half-Wave Potential Increments
($\Delta E_{1/2}$)

Substituent	σ_{Pt-X}	$\Delta E_{1/2}$
2-NH$_2$	−0.03	−0.02
2-OH	−0.02	—
4-OH	0	−0.20
6-OH	−1.11	−0.26
7-OH	−1.72	−0.37
6-CH$_3$	−0.30	−0.04
7-CH$_3$	−0.20	0
6-CH$_2$OH	—	−0.03
6-CHO	+0.30	+0.10
6-COOH	−0.44	+0.04
6-Fo[a]	−0.34	−0.08

[a] Folic acid residue.

4. Effect of Substituents in the Benzene Ring and in Both Rings on the Reduction of the Heterocyclic Ring

Substituent constants for Reaction Series No. 9–21 were chosen by a trial-and-error method. For quinolines (No. 9) as a first approximation the constants σ_{m-X} can be used for substituents in positions 3, 5, and 7, and the constants σ_{p-X} can be used for substituents in positions 2 and 4 (Fig. VII-1). Similarly, for coumarins (No. 12 and 13) for substituents in positions 6 and 8 values of σ_{m-X} and for those in position 7 values of σ_{p-X} can be used. For phenothiazines substituted in position 2 (No. 20 and 21), the application of $(\sigma_m + \sigma_p)/2$ was most satisfactory. In all other reaction series, the application of σ_{p-X} was found to give the best linear $E_{1/2}-\sigma$ relations. This was found to be true even for quinoxalines (No. 10), for which the authors[13] who carried out the half-wave potential measurements recommended the application of values of $(\sigma_m + \sigma_p)/2$. The usefulness of σ_{p-X} shows that mesomeric interactions occur between the substituent and the ring containing it. The change in electron density resulting from such interaction then is transmitted to the electroactive heterocyclic ring and causes a polarization of the electroactive bond. In none of the series No. 9–21 were half-wave potential values available for those substituents which would enable a distinction to be made between the suitabilities of σ_{p-X} and σ^-_{p-X}.

For 7-substituted quinoxalines (No. 10 and 11) the value of the reaction constant $\rho_{\pi,\text{quinoxaline}}$ is only slightly dependent on pH. Whereas, for 5-hydroxyquinoxalines (No. 10 and 11) and coumarins (No. 12) one reaction constant $\rho_{\pi,\text{het}}$ could be used for substituents in various positions, for acridinizinium (benzo [b] quinolizinium) salts (No. 15–17) the situation is different. In this reaction series the changing values of $\rho_{\pi,\text{het}}$ show that various positions are susceptible to substituent effects in different degrees (Fig. VII-3). For substituents in positions 6 and 11, too few substituents (only H and CH_3) were measured to enable a determination of the reaction constant to be made. Tentative values of $\rho'_{\pi,\text{het}} = \Delta E_{1/2}/\sigma_{p-CH_3}$ can be computed. As compared with the treatment of azulenes (see p. 232), the present treatment is even less reliable, because it would be necessary to prove that also in positions 6 and 11 the application of σ_{p-X} gives the best correlation. The negative value of $\rho_{\pi,\text{het}}$ for position 11 gives rise to doubts, at least for this position. The values of the reaction constants obtained for both waves can be schematized:

1st wave   (0.82) (0.84)   2nd wave   0.25

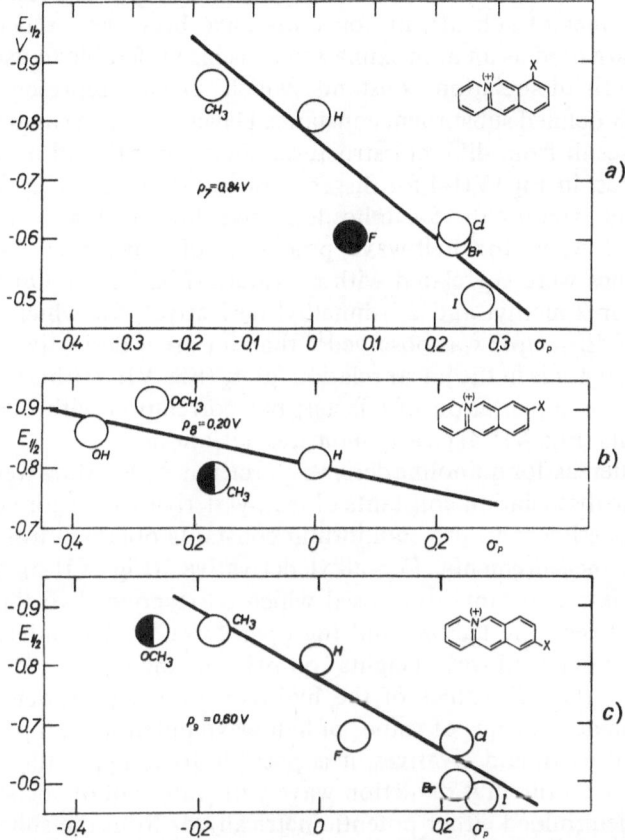

(−0.18)   (0.01)

The values for the first electron uptake are approximately three times as high as those of the second half-wave potential.

Fig. VII-3. Relation of the half-wave potentials for the first reduction waves of acridinizinium (benzo [b] quinolizinium) derivatives to Hammett substituent constants $\sigma_{p\text{-}x}$. a) Substituents in position 7, full point deviates, Reaction Series No. 15a, Table VII-2; b) substituents in position 8, halved point: authors[15] suggest that the specimen is doubtful, possibly a mixture of 8-methyl and 10-methyl derivatives, Reaction Series No. 16a, c) substituents in position 9, halved point: authors[15] suggest that 9-methoxy derivative is doubtful, Reaction Series No. 17a.

For 8-hydroxyquinolines[12] (Fig. VII-4), coumarins[32] (Fig. VII-5), and acridines a better correlation of half-wave potentials with pK values than with σ_X values was found.

For benzene derivatives it cannot be generally recommended that half-wave potentials be compared with the logarithms of rate or equilibrium constants in a given reaction series,* and the application of average and proved substituent constants σ_X seems to be most useful. On the other hand, for heterocyclic compounds for which no exact values of substituent constants have been determined and σ_X values are used as an approximation it seems preferable to use correlations with dissociation constants, which, in fact, represent a set of specially defined substituent constants. Deviations then can be detected which result from different structural effects on pK and on $E_{1/2}$.

Thus, in Fig. VII-4 for dissociation constants a linear relation is observed even for the 5-sulpho derivative, for which a deviation was observed when the half-wave potentials of substituted 8-hydroxyquinolines were correlated with σ_X values (Fig. VII-1). On the other hand, for 2-methyl and 2,3-dimethyl derivatives, for which a greater value of $dE_{1/2}/dpH$ was observed[11] than for other derivatives, the half-wave potentials fit the linear relation for σ_X (Fig. VII-1) (this is probably due to a compensation of effects); but correlation with dissociation constants (Fig. VII-4) clearly indicates deviations.

Whereas, for quinoline derivatives reference pK values were chosen from the dissociation constants of methyl derivatives,[33] for coumarins it was possible[32] to use equilibrium constants obtained from polarographic measurements. For alkyl derivatives (Fig. VII-5), values of equilibrium constants were used which corresponded to the equilibrium between the lactone and the open form, as determined[32] from the dependence of wave heights[1] on pH. For the hydroxy derivatives, however, the pK values of the hydroxy groups were determined[32] from the dependence of values of half-wave potential[1] on pH.

For acridine derivatives, it is possible to compare the half-wave potentials of the first reduction wave with values of dissociation constants determined either potentiometrically or from the shifts in half-wave potential with pH.[14] Values of half-wave potentials of the second

---

* The measured rate or equilibrium constants can be influenced by effects that are specific for the given reaction series either similarly to the half-wave potentials or in a different way. Without comparison with σ_X, it is difficult to distinguish whether a deviation from the $E_{1/2}$–log k or $E_{1/2}$–log K linear relation is due to an anomaly in the course of the polarographic reduction or to an anomaly in the course of the chemical reaction.

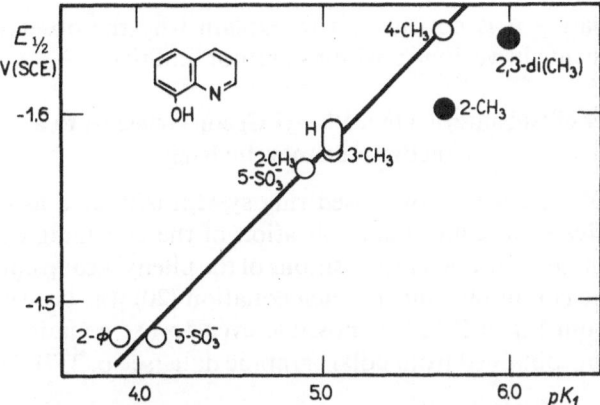

Fig. VII-4. Relation of half-wave potentials for the reduction of substituted 8-hydroxy-quinolines to pK_1-values corresponding to the protonation of the heterocyclic nitrogen. Half-wave potentials from Ref. 11, 12, full points deviate.

wave, corresponding to a radical reduction, cannot be correlated either with the dissociation constants of the oxidized or reduced form, or with the stability of the radical. In cases in which the half-wave potentials of acridine derivatives have been measured,[9,10] only at one or

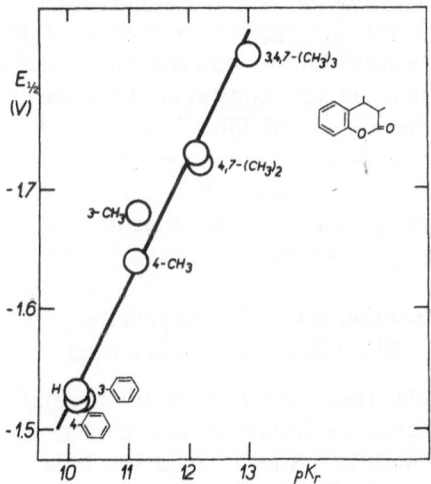

Fig. VII-5. Relation of half-wave potentials for the reduction of substituted coumarins to logarithms of equilibrium constant for the equilibrium between the lactone and acid forms. Experimental data from Ref. 1.

two pH values it is not difficult to explain why the observed effects of substituents in various positions are not additive.

5. Effect of Substitution in a Phenyl Group Attached to an Electro-active Heterocyclic Ring

In these cases the condensed ring system is treated as one single electroactive group, and the application of the constants σ_X for substituents in the *meta* and *para* positions of the phenyl group corresponds fully to the conditions under which equation (20) was derived. As can be seen from Fig. VII-2, it is possible even to use equation (51) with values of σ_B^- obtained from polarographic data (see p. 237).

6. Correlation of the Reactivities of Various Heterocyclic Compounds by the Use of the Dimroth Equation

For a number of compounds which form "a symmetric, dynamically homologous series," Dimroth[34] derived the equation

$$\log k_1 - \log k_2 = b(E_1^{\circ} - E_2^{\circ})$$

where k_1 and k_2 are rate constants for two members of the dynamically homologous series (with the same or similar electroactive groups) and E_1° and E_2° are their oxidation—reduction potentials.

This equation can be applied in polarography for reversible systems, and it has used for the correlation of half-wave potentials with the logarithms of the rate constants of the reaction with phenyl-hydrazine for Methyl Green, Meldola Blue, Cresyl Blue, Capri Blue, Methylene Blue, Thiocarmine, Thionine Blue, New Methylene Blue, and Nile Blue.[35,36] A linear relation was found, and it can be deduced that these oxazines, phenazines, and thiazines form a "symmetric, dynamically homologous series."

7. Effect of Substitution in the Heterocyclic Ring on Reduction in a Side Chain on the Same Ring

For heterocyclic rings which have no aromatic character the participation primarily of inductive effects can be expected. The validity of this assumption was verified for N-substituted phthalimides (No. 28 and 29) (Fig. VII-1), naphthalimides (No. 30), and phthalonimides (No. 31). These reaction series show that, in principle, the polar effects of substituents attached to nitrogen may be analogous to those resulting from the substitution of the same groups on carbon. The difference in the values of the reaction constants in No. 28 and 29 may be due to a change in pH, but also may be due to change in the ethanol content. At pH > 7, at which the reduction of phthalimides occurs in two steps and follows another mechanism, the half-wave potentials are only slightly dependent on the nature of substituent X (compare the small value for $\rho_{\pi,\text{het}}$ in Reaction Series No. 30). For naphthalimides in acid solution no simple correlation was found, and even the sequence of substituent effects differs from that observed in alkaline media.

On the other hand, for 9-substituted acridine N-oxides (No. 32–35) the application of $\sigma_{p\text{-X}}$ values was found to be best, as in the case of the parent acridines (p. 260), indicating the participation of mesomeric interactions. Correlation was found for the first wave at pH 9.25, but not at pH 4.5 in an acetate buffer. No correlation was found in either of these buffers for the second, more negative wave. The identical values of the reaction constants for the pair No. 32 and 33, on one hand, and the pair No. 34 and 35, on the other, and the differing values of the first and second pairs show the effect of the 7-methoxy group on the reaction constant and indicate the nonadditivity of the substituent effects. Revision of the experimental data would be of interest.

8. Effect of Substitution in the Benzene Ring on Reduction in a Side Chain on the Heterocyclic Ring

Here, as with the type (Ab), the unavailability of the special substituent constants makes it difficult to choose the proper kind of σ constant to enable us to make a first approximation. Sometimes it is possible to make a deduction based on the number of atoms between the atom carrying the substituent and that carrying the electroactive element or group. But more often we have to depend on a trial-and-error approach.

Thus, for phthalimides substituted in position 4 (No. 36–42) Tirouflet[21] suggested the application of values of $(\sigma_{m\text{-X}} + \sigma_{p\text{-X}})/2$ on the grounds that in this position the substituent is *meta* relative to one carbonyl and *para* relative to the other. Implicitly, the probability of reduction may be expected to be the same for both carbonyl groups. Nevertheless, good correlation of half-wave potentials with $\sigma_{m\text{-X}}$-values was found (Table VII-2), showing deviation only for the OH group), but correlations with $\sigma_{p\text{-X}}$ were only slightly inferior. The substituents chosen do not allow us to make a decision between these possibilities. The choice of the most appropriate substituent constants for use as a first approximation would be possible if, for example, methoxy derivative were included in the reaction series. With the methoxy group the values of $\sigma_{p\text{-}X}$ and $\sigma_{m\text{-}X}$ are even different in sign and the possibility of complicating mesomeric interaction with the electroactive group via the benzene ring is less, for example, than with a hydroxy or amino group. Values of reaction constants for $\sigma_{m\text{-}X}$ were selected for Series No. 36–42 because their dependence on pH is represented by a smooth curve (nevertheless, it is debatable whether these values should be selected). On the other hand, for substitution in position 3 the values of $\rho_{\pi,R}$ in Table VII-2 were cited from Ref. 22, where the correlation was carried out by the use of $(\sigma_m + \sigma_p)/2$ values.

Tirouflet[21] mentioned that for phthalimides substituted in position 3 identical values of both half-wave potentials and reaction constants $\rho_{\pi,R}$ were found as for substitution by the same groups in position 4. This can be understood, because substitution in position 4 can be regarded as equivalent to a substitution in the *meta* and *para* positions, whereas, substitution in position 3 corresponds to substitution in the *ortho* and *meta* positions. In the absence of steric interactions in position 3, favored by the valence angles in the five-membered ring, the polar contributions from positions 3 and 4 can be equal, accepting identical polar contributions from *ortho* and *para* positions.

For phthalonimides[20] substituted in the benzene ring no correlation could be found for H, NO_2, NH_2, and Cl substituents.

Among other reaction series, it proved useful for isatin derivatives (No. 45) to use $\sigma_{p\text{-}X}^{-}$ for substituents in positions 4 and 6 and $\sigma_{m\text{-}X}$ for substituents in position 5 (Fig. VII-2). Here again, the effect in position 4 (equivalent to *ortho*) is equal to that observed for position 6 (corresponding to *para*).

On the other hand, for xanthones (No. 46) substituents in position 3 need $\sigma_{p\text{-}X}$ values and those in position $2\sigma_{m\text{-}X}$ values, but substituents in position 4 show a pronounced negative *ortho* effect ($\Delta_o = -40\,\text{mV}$). The effect of steric hindrance to coplanarity cannot be ruled out.

Again, acridine derivatives (No. 47a and 47b) show irregular behavior—both for acridine[10] and acridine N-oxide[9,10,24]—demonstrated in the nonadditivity of substituent effects and the necessity for the application of $\sigma_{p\text{-}X}$ for the 2-and of $\sigma_{m\text{-}X}$ for the 3-position for

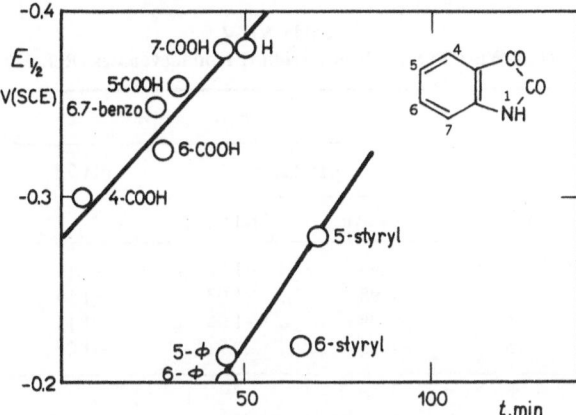

Fig. VII-6. Relation of half-wave potentials for the reduction of substituted isatins to the time needed for decolorization in reactions with Methylene Blue. Half-wave potentials from Ref. 25.

the first wave, but σ_{m-X} for the 2-and σ_{p-X} for the 3- or 7-position for the second wave.

Finally, for isatins the half-wave potentials* were correlated with the times needed for the decolorization of Methylene Blue solutions under defined conditions.[25] A linear relation was found for carboxylic acids (Fig. VII-6). Values of the half-wave potentials of phenyl derivatives, showing deviations in Reaction Series No. 45, and also of styryl derivatives fitted another approximately linear relation (Fig. VII-6).

9. Effect of the Position and Nature of the Heterocyclic Ring on Reduction in the Side Chain

Comparisons between heterocyclic compounds with the electro-active group in various positions in the same heterocyclic ring are restricted by our limited knowledge of the course of reduction of such compounds. More specifically, evidence that the mechanism of the electrode process (e.g., for quinoline aldehydes[37]) is the same for all members of the series is often lacking. Among the examples available, we cite values measured[31] for acridinyl isothiocyanates (Table VII-4).

When the half-wave potentials of the second wave (which for acridine derivatives again show irregular behavior) are neglected, the

---

* They were measured from figures in Ref. 25.

Table VII-4
Half-Wave Potentials of Acridinyl Isothiocyanates (Ref. 31)

Position of NCS	$E_{1/2}$		pH 9.1
	pH 2.2		
	1st wave	2nd wave	
2	−0.93	−1.02	−1.14
3	−0.98	−1.03	−1.15
4	−0.98	−1.05	−1.15
5	−0.83	−0.91	−1.07

effect of the acridine ring follows the sequence

Similarly, 8-nitroquinoline is reduced at considerably more positive potentials (by about 0.1 V) than 6-nitroquinoline.[30] The half-wave potentials of 2- and 3-bromobenzo [b] thiophene 1,1-dioxides[28] are practically identical.

On the other hand, the nature of the hetero atom can have a pronounced effect on polarographic behavior. Thus, the exchange of the NH group in phthalimide for a sulphur atom[22] results in a shift in the half-wave potential of about +0.2 to +0.3 V (according to the pH value) toward more positive values. Also, the exchange of the =CH— group for =N— in the comparison of quinoline N-oxide with quinoxaline N-oxide and of acridine N-oxide with phenazine N-oxide[24] results in a shift of about +0.5 V toward more positive potentials.

(II)

When the change occurs in the side chain rather than in the skeleton, as in (arylmethylene)homophthalimides[20] (II), the effect of the

Table VII-5
Approximate Values of Substituent Constants for the Annelled Heterocyclic Ring of Phthalide

Substituent constant	pH 0	pH 5.4	Average
$\sigma_{2\text{-}CH_2OCO\text{-}3}$	$+1.3_7$	$+1.2_9$	$+1.3$
$\sigma_{3\text{-}CH_2OCO\text{-}4}$	$+1.2_5$	$+1.5_0$	$+1.4$
$\sigma_{4\text{-}CH_2OCO\text{-}3}$	$+1.3_7$	$+1.1_5$	$+1.2_5$

aryl B is negligible at pH 2.0 and 5.0 and is small at pH 7.9, when it increases in the sequence B = furyl > phenyl > thienyl.

The effect of the position of the reducible group in the benzene ring, relative to the annelled heterocyclic ring, was followed for nitrophthalides (III).[29] When the value of the reaction constant

(III)

ρ_{π,NO_2} is put equal to the value for other nitrobenzene derivatives in the supporting electrolyte used, it was possible to calculate approximate values of the substituent constants characterizing the annelled heterocyclic ring (Table VII-5).

References

[1] R. Patzak and L. Neugebauer, *Monatsh. Chem.* **83**: 776 (1952).

[2] M. Deželič and M. Trkovnik, *Croat. Chem. Acta* **33**: 209 (1961).

[3] K. Györbiró, *J. Electroanal. Chem.* **2**: 259 (1961).

[4] G. Sartori and C. Furlani, *Ann. Chim.* **45**: 251 (1955).

[5] R. Curti and S. Locchi, *Ann. Chim.* **45**: 1244 (1955).

[6] J. Komenda, *Collection Czech. Chem. Commun.* **26**: 212 (1961).

[7] Ju. S. Rozum, *Biokhimiya* **20**: 66 (1955).

[8] P. Kabasakalian and J. McGlotten, *Anal. Chem.* **31**: 431 (1959).

[9] O. N. Nečajeva and Z. V. Puškareva, *Zh. Obshch. Khim.* **28**: 2693 (1958).

[10] L. B. Radina, Z. V. Puškareva, N. M. Voronina, and N. M. Chvorova, *Zh. Obshch. Khim.* **30**: 3480 (1960).

[11] Q. Fernando and J. P. Phillips, *J. Am. Chem. Soc.* **74**: 3103 (1952).

[12] G. Bocquet and R. A. Paris, *Anal. Chim. Acta* **15**: 492 (1956).

[13] M. P. Strier and J. C. Cavagnol, *J. Am. Chem. Soc.* **80**: 1565 (1958).

[14] R. C. Kaye and M. L. Stonehill, *J. Chem. Soc.* **1951**: 2638.

[15] J. G. Frost and J. H. Saylor, *Rec. Trav. Chim.* **83**: 340 (1964).

[16] A. P. Kottenhahn, E. T. Seo, and H. W. Stone, *J. Org. Chem.* **28**: 3114 (1963).

[17] J. Volke and V. Szabó, *Chem. Listy* **50**: 1095 (1956).

[18] Y. Oshima, M. Kijutani, and K. Ueda, *J. Agr. Chem. Soc. Japan* **27**: 98 (1953).

[19] R. Dabard, *Compt. Rend.* **244**: 1651 (1957).

[20] R. Dabard, Thesis, Dijon, 1962.

[21] J. Tirouflet, R. Robin, and M. Guyard, *Bull. Soc. Chim. France* **1956**: 571.

[22] J. Tirouflet and R. Dabard, *Compt. Rend.* **242**: 2839 (1956).

[23] C. Furlani, *Gazz. Chim. Ital.* **85**: 1646, 1668 (1955).

[24] L. V. Varjuchina and Z. V. Puškareva, *Zh. Obshch. Khim.* **26**: 1740 (1956).

[25] H. Cassebaum, *Z. Elektrochem.* **58**: 515 (1954).

[26] W. E. Whitman and L. A. Wiles, *J. Chem. Soc.* **1956**: 3016.

[27] R. Curti, S. Locchi, and U. Landini, *Rend. Accad. Naz.* **18**: 78 (1955).

[28] P. Smith, H. G. Sprague, and O. C. Elmer, *Anal. Chem.* **25**: 793 (1953).

[29] J. Tirouflet, *Bull. Chem. Soc. France* **1956**: 27.

[30] R. Zahradnik and K. Boček, *Collection Czech. Chem. Commun.* **26**: 1733 (1961).

[31] D. Vlachová, R. Zahradník, K. Antoš, P. Kristián, and A. Hulka, *Collection Czech. Commun.* **27**: 2826 (1962).

[32] P. Zuman, *Chem. Listy* **48**: 94 (1954).

[33] R. Riccardi and M. Bresesti, *Ann. Chim.* **48**: 826 (1954).

[34] O. Dimroth, *Angew. Chem.* **46**: 511 (1933); *Chemiker-Zt.* **62**: 454 (1958).

[35] K. Schwabe and H. Berg, *Z. physikal. Chem. (Leipzig)* **204**: 18 (1955).

[36] H. Berg, *Chem. Tech.* **85**: (1956).

[37] V. Preininger, H. Potěšilová, and F. Šantavý, *Chem. Listy* **52**: 25 (1958).

VIII: Effects of Substituents in Quinonoid Compounds

It was shown in Chapter II that a measure of the influence of the substituent on the redox potential of a reversible electrode system is provided by the change in the increment of the standard free energy in the interconversion of the oxidized and reduced forms. The shift in the redox potential (or half-wave potential, denoted by $\Delta E^{\circ}_{1/2}$), defined as $\Delta E^{\circ}_{1/2} = (E^{\circ}_{1/2})_X - (E^{\circ}_{1/2})_H$ (where the index X denotes the compound containing the substituent X and the index H the parent reference compound) is proportional to the difference of the logarithms of the equilibrium constants K_X and K_H of the corresponding reversible process

$$\Delta E^{\circ}_{1/2} \sim \ln \frac{K_X}{K_H} \tag{58}$$

It is obvious from equation (58) that the free energy relations that have been applied in preceding chapters mainly to the half-wave potentials of irreversible systems can also be used for the redox potentials of reversible systems. Some examples of applications of modified forms of equation (1) for quinonoid systems are presented here.

1. General

In the following, oxidation-reduction systems will be denoted as quinonoid when the oxidized form is a *para*, *ortho*, or *amphi* quinone, a quinonimine, or a quinonediimine. Compounds of this type hold a

273

special position in the polarography of organic substances because they undergo rapid electrode processes (described as reversible). It is this rapid establishment of equilibrium conditions at the electrode that allows the thermodynamic discussion of measured potentials as functions of the difference in standard free energy between the oxidized and reduced forms. A further exceptional feature of quinonoid systems is that the net result of what is usually a two-electron oxidation–reduction process is a simultaneous change in two reactive groups in the molecule. This change is accompanied by a pronounced change in the aromaticity—from the oxidized form, resembling, an α,β-unsaturated ketone in its behavior, to the aromatic reduced form. In a detailed discussion of a substituent effect one has to consider the influence of the substituent on electron density in both the oxidized and reduced forms. It is supposed that the substituent effect, and especially the mesomeric effect, predominates in the aromatic reduced form.

The measurement of redox potentials can be performed[1] by the use of potentiometry, spectrophotometry, or polarography. Polarography has the advantage in that the measurements of half-wave redox potentials can be performed even when one or both members of the oxidation–reduction couple are unstable, provided that their lives in the solution in question are several times as great as the drop time (usually about 3 s). When the electrode system is polarized by a periodically changed voltage (as realized in the commutator method of Kalousek[2]), even more unstable systems can be followed. Moreover, in polarography the addition of an oxidizing or reducing agent to the solution under study is unnecessary, and interfering side reactions (e.g., with hydrosulfite) are avoided. It is an advantage in work with rare organic substances that only very small amounts are necessary: the complete determination of the pH-dependence of the redox potential of a quinonoid system can be made with as little as 10 mg, and the most important measurements can be made with 2 mg. On the other hand, polarographic measurements carried out by normal techniques are rarely accurate within less than ± 0.002 V, but this is usually small enough for the study of structural effects.

The values of the diffusion coefficients of the oxidized and reduced forms are usually similar, and the half-wave potentials can therefore be treated as equivalent to the potentiometric or spectrophotometric values. Thus, in the following discussion values obtained by all three methods are included without distinction.

It should be stressed that the comparison of structural effects is permissible only in cases in which the electrode processes have the same or similar courses. For quinonoid substances, this requirement is fulfilled if the degrees of dissociation of both the oxidized and reduced forms, and hence the slopes $dE^\circ_{1/2}/d$pH, are identical for all the compounds compared in the reaction series. This principle has often been ignored, and inadmissible values of redox potentials have been compared. The extrapolation of values of redox potential to pH 0 or the choice of pH 7.0 as a reference condition for the comparison of redox potentials is quite arbitrary and theoretically unfounded. Several difficulties encountered in the discussion of structural effects in quinonoid series are probably caused by insufficient care in the interpretation of experimental data. The theoretically preferred procedure would be to make the comparison of values of redox potential in a pH range in which both the oxidized and reduced form are totally dissociated, i.e., in which $dE^\circ_{1/2}/d$pH $= 0$. This pH range is experimentally inaccessible for several substances. However, it is sometimes possible to determine the redox potential polarographically in this pH range, especially when the reduced form (the hydroquinone) is added to the polarographed solution, even when both forms are too unstable to allow potentiometric measurements. In alkaline solution the oxidized forms of most quinones undergo side reactions accompanied by destruction of the quinonoid system.† For certain compounds, these reactions are slow compared with the recording speed of the polarographic curves. But for other quinones decomposition already occurs at pH 9 or 10 at such a rate that measurements of redox potential can be made only in a pH range in which $E^\circ_{1/2}$ is a function of pH. It is then necessary to confirm that in the chosen pH range the value of $dE^\circ_{1/2}/d$pH remains the same for all compounds compared.

It should be remembered that within this pH range the measured values of redox potentials are a function not only of the equilibrium constant of the oxidation–reduction process, but also of the dissociation constants of the oxidized and reduced form. Thus, for a quinonoid system containing one substituent, e.g., an OH group which can undergo dissociation, the value of the redox potential at pH 0.0 is given

† Nucleophilic attack by hydroxyl ions on the double bond is usually the first step in this degradation process. Thus, those quinonoid compounds in which no hydrogen at a double bond is available (such as in chloranil, phthiocol, or anthraquinones) can be stable even in strongly alkaline media.

by

$$E_{\text{pH}=0} = E^\circ + \frac{RT}{n\text{F}} \ln \frac{[\text{QOH}]}{[\text{H}_2\text{QOH}]}$$

$$= \frac{RT}{n\text{F}} \ln \frac{1 + K_{\text{OX}}}{1 + K_{r_1} + K_{r_1}K_{r_2} + K_{r_1}K_{r_2}K_{r_3}} \tag{59}$$

where QOH is the oxidized form of the quinonoid system containing an undissociated substituent, H_2QOH is the reduced form of this system with the undissociated substituent, K_{OX} is the dissociation constant of the oxidized form, and K_{r_1}, K_{r_2}, and K_{r_3} are the dissociation constants of the corresponding reduced form. The values of most of these dissociation constants usually are not known with sufficient accuracy to allow the computation of E°. Thus, when structural effects on redox potentials are compared in a range in which $dE^\circ_{1/2}/d\text{pH} \neq 0$, it should be kept in mind that the structural effects both on oxidation–reduction equilibria and on dissociation constants are being compared simultaneously. At pH $= 0$ the last term can be neglected, for dissociation constants are usually small compared with unity; but at higher pH this term (which is then more complex) should not be neglected.

The structural effects that are reflected in the values of the redox potentials of quinonoid systems are principally of three types: (*i*) the effect of substituents; (*ii*) the role of the type of quinone (comparison of *ortho*, *para*, *ana* and *amphi* quinones or of quinones, quinonimines, and quinonediimines with the same number of condensed rings); and (*iii*) effect of the number of condensed rings (for quinones of the same type, e.g., for *para* quinones and quinonimines). The two latter types represent a change of the reactive group and are not suitable for treatment by means of linear free energy relations. Attempts have been made to treat these structural effects by quantum-mechanical methods, but only the effects of substituents will be discussed here.

Several authors, observing the shifts in half-wave potential due to the introduction of substituents, "have made some comment, if only to call attention to the fact that some substituents cause a positive, others a negative, shift of potential from that of the parent system."[1] Our objective here is the quantitative treatment of these shifts.

2. Classification of Reaction Series

In quinonoid substances substitution can occur either in the ring, on the nitrogen in the case of quinonimines and quinonediimines, or on an annelled benzene (or heterocyclic) ring in polycyclic compounds. For compounds belonging to the first two groups containing substituents X with no pronounced steric or mesomeric effects, it is possible to restrict considerations to the inductive effects of substituents (or so-called polar effects in the case of substituents derived from methyl). In such cases, the whole quinonoid system is treated as one polarographically active group R attached directly to the substituent X, and the following modified[1,3] Taft equations can be applied:

$$\Delta E^{\circ}_{1/2} = \rho^{I}_{\pi,Q}\sigma^{I}_{X} \tag{60}$$

$$\Delta E^{\circ}_{1/2} = \rho^{*}_{\pi,Q}\sigma^{*}_{CH_2X} \tag{61}$$

The definitions of the reaction constants $\rho^{I}_{\pi,Q}$ and $\rho^{*}_{\pi,Q}$, as well as those of substituent constants, which are tabulated by Taft,[3] remain the same as in aliphatic series (cf. Chapter II).

For polysubstituted alkyl quinones two possibilities exist. Either all positions in which substituents are placed are equivalent, and hence, the susceptibility to substituent effects (and the value of the reaction constant) will be the same for all compounds involved; these effects will be then described by

$$\Delta E^{\circ}_{1/2} = \rho^{*}_{\pi,Q}\Sigma\sigma^{*}_{CH_2X} \tag{62}$$

Or, the susceptibility to substituent effects (and hence, the value of the reaction constant) is different according to the position involved, and the following equation ought to be used:

$$\Delta E^{\circ}_{1/2} = \Sigma\rho^{*}_{\pi,Q}\sigma^{*}_{CH_2X} \tag{63}$$

For bulky substituents the next two equation can be tried:

$$\Delta E^{\circ}_{1/2} = \rho^{*}_{\pi,Q}\sigma^{*}_{CH_2X} + \delta_{\pi,Q}(E_S)_{CH_2X} \tag{64}$$

$$\Delta E^{\circ}_{1/2} = \rho_{\pi,Q}\sigma_{o\text{-}X} + \delta_{\pi,Q}(E_S)_{o\text{-}X} \tag{65}$$

The symbols for polar reaction constants $\rho^{*}_{\pi,Q}$ and total polar reaction constants $\rho_{\pi,Q}$, for steric reaction constants $\delta_{\pi,Q}$, for substituent constants σ, and for steric substituent constants E_S have the same meaning as in previous chapters. The reaction constant $\rho_{\pi,Q}$ expresses the ability of the quinonoid system to transmit the effects of substituents.

Finally, for substituents which exert a considerable mesomeric effect equations (66) and (67) are suggested:

$$\Delta E^{\circ}_{1/2} = \rho_{\pi,Q}\sigma^{-}_{p\text{-}X} \tag{66}$$

or better,

$$\Delta E^{\circ}_{1/2} = \rho^{I}_{\pi,Q}\sigma^{I}_{X} + \rho^{R}_{\pi,Q}\sigma^{R}_{X} \tag{67}$$

Here the substituent constant σ^{-}_{X} includes the mesomeric interaction of the polarographically active group and the substituent. Superscript I represents the polar inductive effect and superscript R the polar resonance effect[3] (the values of these constants for quinonoid systems in principle, can be different from those derived for benzene derivatives).

On the simplifying assumption that $\rho^{I}_{\pi,Q} \approx \rho^{R}_{\pi,Q}$, equation (67) takes the form

$$\Delta E^{\circ}_{1/2} = \rho_{\pi,Q}(\sigma^{I}_{X} + \sigma^{R}_{X}) \tag{68}$$

On the supposition, mentioned above, that the mesomeric interactions predominantly influence the aromatic reduced form, we can consider as a first approximation that $(\sigma^{I}_{X} + \sigma^{R}_{X}) \approx \sigma_{p\text{-}X}$, and thus, for the shifts in redox potential we can use the modified Hammett equation in the form

$$\Delta E^{\circ}_{1/2} = \rho_{\pi,Q}\sigma_{p\text{-}X} \tag{69}$$

The contribution of the polar resonance effects of substituents to the shifts in redox equilibria explains why better correlations of shifts in redox potential can be obtained with $\sigma_{p\text{-}X}$ (which includes both σ^{I}_{X} and σ^{R}_{X}) than with either $\sigma_{m\text{-}X}$ (which includes almost only σ^{I}_{X}) or $(\sigma_{m\text{-}X} + \sigma_{p\text{-}X})/2$. This is true even when the substitution on the quinonoid ring is in the *ortho* position (equivalent to *para*) to one, but in the *meta* position to the other C=O or C—OH group.

Although the Hammett equation has been applied for the correlation of various equilibrium constants, rate constants, and physical quantities,[4,5] the possibility of applying equations similar to (69) to oxidation–reduction potentials has been only briefly mentioned by Evans and Heer in their review.[6] Apparently, the work by Carter mentioned as a private communication in this review[6] was never published. Only recently, in a few papers,[7-11] linear free energy relations have been used for the treatment of oxidation–reduction potentials.†

† When preparing the translation of his paper[7] the present author received a manuscript of a paper by Dr. M. Charton, which was to be presented at a meeting of the American

A quinone system with a phenyl substituent can be treated as a single, polarographically active group attached to the phenyl ring, and we may examine how the electroreduction of the active group is influenced by substituents in the phenyl ring. Thus, the simple modified form of the Hammett equation can be used:

$$\Delta E^\circ_{1/2} = \rho_{\pi,R}\sigma_X \tag{70}$$

Here the reaction constant $\rho_{\pi,R}$ expresses how the phenyl ring transmits the effects of the substituent X to the polarographically active group R (which is represented by the whole quinone ring).

Substitution in a benzene ring in a polycyclic quinonoid compound can be treated by

$$\Delta E^\circ_{1/2} = \rho_{\pi,pQ}\sigma_{pQ-X} \tag{71}$$

To use equation (69), it would be necessary to know the values of the substituent constants σ_{pQ-X} characterizing the effects of substituents in particular positions in the annelled benzene ring, but these are unknown. As a first approximation, by the method of trial and error, the effect of the substituent on one or other of the carbonyl or hydroxyl groups can be considered, or a mean value of the effects on both carbonyl (or hydroxyl, respectively) groups can be considered. We use the modified Hammett equation in the form

$$\Delta E_{1/2} = \rho_{\pi,pQ}\sigma_X \tag{69'}$$

In this equation σ_m, σ_p or $(\sigma_m + \sigma_p)/2$ are used as approximate substituent constants values, and are chosen by trial and error.

The reaction series are further classified into series of monocyclic compounds (where only substitution in the quinonoid ring, or on the nitrogen in a quinonimine system, is possible), bicyclic derivatives (here there is the possibility of substitution both in the quinonoid ring and in the benzene or annelled ring), and polycyclic derivatives (with possible substitution in benzene rings only and/or on the nitrogen of the quinonimine system).

Chemical Society.[8] In the treatment of total polar substituent effects on the oxidation–reduction potentials of benzoquinones and naphthoquinones, very good agreement in the way of treatment and in the numerical results was observed between these two papers[7,8] in the parts which overlapped. The present author wishes to express his thanks to Dr. Charton for kindly sending him the unpublished manuscript.

The principles determining the choice of substituent constants and their treatment are the same as in previous chapters. In most instances only pH-dependent values of redox potentials are available for these correlations. As mentioned on p. 276, these values are functions of dissociation constants, and the measured shift in the redox potential of a substituted quinonoid compound relative to the parent compound can be affected both by change in the equilibrium constant of the oxidation–reduction process and by a change in the dissociation constants. The linear correlations observed arise because the principle of linear free energy relations hold both for the redox equilibria and for the corresponding dissociation constants. The dependence of the pK and $E°$ values on the kind of substituent is reflected in the value of the reaction constant ρ. It should always be remembered that, when redox potentials are compared at a lower, conventionally chosen, pH-value, the observed dependence on structure is a complicated one.

3. Monocyclic Quinonoid Systems

The number of monoalkyl-substituted quinonoid compounds is too limited to permit discussion of the application of equation (61). Hence, equations (62) and (63) for polysubstituted quinonoid compounds have been applied to values of half-wave potentials of poly-alkyl-hydroquinones[10] and -pyrocatechols[11] measured in acetate buffers containing 50% ethanol, extrapolated to pH 0.

For alkylhydroquinones, all substituent positions are equivalent, being *ortho* relative to one oxygen-containing group and *meta* relative to the other. Hence, the same reaction constant and equation (62) can be applied. The linear relation of $(\Delta E_{1/2}^0)_H$, expressed against unsubstituted hydroquinone as standard, to the sum $\Sigma\sigma_X^*$ (Fig. VIII-1) with a value of 0.10_2 V for $\rho_{\pi,Q}^*$ proves the validity of the assumption made. Deviations for the 2,6-t-butyl derivative toward more negative potentials (and, in a lesser degree, for the 2-methyl-6-t-butyl derivative) indicate the possibility of the operation of a steric effect, e.g., steric hindrance to coplanarity.

The situation is more complicated for substituted pyrocatechols, where substituents in position 3 or 6 exert a greater effect $(\rho_{\pi,Q_{3,6}}^* = 0.09$ V$)$ than the same substituents in position 4 or 5 $(\rho_{\pi,Q_{4,5}}^* = 0.05$ V$)$. This is understandable when it is remembered that in 3- or 6-substituted compounds the substituent is in the *ortho* position, whereas, in 4- or 5-substituted compounds it is in the *meta* or *para*

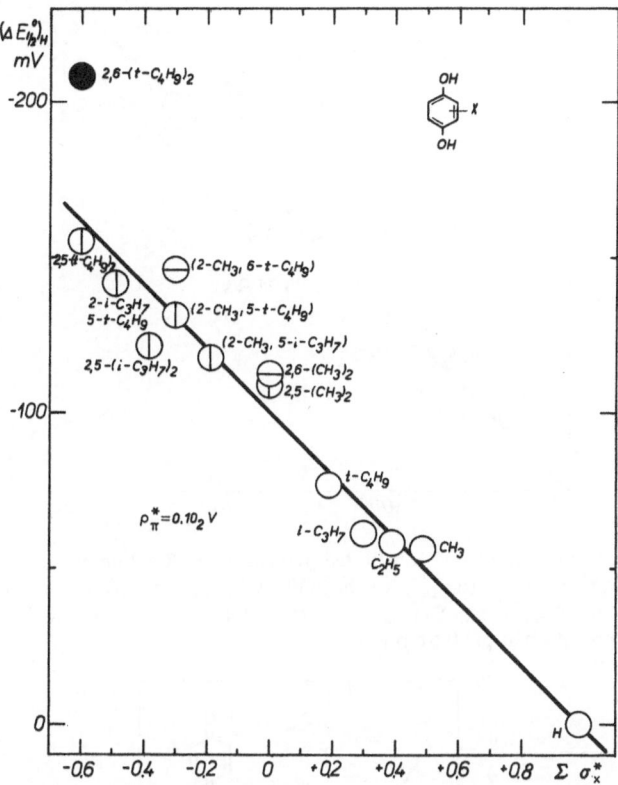

Fig. VIII-1. Relation of shifts in half-wave potentials for the oxidation of alkyl-substituted hydroquinones to the sum of Taft polar substituent constants $\Sigma\sigma_X^*$. ○ Monoalkyl; ⊕ 2,5-dialkyl; ⊖ 2,6-dialkyl derivatives; ⊙ deviates. Half-wave potentials at pH 0 from Ref. 10.

position relative to one hydropyl group. Hence, the application of equation (62) is not recommended, and the following equation usually is preferred:

$$(\Delta E_{1/2})_{\text{H,calc}} = \rho^*_{\pi,Q_{3,6}}\Sigma\sigma^*_{3,6\text{-X}} + \rho^*_{\pi,Q_{4,5}}\sigma^*_{4,5\text{-X}} \tag{63}$$

or

$$(\Delta E_{1/2})_{\text{H,calc}} = 0.09\Sigma\sigma^*_{3,6\text{-X}} + 0.05\Sigma\sigma^*_{4,5\text{-X}} \tag{63a}$$

Values calculated in this way give a linear plot (Fig. VIII-2, showing 10-mV limits for the scattering) against measured shifts of half-wave potential $(\Delta E_{1/2})_{\text{H,exp}}$, and it has the theoretical slope of

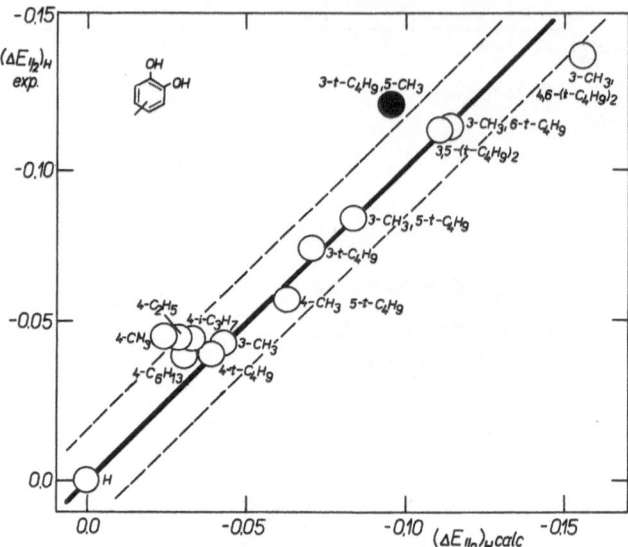

Fig. VIII-2. Oxidation of alkyl-substituted pyrocatechols. Relation of the shifts in meas-ured half-wave potential $(\Delta E_{1/2})_H$ to the shift $(\Delta E_{1/2})_{H\ calc}$ calculated according to the equation $(\Delta E_{1/2})_{H,calc} = \rho^*_{3,6}\Sigma\sigma^*_{3,6-x} + \rho^*\Sigma\sigma^*_{4,5-x}$ for $\rho^*_{3,6} = 0.09$ V and $\rho^*_{4,5} = 0.05$ V. Half-wave potentials at pH 0 from Ref. 11.

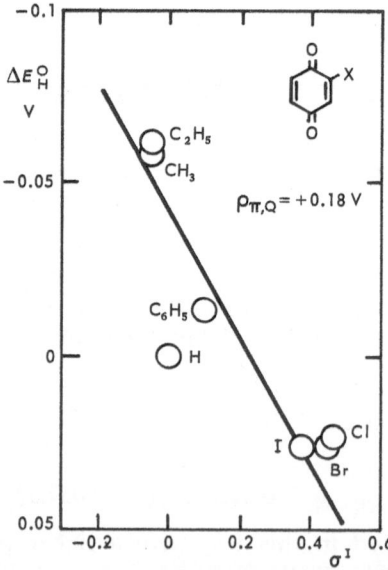

Fig. VIII-3. Reduction of 2-substituted p-benzoquinones. Relation of shifts in oxidation-reduction potential relative to the unsubstituted parent compound ΔE_H to inductive substituent constants σ^I_X. Oxidation-reduction potentials in benzene solutions from Ref. 12.

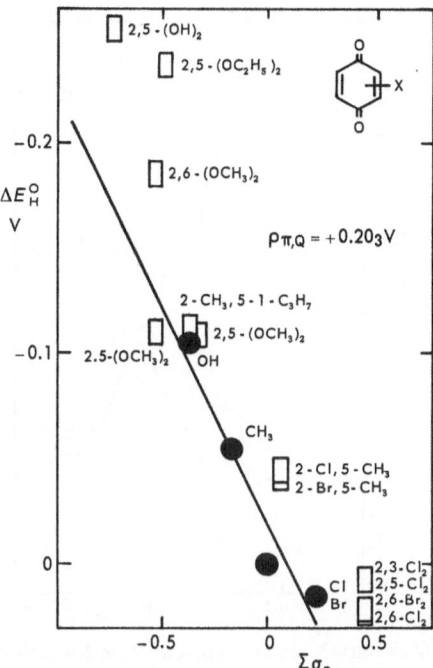

Fig. VIII-4. Relation of oxidation–reduction potentials of mono- and di-substituted
p-benzoquinones in aqueous solution to the sum of the Hammett substituent constants
$\Sigma\sigma_p$. Oxidation–reduction potentials expressed as shifts, ΔE_H, relative to the unsubstituted parent compound. Full points: monosubstituted compounds; squares: disub-
stituted compounds. Potentials for H, 2-Cl, 2-Br, 2-CH$_3$, 2,5-di-Cl, 2-Cl-5-CH$_3$, 2,6-di-Br,
and 2-Br-5-CH$_3$ from Ref. 13; for 2-OH, 2,5-di-OC$_2$H$_5$, and 2,5-di-OH from Ref. 15;
for 2,3-di-Cl and 2,6-di-Cl from Ref. 14; for 2,5-di-CH$_3$, 2-CH$_3$-5-i-C$_3$H$_7$, 2,5-di-OCH$_3$,
and 2,6-di-OCH$_3$ from Ref. 16.

1.00. For unsubstituted pyrocatechol, the value $(\Delta E_{1/2})_{H,calc} = 0.09$
$\times 2(0.49) + 0.05 \times 2(0.49)$ was used with respect to the highest
number of substituted hydrogen atoms. Only the value for the 3-t-
butyl-5-methyl derivative deviates. Hence, even in this case the predominant polar effect of the substituents was proved.†

For alkyls and halogens,[12] the applicability of equation (60) has
been demonstrated, showing that for these groups the inductive effect
predominates. As expected, values for 2-hydroxy- and 2-methoxy-p-
benzoquinones show no correlation according to equation (60).

† The authors of the original papers[10,11] used a somewhat different treatment and stressed
  primarily the role of steric effects.

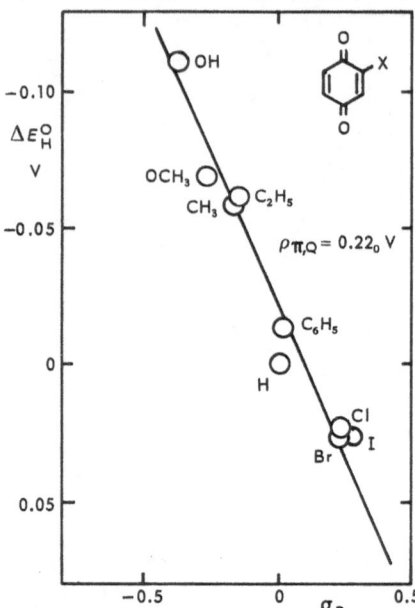

Fig. VIII-5. Relation of oxidation–reduction potentials of 2-substituted p-benzoquinones in benzene solutions to Hammett substituent constants σ_p. Oxidation–reduction potentials from Ref. 12 are expressed as shifts, ΔE_H, relative to the unsubstituted parent compound.

The effect of σ_X^R is pronounced in these two cases. A shift of the whole linear relation to more negative potentials, when compared with the parent unsubstituted compound (Fig. VIII-3; and also observed in the plots in Figs. VIII-4 and VIII-5) can involve an additive steric effect, already present when one substituent is bound to the quinone ring (see p. 286).

The main difficulty in testing the approximate equation (69) lies in the lack of data for reaction series measured under identical conditions. This is especially true for ethanolic solutions, for which several ethanol concentrations and ionic strengths have been used by various authors.[1] Since the effects of ethanol concentration and of ionic strength on the oxidation–reduction potentials are not very pronounced,[1] values measured[13–16] in aqueous ethanolic solutions (Fig. VIII-4) and in benzene solutions[12] (Fig. VIII-5) also have been compared.

Whereas, for monosubstituted p-benzoquinones the correlation of oxidation-reduction potentials in hydroxylic solvents in acid media was good (Fig. VIII-5, full points in Fig. VIII-4), a rather random distribution of values was obtained for disubstituted quinones (Fig. VIII-4 squares). Most of the deviations are above the regression line, which seems to indicate that a steric factor is involved. This is even more clearly demonstrated in Table VIII-1, showing the nonadditivity of the shifts caused by individual substituents. Values of $\Delta\Delta E_H$, expressing the difference between the measured shifts and shifts calculated on the basis of additivity, are all negative (in Table VIII-1). This means that polysubstituted p-benzoquinones have always more negative potentials than they would have if the effects of substituents were independent. Several types of *ortho* effect can be effective here. Two main possibilities exist: in addition to the polar and resonance effects of substituents

Table VIII-1

Nonadditivity of the Shifts of Oxidation–Reduction Potentials at $25°C$ Due to Substituents in Polysubstituted p-Benzoquinones

Substituents	HCl, M	C_2H_5OH, %	$E°(V)$	$\Delta E_H(V)$[a]	Calc.[b]	$\Delta\Delta E_H(V)$ Exp.[c]	Ref. on $E°$
H	0.2	95	0.709	0	—	—	14
2-Cl	1.0	95	0.735	+0.026	—	—	14
2,3-Cl$_2$	0.5	95	0.739	+0.030	+0.052	−0.022	14
2,5-Cl$_2$	0.5	95	0.740	+0.031	+0.052	−0.021	14
2,6-Cl$_2$	0.5	95	0.747	+0.036	+0.052	−0.016	14
2,3,5-Cl$_3$	0.5	95	0.731	+0.022	+0.078	−0.056	14
2,3,5,6-Cl$_4$	0.5	95	0.695	−0.014	+0.104	−0.118	14
CH$_3$	0.2	50	0.657	−0.052	—	—	17
2,3-(CH$_3$)$_2$	d	50	0.588	−0.121	−0.104	−0.017	18
2,3,5-(CH$_3$)$_3$	0.5	50	0.529	−0.156	−0.156	0.024	14
2,3,5,6(CH$_3$)$_4$	0.5	50	0.466	−0.243	−0.208	−0.035	14
2-CH$_3$, 5-Cl	0.5	—	0.679	−0.030	−0.026	−0.004	19
2-CH$_3$, 6-Cl	0.5	—	0.681	−0.028	−0.026	−0.002	19
2-CH$_3$, 3 Cl	0.5	—	0.676	−0.033	−0.026	−0.007	19
2-CH$_3$, 3,5-Cl$_2$	0.5	—	0.691	−0.018	0.0	−0.018	19
2-CH$_3$, 5,6-Cl$_2$	0.5	—	0.678	−0.031	0.0	−0.031	19
2-CH$_3$, Cl$_3$	0.5	—	0.681	−0.028	+0.026	−0.054	19

[a] The difference between the oxidation–reduction potentials of the substituted and the parent benzoquinone.
[b] The shifts calculated on the assumption of additivity.
[c] The difference between the measured and calculated shifts.
[d] The buffer of pH 4.7.

on the quinonoid system, substituents exhibit either mutual interaction or direct, possibly steric, interactions with the carbonyl (or hydroxy) groups; and secondly, most substituents, at least in the examples given in Table 1 and Fig. VIII-2, show a tendency to shift potentials in one direction to more negative potentials, as already mentioned. This would be hardly expected to arise from a mutual interaction of substituents. Furthermore, if the mutual interaction of substituents were the predominating factor, the difference between the theoretical and observed shift $\Delta\Delta E_{H}$ would be more pronounced for 2,3-derivatives than for other isomers. However, this is not found: the values in Table VIII-1 show no substantial difference between 2,3-, 2,5- and 2,6-isomers. We assume, therefore, that the nonadditivity of the effects of substituents is caused predominantly by interaction between substituents and the adjacent carbonyl or hydroxy group. Such an effect seems to exist in a lesser degree also in monosubstituted p-benzoquinones, where it causes the deviation of the value for the present compound in the correlations in Figures VIII-3 to -5. One feature that could explain the nonadditivity of the effects is steric hindrance by the substituents to coplanarity. A shift of the carbonyl or hydroxyl group from the position coplanar with the ring would decrease the conjugation and result in a shift of oxidation–reduction potentials to more negative values, in agreement with observations.

The shifts $\Delta\Delta E_{H}$ depend on the solvent used, which can affect both their magnitude and sign. This could be explained by solvation effects on the quinonoid systems and on the substituents. For example, half-wave potentials of polymethylbenzoquinones,[20] measured in a mixture of CH_3COOH (0.1 M) CH_3COONa (0.1 M), and methanol (50%) ('pH' 5.4) show additivity of the substituent effect. Similarly, the values of the half-wave potentials[21] of the first, more negative anodic wave corresponding to a one-electron transfer in a 0.1 M solution of NaOH in 75% ethanol show additivity for dichloro and trichloro derivatives. Nevertheless, in this case, considerable deviations from additivity were observed for tetrachloro- and tetrabromo-p-benzoquinones.

The additivity of total polar substituent effects and the applicability of the substituent constants σ_p in equation (69) have been proved also for oxidation–reduction potentials measured in acetonitrile,[21] in which a different electrode mechanism is followed: in this case, as in some other aprotic solvents, the reduction of the quinone and the oxidation of the hydroquinone occur in two one-electron steps with

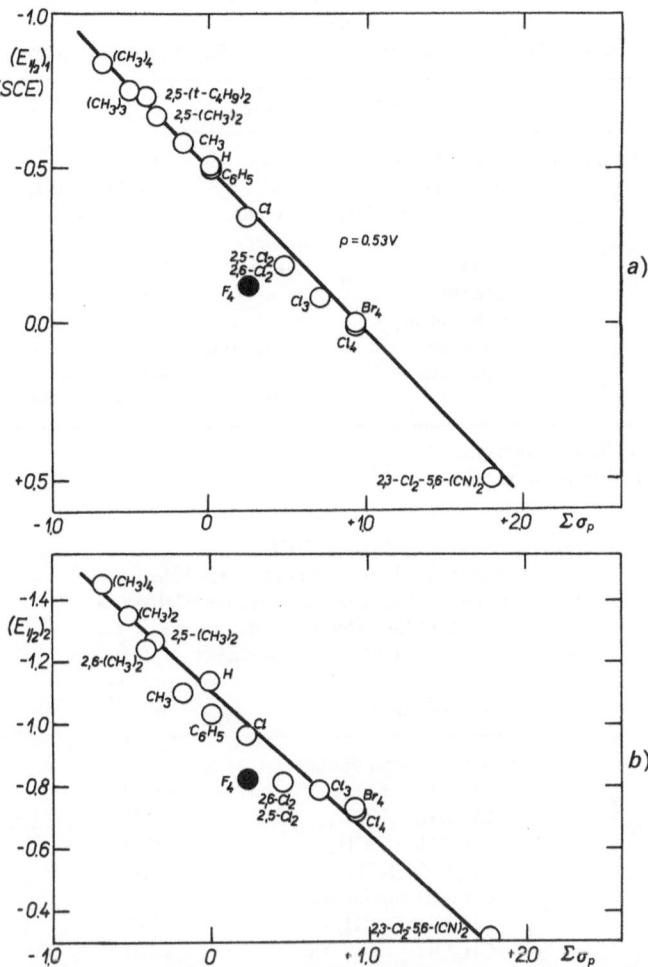

Fig. VIII-6. Relation of half-wave potentials for the reduction of substituted p-benzoqui-nones in acetonitrile solution to the sum of the Hammett substituent constants $\Sigma\sigma_{p-X}$. Half-wave potentials from Ref. 21; supporting electrolyte 0.1 M $N(C_2H_5)_4ClO_4$. Value for tetrafluoro derivative (full point) deviates. a) first wave; b) second wave.

formation of the semiquinone ion as intermediary. The half-wave potentials of both the first and the second waves are linear functions of the sum of the substituent constants $\Sigma\sigma_p$ even when tri- and tetra-substituted compounds are included, and a range of more than 1.3 V is covered (Fig. VIII-6).

Table VIII-2

Values of Reaction Constants $\rho_{\pi,Q}$ Found for Substituted Benzoquinones

Compound	Solvent	Medium	$\rho_{\pi,Q}$, V	r^a	Ref.
p-Benzoquinone	EtOH-H$_2$O	acid	$+0.20_3$	—	7
	EtOH-H$_2$O	acid	$+0.20_7$	0.956	8
	50% MeOH-H$_2$O	acetate buffer	$+0.35$	—	7
	H$_2$O	alkaline	$+0.30$	—	—
	Benzene		$+0.22_0$	—	7
	Acetonitrile				
	1st wave	NEt$_4$ClO$_4$	$+0.53$	—	—
	2nd wave		$+0.46$	—	—
o-Benzoquinone[b]	—	buffers	0.26_9	0.962	8

[a] r = correlation coefficient.
[b] Substituent in position 4.

Table VIII-3

Approximate[a] Values of Total Polar Substituent Constants, $\sigma_{p\text{-}X}$, Derived from the Half-Wave Potentials of Substituted p-Benzoquinones[20] and p-(Diethylamino)anilines[20]

Substituent X	$\sigma_{p\text{-}X}$
From p-Benzoquinones	
CH(OH)CH$_3$	-0.14
CH(OH)CH$_2$CH$_3$	-0.16
CH$_2$CH(OH)CH$_3$	-0.17
CH$_2$C(OH)(CH$_3$)$_2$	-0.17
From p-(Diethylamino)anilines	
CH$_2$NHSO$_2$CH$_3$	-0.00_5
CH$_2$OH	-0.01
CH$_2$CH$_2$NH CO CH$_3$	-0.05_5
CH$_2$CH$_2$OH	-0.05_5
CH$_2$CH$_2$NH$_2$	-0.06
CH$_2$CH$_2$NH SO$_2$CH$_3$	-0.11
CH$_2$CH$_2$N SO$_2$CH$_3$ $\quad\mid\quad$ CH$_3$	-0.10

[a] The approximate character of the tabulated data arises from the approximate character of equation (69) for the system studied and from the fact that substituent constants were obtained from only one reaction series.

The values of the reaction constants $\rho_{\pi,Q}$ summarized in Table VIII-2 show that the effect of the types of solvent and supporting electrolyte used is more pronounced than that of the type of quinone. The replacement of water by another hydroxylic solvent has very little effect, and surprisingly, neither does the replacement of water by benzene. On the other hand, half-wave potentials in acetonitrile[21] are markedly more sensitive to substituent effects. It is assumed that in these solutions, even when unbuffered, the measured half-wave potentials are thermodynamic quantities. If this assumption is correct, the observed susceptibility to substituent effects can be interpreted directly as a property of the electron-transfer process, whereas, in other cases the value of $\rho_{\pi,Q}$ can include the effect on the dissociation constant as well. An idea about the magnitude of the participation of the latter effect can be gained from a comparison of $\rho_{\pi,Q}$ values in acid and alkaline media, since in the latter case the effect of changes of dissociation constants must be absent.

By the use of the value $\rho_{\pi,Q} = 0.35$ V for acetate buffers in 50% methanolic solutions, some approximate values for substituent constants were calculated, and these are given in Table VIII-3. The values differ only a little from the value for the p-methyl group $\sigma_{p\text{-}CH_3} = -0.17$, which is in agreement with the well-known small effect of structural changes in a part of the molecule remote from the ring.

For the oxidation–reduction potentials[12] of 2-phenyl-p-benzoquinones, substituted in the *para* or *meta* position of the phenyl ring, the simple, modified Hammett equation (70) fits well (Fig. VIII-7). The only deviation observed was for the biphenylyl quinone. The small value of the reaction constant $\rho_{\pi,R} \approx +0.03$ V is in agreement with the positive value of the potential for unsubstituted 2-phenyl-p-benzoquinone.

As discussed in Chapter VII, Dimroth[22] derived an equation for a linear relation between the oxidation–reduction potentials of quinones and the logarithms of the rate constants of homogeneous chemical reactions of these substances in groups of related compounds. This treatment was extended later by Braude.[46] Musso[47] has since shown that an analogous treatment can be applied to heterogeneous catalytic hydrogenation is glacial acetic acid. The relation of $\log(1/t_{1/2})$ to E° was found to be linear. It remained linear in the presence of a catalytic poison: the plot was merely shifted. All t-butyl-substituted quinones show deviations from the linear relation, being hydrogenated more slowly than would be predicted from the free energy relations of related

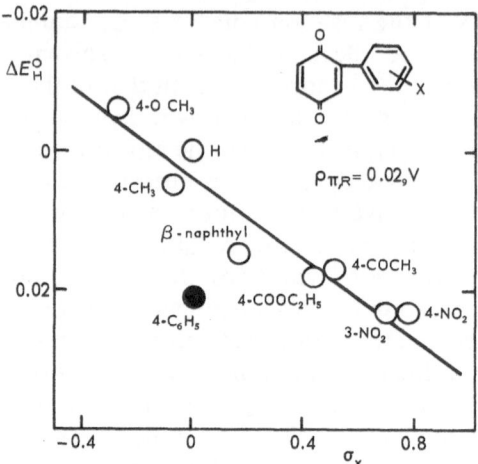

Fig. VIII-7. Relation of oxidation–reduction potentials of 2-phenyl-p-benzoquinones substituted in the *meta* or *para* position of the phenyl group in benzene solution to the Hammett substituent constants σ_X. Oxidation–reduction potentials from Ref. 12 are expressed as shifts, ΔE_H, relative to the unsubstituted parent compound. Full point deviates.

compounds. Berg[23] recently reported a similar relation for micro-heterogeneous reactions, namely, for reductions of substituted ben-zoquinones in the presence of palladium sols at the surface of a mercury electrode. Whereas, for homogeneous reactions in the relation $\Delta \ln k = B \Delta E_{1/2}$, $B \geq 17.2$, for microheterogeneous reductions of substituted p-benzoquinones $B = 1.9 \, V^{-1}$ (in phosphate buffer of pH 7.0). For anthraquinones, which are correlated by another line, $B = 2.7 \, V^{-1}$ was found. The microheterogeneous reduction therefore, is much more susceptible to substituent effects than homogeneous reactions.

I   II   III

For substituted p-(diethylamino)anilines of Types I-III the half-wave potentials of the oxidation waves were measured[24] at pH 11.0.

It is not stated whether the half-wave potentials of these compounds are pH-dependent. However, from the known dissociation constants of substances of this type, it can be assumed that a range was reached in which the half-wave potentials are pH-independent. For compounds of Type I, in which the substituent is in the *ortho* position relative to the NH_2 group, equations (60) and (61) can be used for correlations of the half-wave potentials of substances for which $X = CH_3$, C_2H_5, CH_2OH, and Cl. For all other values measured, the mesomeric interaction of the substituent with the ring plays an important role, and equation (69) is to be used for correlation. In cases in which values for *ortho*-substituent constants[3] were known, they fitted the correlation better than $\sigma_{p\text{-}X}$ constants. Significant deviations were observed for $X = NH_2$, $NHCH_3$, and OH; and a change in the mechanism of the oxidation process, with preferential formation of the *ortho* quinonoid system (due to the presence of two alkyl groups on the nitrogen in *para* position), is assumed in explanation. The value of the reaction constant $\rho_{\pi,Q} = +0.20$ V actually characterizes the oxidation–reduction properties of the systems, if the assumption concerning the pH-dependence is justified. By the use of equation (69), it has been possible to compute some approximate values of the substituent constants $\sigma_{p\text{-}X}$ (Table VIII-4).

Even for bulky substituents X adjacent to the primary amino group in substances of Type I, no shifts were observed that could possibly be ascribed to proximity steric effects. The effective volume of the NH_2 group seems to be so small that interaction does not occur.

Table VIII-4
The Values of Reaction Constants for Some Oxidation Potentials of Series of p-(Diakyl-amino)anilines H_2N—C_6H_4—NR^1R^2 (Values of half-wave potential measured by Bent and co-workers[24] at pH 11.0)

R^1	R^2	$\rho_{\pi,Q,V}$	$\delta_{\pi,Q,V}$
CH_3	X^a	0.04_8	0.01_2
C_2H_5	X^a	0.08_0	0.01_3
C_3H_7	X^a	0.05_2	0.04_2
X^a	X^a	$0.04_4{}^b$	0.02_9

[a] The alkyl changed.
[b] $E_{1/2} = \sum [\rho^*_{\pi,Q}\sigma^*_X + \delta_{\pi,Q}(E_S)_X]$.

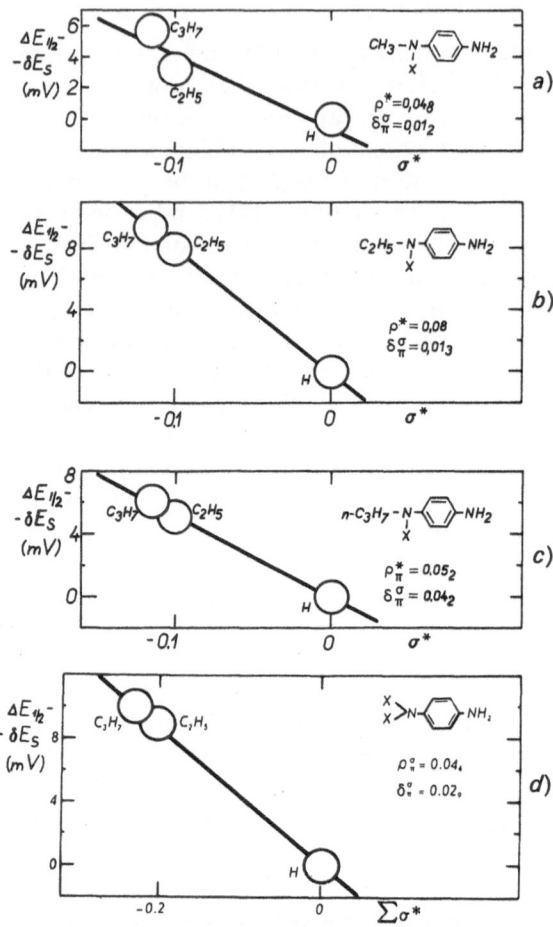

Fig. VIII-8. Relation of shifts in half-wave potential [corrected for the steric contribution $(\Delta E_{1/2} - \delta_{\pi,Q}E_S)$] for N-p-(N-substituted alkylamino) anilines to the Taft substituent constants σ^*. Shifts in half-wave potential are expressed relative to the value for the unsubstituted parent compound. Values of $\rho^*_{\pi,Q}$ and $\delta_{\pi,Q}$ are given in Table VIII-4. Half-wave potentials from Ref. 24. a) $CH_3NXC_6H_4NH_2$; b) $C_2H_5NXC_6H_4NH_2$; c) $C_3H_7NXC_6H_4NH_2$; d) $X_2NC_6H_4NH_2$ ($\Sigma\sigma^*$ in this case).

A different situation was observed for substances of Type II, containing an alkyl or alkoxy group in the *ortho* position relative to the tertiary nitrogen. Here there is no linear relation between the half-wave potentials and the substituent constants σ_{p-X} or σ_{o-X}, and equation (65) could possibly be applied. The available half-wave potentials[24] do not allow the computation of the constant $\delta_{\pi,Q}$. The trend of the half-wave potentials seems to indicate the possibility of the steric hindrance of the optimum orientation of the $-N(C_2H_5)_2$ or $=\overset{+}{N}(C_2H_5)_2$ group relative to the ring and, hence, a hindrance of the π-p conjugation.

Also, substitution in the side chain in substances of Type III involves steric interaction, and the simple equation (61) is not obeyed. For alkyl derivatives, for which steric substituent constants have been tabulated,[3] it has been possible to calculate values of the steric reaction constants $\delta_{\pi,Q}$ in equation (64). Shifts in half-wave potential corrected for the steric contribution (i.e., $\Delta E_{1/2} - \delta_{\pi,Q}E_S$) show substantially better correlations with σ_X^* than uncorrected $\Delta E_{1/2}$ values (Fig. VIII-8). The observed values of the reaction constants $\rho_{\pi,Q}^*$ and $\delta_{\pi,Q}$ are gathered in Table VIII-4.

Even when the absolute value of the steric reaction constant $\delta_{\pi,Q}$ is small, the value of the product $\delta_{\pi,Q} E_S$ is still comparable with that of $\rho_{\pi,Q}^*\sigma_X^*$.

Among other empirical relations in this group of compounds, the linear relation of the half-wave potentials of p-(diethylamino)anilines of Type I to the oxidation–reduction potentials[24] of 1,4-naphthoquinones substituted in position 2, and to 1,2-naphthoquinones substituted in position 4 can be mentioned. This is now explained by the fact that for both p-(diethylamino)anilines and naphthoquinones polar and mesomeric effects of substituents influence the redox properties, and in both types of reaction series correlations using σ_p constants in equation (69) can be used as an approximation. In the correlation of the half-wave potentials of p-(diethylamino)anilines with the oxidation-reduction potentials of naphthoquinones deviations for the amino and ethylamino derivatives of Type I is found, as was observed in the correlation of the former values with σ_{p-x} (see Chapter III). This shows that with 1,2-naphthoquinones no change in mechanism occurs, such as is found in 2-amino-4-(diethylamino)anilines, and the contribution of the p-quinonimine form in 4-amino-1,2-naphthoquinones is thus not a substantial one.

Similarly, in the correlation of the half-wave potentials of p-(diethylamino)anilines of Type I with the oxidation–reduction potentials of 2-substituted 1,4-naphthoquinones a deviation was observed for the hydroxy derivative, for which the contribution of the 1,2-naphthoquinone form in the potential is negligible (*cf.* discussion in Ref. 25). This implies a change of mechanism in 2-hydroxy-4-(diethylamino)aniline. On the other hand for 2-amino-1,4-naphthoquinone the deviation is just compensated by that found for the amino derivative of Type I. This permits the assumption of a contribution from the naphthoquinonimine form in both compounds.

The half-wave potentials of *p*-(dialkylamino)anilines were compared by Bent and his co-workers[24] with the rates of their autoxidation and with the rates of photographic development by means of these substances, and a linear relation was found (see Ref. 26) for several compounds.

Finally, a linear relation was found between the half-wave potentials and absorption maxima given by Eggersten and Weiss[27] for substituted *p*-phenylendiamines, 4-aminodiphenylamines, and benzidines. On this relation similar limitations are imposed as on other correlations involving ultraviolet spectra.

4. Bicyclic Quinonoid Systems

(a) Substitution in the Quinonoid Ring.   When the oxidation–reduction potentials[15,28,29] of 1,4-naphthoquinones containing a substituent without a pronounced mesomeric effect were compared

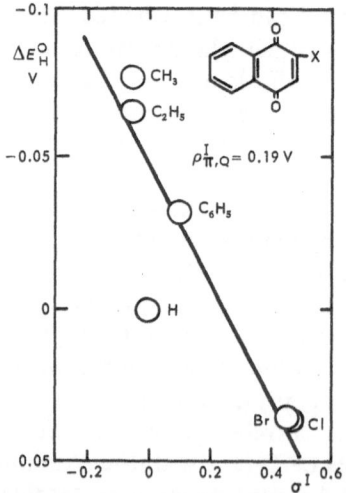

Fig. VIII-9. Reduction of 2-substituted 1,4-naphthoquinones. Relation of shifts in oxidation–reduction potential relative to the unsubstituted parent compound ΔE_H to inductive substituent constants σ^I. Oxidation–reduction potentials in 0.1 *N* HCl, 0.2 *N* LiCl, 50% ethanol for H and C_2H_5 and in 0.2 *N* HCl, 0.2 *N* LiCl, 70% ethanol for CH_3 and C_6H_5 from Ref. 28 and in 0.5 *N* HCl 95% ethanol for Cl and Br from Ref. 15.

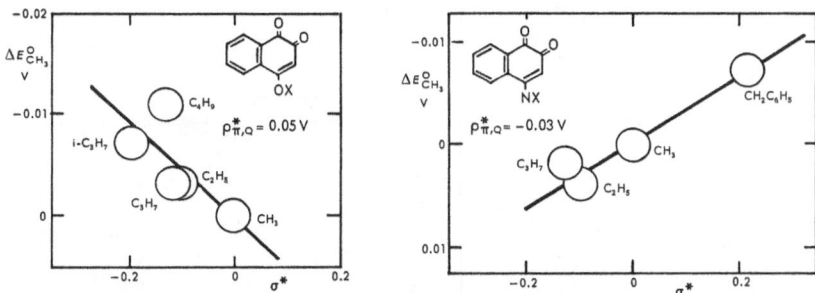

Fig. VIII-10. Relation of oxidation–reduction potentials of substituted 1,2-naphthoqui-nones to Taft polar substituent constants σ_X^*. Oxidation–reduction potentials measured in 0.047 M KH_2PO_4 and 0.047 M Na_2HPO_4 with 37% ethanol at 25°C expressed as shifts ($\Delta E_{CH_3}^o$) relative to the value for the methyl-substituted compound. a) 4-Alkoxy-1,2-naphthoquinones, half-wave potentials reported in Ref. 31. b) 4-alkylamino-1,2-naphtho-quinones, half-wave potentials from Ref. 28.

with the inductive sub-substituent constants σ_X^I, shifts corresponding to equation (60) were observed (Fig. VIII-9). When the effects of alkyl groups in position 2 in 1,4-naphthoquinones[28] and in position 3 in 2-hydroxy-1,4-naphthoquinones[30] were compared, only small shifts of redox potentials were observed. Thus, the value of $\rho_{\pi,Q}^*$ in equation

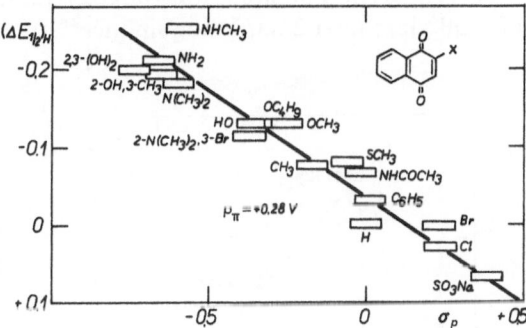

Fig. VIII-11. Relation of selected values of oxidation–reduction potentials of 2-substituted 1,4-naphthoquinones to Hammett substituent constants σ_p. Oxidation–reduction potentials expressed as shifts ($\Delta E_{1/2}$)$_H$ relative to the value for the unsubstituted parent compound. Potentials for $NHCN_3$, NH_2, $N(CH_3)_2$, OCH_3, CH_3, $NHCOCH_3$, C_6H_5, $OCOCH_3$, and SO_3^- from Ref. 28, for OH, Cl, and Br from Ref. 15, for C_2H_5 from Ref. 31, and for SCH_3 Ref. 32.

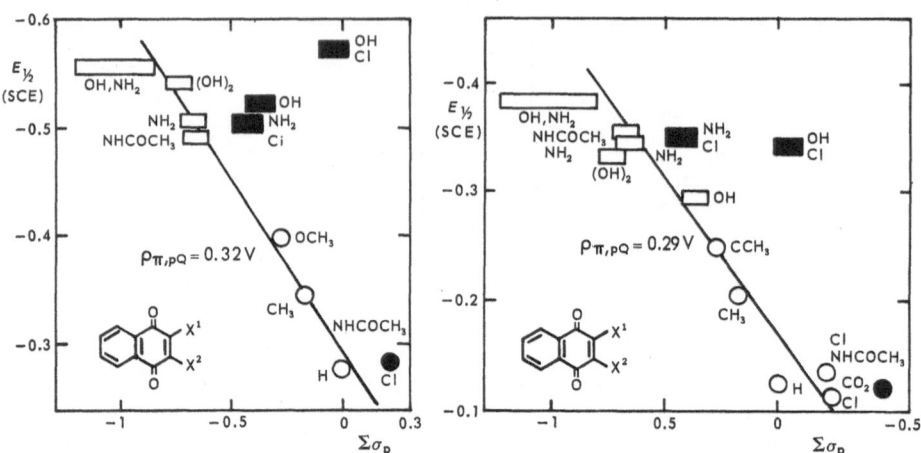

Fig. VIII-12. Relation to half-wave potentials of 2- and 2,3-substituted 1,4-naphtho-quinones to the sum of the Hammett substituent constants $\Sigma\sigma_{p\text{-}x}$. Half-wave potentials measured in a) 0.1 M CH$_3$COOH, 0.01 M CH$_3$COONa, and b) 0.1 M NH$_3$, 0.1 M NH$_4$Cl, both with 75% ethanol. Half-wave potentials from Ref. 33. Full points deviate.

(61) is small. The effect of alkyl groups attached to oxygen or nitrogen in position 4 of 1,2-naphthoquinones is more pronounced. The shifts in the oxidation–reduction potentials of 4-alkoxy-1,2-naphthoquin-ones[31] and of 4-alkylamino-1,2-naphthoquinones[28] fit equation (61)

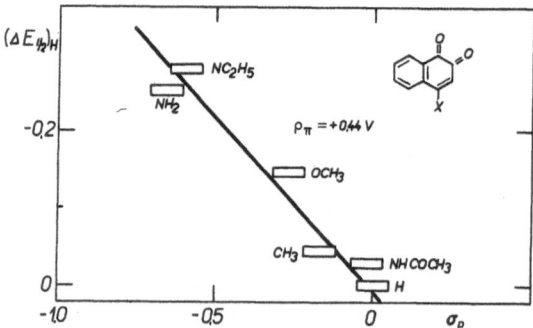

Fig. VIII-13. Relation of oxidation–reduction potentials of 4-substituted 1,2-naphtho-quinones to Hammett substituent constants $\sigma_{p\text{-}x}$. Oxidation–reduction potentials from Ref. 28 expressed as shifts $(\Delta E_{1/2})_H$ relative to the value for the parent compound.

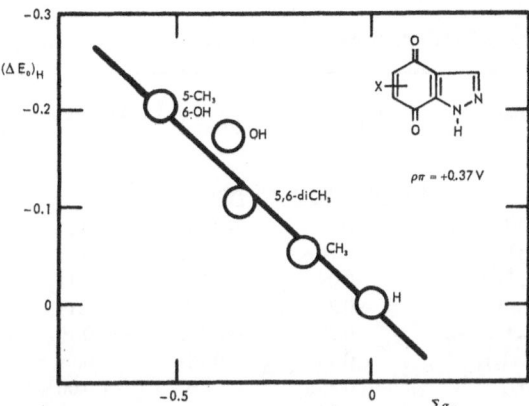

Fig. VIII-14. Relation of half-wave potentials of 5- and 6-substituted 4,7-indazole-quinones to the sum of the Hammett substituent constants $\Sigma\sigma_{p-x}$. Shifts in half-wave potential from Ref. 34 in the region pH 2–8 (relative to the parent compound).

satisfactorily (Fig. VIII-10). The opposite slopes of these relations is remarkable. The contribution of the p-benzoquinonimine form cannot be excluded in the case of amino derivatives.

Taking both polar and mesomeric interactions of substituents with the quinonoid ring into consideration, we tested equation (69) for 2-substituted 1,4-naphthoquinones. Fig. VIII-11 shows the relation obtained with selected values for the shifts in oxidation–reduction potential.[15,28,29,32] Fig. VIII-12 shows the half-wave potentials measured[33] in a mixture of CH_3COOH (0.1 M), CH_3COONa (0.1 M), and ethanol (75%). The good correlations observed demonstrate that the approximations made in the derivation of equation (69) are justified. In a similar manner, it was possible to use equation (69) to correlate the shifts in the oxidation–reduction potentials of 4-substituted 1,2-naphthoquinones[28] (Fig. VIII-13) and in the half-wave potentials of 4,7-indazolequinones[34] substituted in position 5 or 6 (Fig. VIII-14).

The additivity of potential shifts for substances containing two substituents in the quinonoid ring has been observed only occasionally (cf. Fig. VIII-12); more generally, deviations of variable magnitude have been observed (Table VIII-5). Apart from the fact that the deviation from additivity is least for dimethyl derivatives, there is no evident correlation between the deviation (expressing the nonadditivity) and the shape, polar or mesomeric properties of the substituents. The simultaneous presence of two substituents may lead to decrease or

Table VIII-5
Deviations from Additivity of Substituent Effects in Bicyclic Quinones

Substituents	ΔE_H^o or Calc.	$[\Delta(E_{1/2})_H]$, V Found	Difference	Source of E^o or $E_{1/2}$
2,3-Disubstituted 1,4-Naphthoquinones				
2-OH, 3-OCH$_3$	-0.26	-0.14	$+0.12$	29
2,3-(SCH$_3$)$_2$	-0.16	-0.04_5	$+0.11_5$	32
2,3-(SC$_6$H$_5$)$_2$	-0.09	$+0.02_5$	$+0.11_5$	32
2-CH$_3$, 3-OCH$_3$	-0.21	-0.10	$+0.11$	29
2-NHC$_6$H$_5$, 3-Br	-0.21	-0.11_5	$+0.10_5$	32
2,3-(OCH$_3$)$_2$	-0.26	-0.18	$+0.08$	29
2,3-(OH)$_2$	-0.26	-0.20	$+0.06$	29
2-SC$_6$H$_5$, 3-Br	-0.03_8	$+0.02_7$	$+0.05_5$	32
2-N(CH$_3$)$_2$, 3-Br	-0.16_5	-0.11_5	$+0.05$	32
2-NHCOCH$_3$, 3-Cl	-0.05_5	-0.00_5	$+0.05$	33
2-SCH$_3$, 3-Br	-0.06_5	-0.02	$+0.04_5$	32
2-CH$_3$, 3-OH	-0.21	-0.18	$+0.03$	29
2-SCH$_3$, 3-Cl	-0.05_5	-0.03_5	$+0.02$	32
2,3-(CH$_3$)$_2$	-0.15	-0.14_5	$+0.005$	28
2-OH, 3-Cl	-0.09_5	-0.13	-0.03_5	15
2-NH$_2$, 3-Cl	-0.18_5	-0.23_5	-0.05	32
2-NH$_2$, 3-Br	-0.19_5	-0.22_5	-0.05	32
2,3-Cl$_2$	$+0.07$	$+0.01$	-0.06	15
5,6-Disubstituted 4,7-Indazolequinones				
5-CH$_3$, 6-OH	-0.23	-0.20_3	$+0.02_7$	34
5,6-(CH$_3$)$_2$	-0.11	-0.10_5	$+0.00_5$	34

Table VIII-6
Values of Reaction Constants $\rho_{\pi,Q}$ Found for Substituted Naphthoquinones

Compound	Solvent	Medium	$\rho_{\pi,Q}$, V	r^a	Ref.
1,4-naphthoquinone	EtOH-H$_2$O	acid	$+0.28$	—	7
	EtOH-H$_2$O	acid	$+0.29$	0.972	8
	75% EtOH-H$_2$O	acetate pH 4.7	$+0.32$	—	7
	75% EtOH-H$_2$O	NH$_3$-NH$_4$Cl pH 9.3	$+0.29$	—	7
1,2-naphthoquinone	EtOH-H$_2$O	acid	$+0.44$	—	7
	EtOH-H$_2$O	acid–neutral	$+0.41_2$	0.961	8
4,7-indazolequinones	EtOH-H$_2$O	pH 2-8	$+0.37$	—	7

^{a} Correlation coefficient.

increase in the effect of either. An explanation is sought in the interaction between the substituents and the quinonoid carbonyls (see p. 286).

When the values of the reaction constants for 1,4-naphthoquinones, 4,7-indazolequinones, and 1,2-naphthoquinones are compared (Table VIII-6), the effect of the heterocyclic nucleus and the differences between *ortho* and *para* quinones are obvious. Nevertheless, we must not forget that values of the ρ-constants can include also the effects of the changes in the dissociation constants (p. 276). On the other hand, the effect of the composition of the medium in which the oxidation-reduction potentials were measured on the values in Table VIII-6 is slight, but the number of known values of half-wave potential in non-hydroxylic solvents is too small for convincing conclusions to be reached.

The value of $\rho_{\pi,Q}$ for 1,4-naphthoquinone was used for the determination of the approximate values of substituent constants given in Table VIII-7.

(b) Substitution in the Annellated Benzene Ring, or in the Attached Phenyl Ring. The comparison of the effects of substituents in the benzene ring of naphthoquinones on their oxidation–reduction potentials is limited by the small number of values available for monosubstituted

Table VIII-7

Approximate[a] Values of Total Polar Substituent Constants, σ_{p-X}, Derived from the Half-Wave Potentials of 2-Substituted 1,4-Naphthoquinones

Substituent	σ_{p-X}	Source of $E_{1/2}$ (or E°)
NHC_6H_5	-0.68	32
	-0.59	28
	-0.56	33
$NHC_{10}H_7$-2	-0.54	33
NHC_6H_4COOH-4	-0.42	33
$NHC_6H_4COO^-$-4	-0.52	33
$NHC_6H_4SO_2NH_2$-4	-0.40	33
$NHC_6H_3Cl_2$-2,4	-0.37	33
$CH(C_6H_5)_2$	-0.05	28
$SO_2C_6H_4CH_3$-4	$+0.56$	28
SC_6H_5	$+0.02$	32

[a] See footnote under Table VIII-3.

Table VIII-8
Additivity of Shifts in Oxidation–Reduction Potential[29] at pH 0
for 1,4-Naphthoquinones Substituted in the Benzene Ring

Substituent	$\Delta E°$, V		Difference
	Calc.	Found	
2-CH$_3$	—	-0.06_6	—
5-OH	—	-0.03_8	—
2-CH$_3$, 5-OH	-0.10_7	-0.10_7	0.00_0
5,8-(OH)$_2$	-0.07_6	-0.10_2	-0.02_6
2-CH$_3$, 5,8-(OH)$_2$	-0.16_8	-0.16_5	$+0.00_3$
2-OH	—	-0.13	—
2,5,8-(OH)$_3$	-0.23_2	-0.24	-0.00_8
2-CH$_3$, 3-OH	—	-0.18_2	—
2-CH$_3$, 4,5,8-(OH)$_3$	-0.28_4	-0.27_8	$+0.00_6$

compounds. From the few data available (Table VIII-8), it follows that
a substituent in the benzene ring has a smaller effect on the shift of the
potential than the same substituent in the quinonoid ring and, furthermore, that the effects of two substituents, one of which is in the benzene
ring and the other is in the quinonoid ring, are approximately additive

Table VIII-9
Computed Shifts of the Oxidation–Reduction Potentials of 1,4-Naphthoquinones
Substituted in the Benzene Ring

Substituent	Shift, V	Computed from $E°$ of	Source of $E°$ (or $E_{1/2}$)
6(7)-CH$_3$	-0.01_5	2-CH$_3$; 2,7-(CH$_3$)$_2$	34
	-0.01_9	2-CH$_3$; 2,6-(CH$_3$)$_2$	28
	-0.02_0	2-OH; (2-OH, 6-CH$_3$)	35
	-0.03_2	2,3-(OH)$_2$; (2,3-(OH)$_2$, 6-CH$_3$)	29
	-0.02_1	(2-CH$_3$, 3-OH); (2,6-(CH$_3$)$_2$, 3-OH)	35
6(7)-Br	$+0.03_2$	2,3-(OH); (2,3-(OH), 6-Br)	29
	$+0.02_2$	2,3,6-(OH)$_3$[a]; (2,3,6-(OH)$_3$, 7-Br)	29
6(7)-OCH$_3$	-0.10_0^0	(2-CH$_3$, 5,8-(OH); (2-CH$_3$, 5,8-(OH)$_2$, 6(7)-OCH$_3$)	29

[a] Computed from 2,3-(OH)$_2$ and 6-OH.

(Table VIII-8). Thus for substituents in the benzene ring (even in the *peri* position) the interaction with the quinonoid carbonyl group is less pronounced than for substituents in the quinonoid ring.

By the use of the additivity principle it has been possible to calculate the increments in the shifts in oxidation–reduction potential caused by some substituents (Table VIII-9).

The mean values of the shifts given in Table VIII-9, together with the few values obtained[29] by direct measurement, show a good correlation by equation (69′) with substituent constants σ_p (Fig. VIII-15), but not with σ_m. The correlation with $(\sigma_m + \sigma_p)/2$ is worse than with σ_p. The smaller value of the reaction constant $\rho_{\pi,pQ}(\rho_{\pi,pQ} = 0.14 \text{ V})$, compared with the value for shifts due to substituents in the quinonoid ring $(\rho_{\pi,Q} = 0.28 \text{ V})$, expresses quantitatively the smaller effect of substituents in the benzene ring.

Even more screened is the effect of a substituent separated from the quinonoid ring by an anilino residue. Thus, for the half-wave potentials[33] of 2-anilino-3-hydroxy-1,4-naphthoquinones substituted in positions 2 and 4 of the anilino group equation (70) fits satisfactorily (Fig. VIII-16), and the insertion of the C_6H_4NH group has reduced the value of the reaction constant $(\rho_{\pi,R} = 0.03 \text{ V})$ about ten fold.

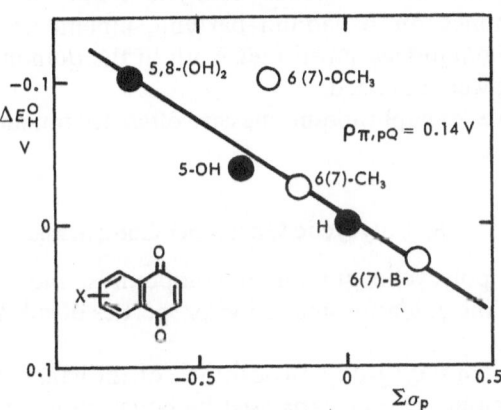

Fig. VIII-15. Relation of oxidation–reduction potentials of 1,4-naphthoquinones substituted in the benzene ring to the sum of the Hammett substituent constants $\Sigma\sigma_{p\text{-}X}$. Oxidation–reduction potentials, expressed as shifts ΔE_H° relative to the value for the parent compound, for H, 5-OH and 5,8-di-OH from Ref. 29; for 6(7)-CH$_3$, 6(7)-Br and 6(7)-OCH$_3$ see Table VIII-9.

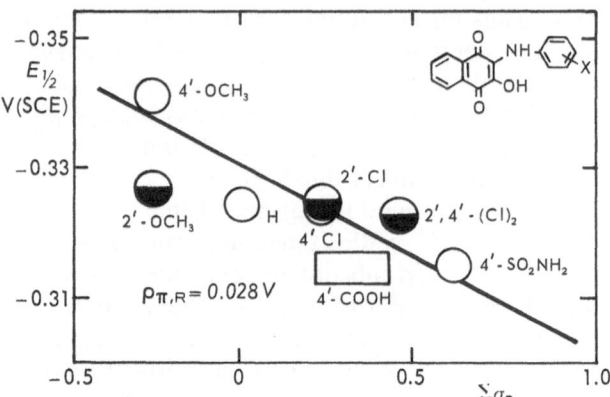

Fig. VIII-16. Relation of half-wave potentials of 2'- and 4'-substituted 2-anilino-3-hydroxy-1,4-naphthoquinones to the sum of the Hammett substituent constants $\Sigma\sigma_{p-X}$. Half-wave potentials in 0.1 M CH$_3$COOH, 0.1 M CH$_3$COONa with 75% ethanol from Ref. 33. Halved points: *ortho* derivatives.

The empirical relation between the half-wave potentials of naphthoquinones and *p*-(diethylamino) anilines was mentioned earlier (p. 293). A linear relation between $\log(1/t_{1/2})$ and E° has also been found[47] for the heterogeneous catalytic reduction of substituted naphthoquinones on palladium–barium sulphate in glacial acetic acid. Naphthoquinones substituted both in the quinonoid and in the benzene ring were included.

The waves of naphthoquinones are often accompanied by adsorption prewaves.[36]

5. Polycyclic Quinonoid Compounds

Among polycyclic quinonoid compounds, most attention has been paid to the oxidation-reduction properties of substituted anthraquinones.

The effect of alkyl groups in position 2 on the half-wave potentials[37] of anthraquinones can be expressed by equation (61) when measurements are carried out in 0.1 N HCl ($\rho^*_{\pi,Q} = +0.02_8$ V for X = H, CH$_3$, C$_2$H$_5$, and i-C$_3$H$_7$). On the other hand, for solutions of the same substances in 0.1 M NaOH two waves and correlation with $\rho^*_{\pi,Q} = 0.04$ V for the more negative one were observed. The more positive wave showed no correlation with σ^*_X. Very small substituent effects,

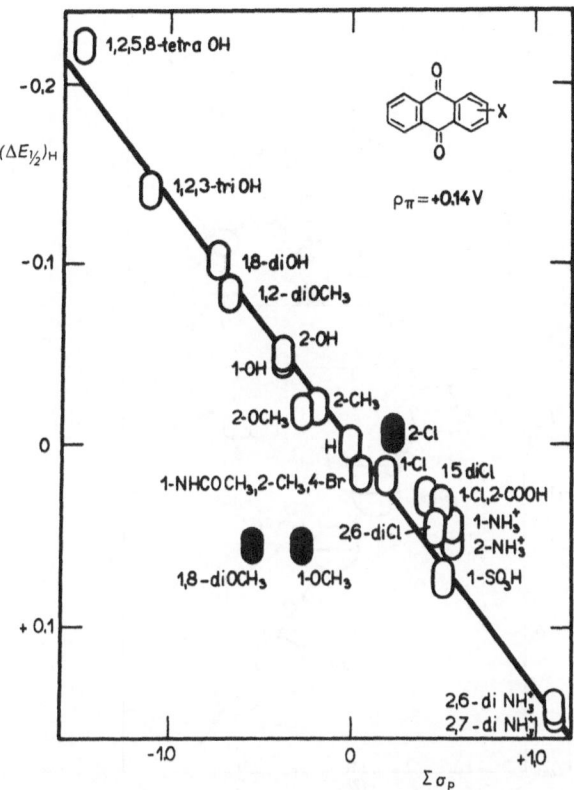

Fig. VIII-17. Relation of half-wave potentials of substituted anthraquinones to the sum of the Hammett substituent constants $\Sigma\sigma_{p\text{-}X}$. Half-wave potentials in 10% H_2SO_4 in glacial acetic acid from Ref. 38 are expressed as shifts relative to the value for the parent compound $(\Delta E_{1/2})_H$. Full points deviate.

which do not allow a more precise determination of the reaction constant $\rho^{*}_{\pi,Q}$, were observed for the two waves of 2-substituted tetrahydro- and octahydro-anthraquinones[37] in alkaline media.

In most other cases, values of half-wave potential are available for substituents for which both inductive and mesomeric effects can play a part. For these substances, the question of the proper choice of substituent constants for use in equation (69a) remains open. For substituents in position 1, the application of the constants $\sigma_{p\text{-}X}$ (or $\sigma_{o\text{-}X}$ when available) has proved useful in all instances so far. The same constants were found to be best for several substituents in position 2

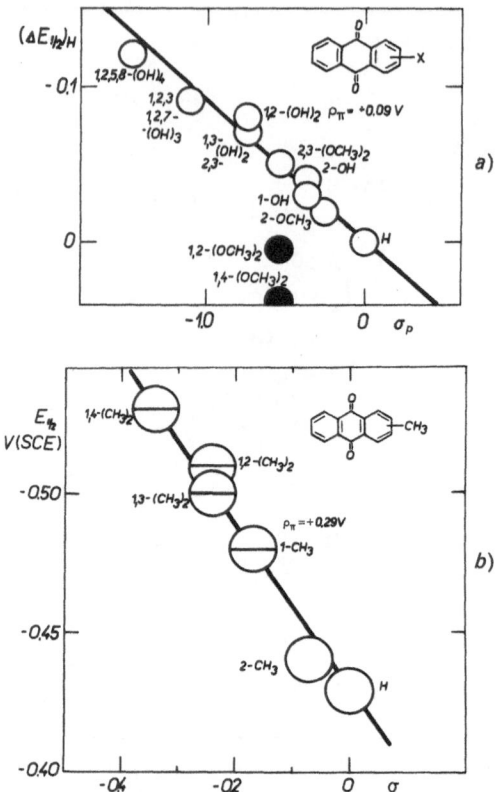

Fig. VIII-18. Relation of half-wave potentials of substituted anthraquinones to the sum of the Hammett substituent constants $\Sigma\sigma_{p\text{-}X}$. a) Shifts in half-wave potential relative to the value for the parent compound $(\Delta E_{1/2})_H$ at pH 1.25 in 70% ethanol from Ref. 39; b) half-wave potentials in 50% 2-propanol from Ref. 40. Circles $\sigma_{m\text{-}X}$, halved circles $\Sigma\sigma_{o\text{-}X}$.

when the half-wave potentials were obtained in acid media-as in glacial acetic acid containing 10% of sulfuric acid[38] (Fig. VIII-17) or at pH 1.25 in 70% ethanol[39] (Fig. VIII-18a). The correlation of half-wave potentials with σ_p was better than with $(\sigma_p + \sigma_m)/2$ and considerably better than with σ_m. Hence, in the expression $\rho_1\sigma_{p\text{-}X} + \rho_2\sigma_{m\text{-}X}$ is $\rho_1 \gg \rho_2$.

On the other hand, for the shifts in half-wave potential of polymethylanthraquinones measured[40] in 0.1 M HCl and in acetate buffers of pH 5.6 containing 50% of isopropyl alcohol (the shifts are reported to be almost identical in these two media) the use of $\sigma_{m\text{-}CH_3}$

for the 2-methyl group was found to be the most satisfactory (Fig. VIII-18*b*). This difference may be influenced by the changes in solvation (see p. 286).

Significant deviations in aqueous solutions and in solutions in glacial acetic acid were observed for the 1-methoxy derivative (Figs. VIII-17 and -18*a*). Wiles[39] tried to explain the observed difference in behavior between the methoxy and hydroxy derivatives by the formation of hydrogen bridges in the latter. The relations given in Figs. VIII-17 and -18*a* show that the behavior of the 1-hydroxy derivatives is normal, involving only polar and resonance effects, but that of the 1-methoxy derivatives is abnormal. Also, the half-wave potentials of 1-hydroxyanthraquinones measured in dimethylformamide in the presence of tetraethylmmonium iodide have been interpreted[41] as proof of intramolecular hydrogen bonding. The limited choice of the type of substituent prevents a more conclusive treatment, but the presently available data show deviations not only for 1-hydroxy, but also for 2-hydroxy and some other derivatives, the correlation being poor for σ_p, for $\sigma_{1\text{-}X} = \sigma_{p\text{-}X}$ and $\sigma_{2\text{-}X} = \sigma_{m\text{-}X}$, and for $(\sigma_m + \sigma_p)/2$.

In anthraquinones with one methoxy group in a *peri* position, a positive *ortho* effect was observed. The introduction of a second methoxy group into the *peri* position on the other side of the same carbonyl group (the 1,8-dimethoxy derivative) did not change the shift, which was the same as for the monomethoxy compound; thus, a regular polar effect of the methoxy group can be excluded. Nevertheless, it cannot be decided whether a field effect or a steric fixing of the carbonyl group into an advantageous position is operating. In anthraquinones with two methoxy groups in *peri* positions relative to different carbonyl groups (the 1,4- or 1,5-dimethoxy derivative), a considerable shift in the opposite direction has been reported, in fact a shift toward such negative potentials that no wave is observed. A steric effect on the orientation of protonated carbonyls can be presumed. Nothing similar has been observed for unprotonated forms, which are reduced[39] in alkaline solutions.

Deviations from the additivity of substituent effects observed in water–ethanol media[39] and in solutions of sulfuric acid in glacial acetic acid[38] can be explained by the influence of substituents on protonation or solution. For some compounds, such as the 1,4- and 1,2-dihydroxy and corresponding diamino derivatives, the observed deviations can be explained by a change in the mechanism of the electrode process, which may involve other quinonoid forms. But such

Table VIII-10

Shifts in the Half-Wave Potentials[42] of Substituted Anthraquinones in 0.1 N NaOH in Water and in 50 % Ethanol–Water

Substituent	$\Delta(E_{1/2})_H$, V	
	Water	50 % Ethanol
1-SO$_3^-$	$+0.01_2$	$+0.01$
2-SO$_3^-$	$+0.07_0$	$+0.01$
1-O$^-$	-0.03_7	-0.05
2-O$^-$	-0.13_7	-0.19
1,2-di-O$^-$	-0.19_3	-0.23
1,4-di-O$^-$	-0.06_2	—
1,8-di-O$^-$	-0.09_2	—
2,6-di-O$^-$	-0.28_0	-0.42

an explanation is inadequate for the nonadditivity of substituent effects in compounds like 1,3,8-trihydroxy-7-methyl- or 1-amino-2,4-dibromo-anthraquinone. The present suggestion of a change in protonation or solvation is in accord with the fact that the half-wave potentials[42] of such compounds in 0.1 N NaOH show additivity of substituent effects.

The marked effect of the solvent on the shifts in half-wave potential is seen from a comparison of Table VIII-10 and -11: not only the values, but also the signs of the shifts are dependent on the solvent. In the comparison of data in Table VIII-11, the drastic changes in acidity must be kept in mind. The striking effect of the kind of buffer used on the half-wave potentials of anthraquinones has been discussed else-

Table VIII-11

Shifts in the Half-Wave Potentials of Substituted Anthraquinones in Different Solvents

Substituent	Glacial acetic acid,[38] 10% H_2SO_4	$\Delta(E_{1/2})_H$, V Dimethylformamide[43]		Dimethylformamide[43]	
		0.1 M-(CH$_3$)$_4$NI		LiCl	
		1st wave	2nd wave	1st wave	2nd wave
1-OH	-0.44_8	$+0.17$	$+0.17$	$+0.20_5$	$+0.02$
1,4-(OH)$_2$	-0.12_2	$+0.26$	$+0.38$	$+0.11_5$	-0.26
1,2,5,8-(OH)$_4$	-0.22_2	—	—	-0.06_5	-0.13

where.[26] The effect of the acidity of the medium should also be kept in mind in a comparison of the reaction constants $\rho_{\pi,pQ}$:

	$\rho_{\pi,pQ}$	Reference for $E_{1/2}$
Glacial acetic acid with 10% H_2SO_4	+0.14 V	38
pH 1.25, 70% ethanol	+0.09 V	39
pH 1.5 to 6.0, 50% 2-propanol	+0.29 V	40

It should be further mentioned that for 1,2-trimethylene- and 1,2-tetramethyleneanthraquinones the shifts in half-wave potential do not correspond to the values[4,5] of $\sigma_{1,2\text{-}(CH_2)_3}$ and $\sigma_{1,2\text{-}(CH_2)_4}$. The observed values do not differ substantially from the half-wave potential of the unsubstituted compound.

Of the other types of empirical correlation the $\ln k$–$E_{1/2}$ correlations[23] for anthraquinones were mentioned earlier (p. 290). Deviations have been observed for anthraquinonesulfonates, which showed also deviations from reversible behavior. Half-wave potentials of quinones also have been correlated with electron affinities[42] and charge-transfer spectra.[43–45]

References

[1] W. M. Clark, "Oxidation–Reduction Potentials of Organic Systems," Williams and Wilkins, Baltimore, 1960.

[2] M. Kalousek and M. Rálek, *Collection Czech. Chem. Commun.* **19**: 1099 (1954); *Chem. Listy* **48**: 808 (1954).

[3] R. W. Taft, Jr., *Separation of Polar, Steric, and Resonance Effects in Reactivity* in "Steric Effects in Organic Chemistry" (M. S. Newman ed.) John Wiley and Sons, New York, 1956.

[4] L. P. Hammett, "Physical Organic Chemistry," p. 184, McGraw-Hill, New York, 1940.

[5] H. H. Jaffé, *Chem. Rev.* **53**: 191 (1953).

[6] M. G. Evans and J. De Heer, *Quart. Rev.* **4**: 94 (1950).

[7] P. Zuman, *Collection Czech. Chem. Commun.* **27**: 2035 (1962).

[8] M. Charton, A paper presented before the Meeting of the American Chemical Society (1961).

[9] T. Hayashi and R. Shibata, *Kogyo Kagaku Zasshi* **63**: 840 (1960); quoted according to T. Hayashi and R. Shibata, *Bull. Chem. Soc. Japan* **34**: 1116 (1961).

[10] O. Ryba, J. Petránek, and J. Pospíšil, *Collection Czech. Chem. Commun.* **30**: 843 (1965).

[11] O. Ryba, J. Petránek, and J. Pospíšil, *Collection Czech. Chem. Commun.* **30**: 2157 (1965).

[12] D. E. Kvalnes, *J. Am. Chem. Soc.* **56**: 667, 2478 (1934).

[13] E. Bilman, A. L. Jensen, and K. O. Pederson, *J. Chem. Soc.* **127**: 199 (1925).

[14] J. B. Conant and L. F. Fieser, *J. Am. Chem. Soc.* **45**: 2194 (1923).

[15] J. B. Conant and L. F. Fieser, *J. Am. Chem. Soc.* **46**: 1858 (1924).

[16] V. K. LaMer and L. E. Baker, *J. Am. Chem. Soc.* **44**: 1954 (1922).

[17] J. B. Conant and L. F. Fieser, *J. Am. Chem. Soc.* **44**: 2480 (1922).

[18] R. T. Arnold and H. E. Zaugg, *J. Am. Chem. Soc.* **63**: 1317 (1941).

[19] E. Scrocco and G. Marmani, *Ann. Chim.* (Rome) **41**: 716 (1951).

[20] L. I. Smith, I. M. Kolthoff, S. Wawzonek, and P. M. Ruoff, *J. Am. Chem. Soc.* **63**; 1018 (1941).

[21] M. E. Peover, *J. Chem. Soc.* **1962**: 4540.

[22] O. Dimroth, *Angew. Chem.* **46**: 571 (1933).

[23] H. Berg, *Z. Chem.* **2**: 237 (1962).

[24] R. L. Bent, J. C. Dessloch, F. C. Duennebier, D. W. Fassett, D. B. Glass, T. H. James, D. B. Julian, W. R. Ruby, J. M. Snell, J. H. Sterrer, J. R. Thirtle, P. W. Vittum, and A. Weissberger, *J. Am. Chem. Soc.* **73**: 3100 (1951).

[25] L. F. Fieser and M. Fieser, "Lehrbuch der organischen Chemie," Verlag Chemie, Weinheim, 1954.

[26] P. Zuman, *Chem. Listy* **48**, 94 (1954).

[27] F. T. Eggertsen and F. T. Weiss, *Anal. Chem.* **28**: 1008 (1956).

[28] L. F. Fieser and M. Fieser, *J. Am. Chem. Soc.* **57**; 491 (1935).

[29] K. Wallenfels and W. Möhle, *Ber.* **76**: 924 (1943).

[30] E. G. Ball, *J. Biol. Chem.* **114**: 649 (1936).

[31] L. F. Fieser, *J. Am. Chem. Soc.* **50**, 439 (1928).

[32] N. Ikeda, *J. Pharm. Soc. Japan* **75**: 1073 (1955).

[33] I. F. Vladimircev and A. G. Stromberg, *Zh. Obshch. Khim.* **27**: 1029 (1957).

[34] P. Zuman, unpublished results.

[35] L. F. Lugg, A. K. Macbeth, and F. L. Winsor, *J. Chem. Soc.* **1936**: 145, 1457.

[36] P. Zuman, *Collection Czech. Chem. Commun.* **19**: 1140 (1954); *Chem. Listy* **48**: 524 (1954).

[37] V. E. Dicent, *Zh. Obshch. Khim.* **29**: 1370 (1959).

[38] L. Stárka, L. Vystrčil, and B. Stárková, *Chem. Listy* **51**: 1440 (1957).

[39] L. A. Wiles, *J. Chem. Soc.* **1952**: 1958.

[40] R. J. Crawford, S. Levine, R. M. Elofson, and R. B. Sandin, *J. Am. Chem. Soc.* **79**: 3153 (1957).

[41] R. Jones and T. McL. Spotswood, *Australian J. Chem.* **15**: 492 (1962).

[42] M. E. Peover, *Nature* **193**, 475 (1962).

[43] M. E. Peover, *Nature* **191**, 702 (1961).

[44] M. E. Peover, *Trans. Faraday Soc.* **58**: 1656 (1962).

[45] M. E. Peover and J. D. Davies, *Trans. Faraday Soc.* **60**: 476 (1964).

[46] E. Braude, L. Jackmann, and R. Linstead, *J. Chem. Soc.* **1954**: 3548.

[47] H. Musso, K. Figge, and D. J. Becker, *Chem. Ber.* **94**: 1107 (1961).

IX: Alicyclic Systems

Generally speaking, the polarography of alicyclic compounds is a little-exploited field to which only a limited number of papers have been devoted so far. In these systems the electroactive group either can be situated in the side chain, or be attached directly to the cyclic system and form a part of it. The structural effects that can exert an influence on the electroactive group can be caused either by a substituent in the cyclic system or by the ring size and the stereochemistry of the system.

Of the four possible combinations of these effects, namely, (*i*) electroactive group in the side chain with a substituent in the ring; (*ii*) electroactive group directly attached to the ring with a substituent in the ring; (*iii*) electroactive group in the side chain affected by the ring size and stereochemistry; and (*iv*) electroactive group directly attached to the ring affected by the ring size and stereochemistry, no examples of the first type have been examined.† The few observed effects on half-wave potentials of the other three types are discussed in next paragraphs. Because of the limited number of compounds for which half-wave potentials have been determined, it has been necessary in most instances to limit the treatment to qualitative discussion, and only in few cases has the application of linear free energy relation been possible. Rather than a review of accomplished goals, this part should

---

† Another division was used in a recent review by the present author.[56]

be regarded as an indication of the possibilities that polarography offers for studies in this group of compounds.

1. Effects of Substituents in the Ring on an Electroactive Group Attached to the Ring

For rethrolones (I) the half-wave potentials in $0.1\ N$ LiOH

$$CH_3 \quad X$$
$$HO \quad O$$
(I)

measured by Krupička[1] satisfactorily fit equation (29) in the form

$$\Delta E_{1/2} = \rho_{\pi,R}^* \sigma_X^* \tag{29}$$

for $X = C_4H_9$, $i\text{-}C_3H_7$, $CH_2C_6H_5$ and C_6H_5 (Fig. IX-1). The value for phenyl fits the linear relation well, so that it can be deduced that this group exerts predominantly a polar effect and that its conjugation is not of primary importance. The determined value $\rho_{\pi,R}^* = +0.10_6$ V allowed computation of the following approximate values of substituent constants:

$\sigma_{CH=CH_2}^*$	$= +0.12$
$\sigma_{furfuryl}^*$	$= +0.29$
$\sigma_{1\text{-cyclohexen-1-yl}}^*$	$= +0.01$
$\sigma_{2\text{-cyclopenten-1-yl}}^*$	$= -0.01$

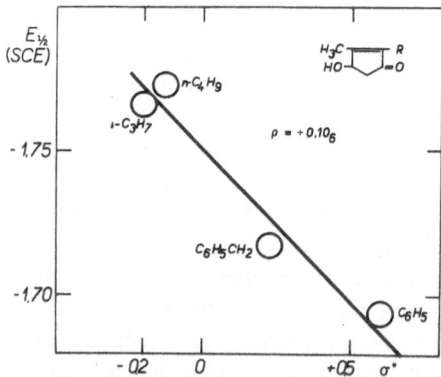

Fig. IX-1. Relation of half-wave potentials of substituted rethrolones to Taft polar substituent constants σ^*. Half-wave potentials in $0.1\ M$ LiOH from Ref. 1.

For naphthylcycloalkenes[2] both the position of attachment in the naphthalene ring and the size of hydroaromatic ring can affect the half-wave potentials. The latter effect is discussed in Section 3, of this chapter. Here, only the effect of exchange of 1-naphthyl in (II) and (IV) for 2-naphthyl in (III) and (V) will be mentioned briefly.

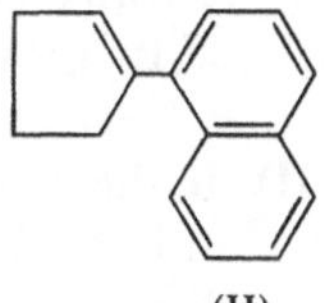

$E_{1/2}$

-2.24 V

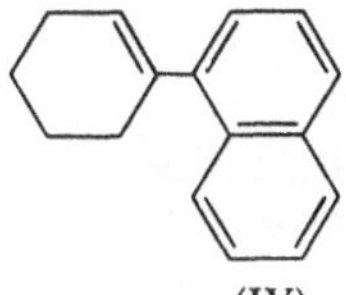

(II)

$E_{1/2}$

-2.42 V

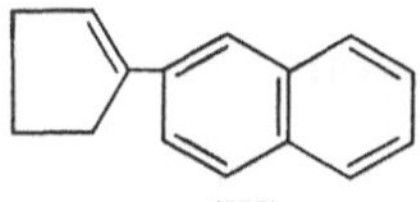

(IV)

-2.25 V

(III) $\Delta - 0.01$ V

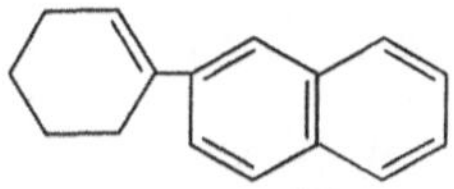

-2.32 V

(V) $\Delta + 0.10$ V

The difference in both magnitude and direction in the shifts due to this exchange for cyclopentene and cyclohexene is caused by steric hindrance to coplanarity, as proved in more detail in Section 3.

The sequence of polarographic reducibility, (II) $\approx$ (III) $>>$ (V) $>$ (IV), is analogous to that observed for the rate of maleic anhydride addition in the Diels–Alder reaction, namely, (II) $\approx$ (III) $>$ (V) $>>$ (IV). It is assumed[2] that this proves that in the electroreduction a primary product, with an exocyclic double bond in the pivotal position between the naphthalene and hydroaromatic rings, is formed by 1,4-addition. The formation of this intermediate is thought to be hindered in the compound (IV).

Apart from effects transmitted along chemical bonds, trans-annular effects also can affect the polarographic behavior.

Examples of this type of behavior can be found in the camphor series (VI). The keto group in unsubstituted camphor is not reducible

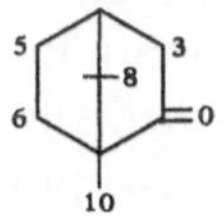

(VI)

in the commonly used buffer solutions before -1.9 V. Similarly, no reduction waves have been observed[3] for 5-oxocamphor. On the other hand, a reduction wave has been recorded for 6-oxocamphor at -1.67 V at pH 8, indicating transannular interaction of the two keto groups.

Also, the effect of the exchange of a methyl group in position 8 for an aldehyde group, causing a shift of the half-wave potential of the C—Br bond in 3-bromocamphor derivatives from -0.66 V to -0.45 V, is unlikely to be transmitted via three carbon atoms. A transannular interaction between the aldehydic group and the COCHBr grouping can be postulated.

Similarly, the reducibility[4] of compounds (VII) in the tropone series (where $X = >\overset{+}{N}(CH_3)_2$, $>N(CH_3)$, $>\overset{+}{S}CH_3$ and $>S$) and (VIII) can be explained only on the basis of transannular interaction

between the hetero atom and the carbonyl group. Sulphur derivatives are more easily reducible than the corresponding nitrogen derivatives, and onium compounds more easily reducible than the corresponding nonquaternary derivatives, which are reduced in the protonated form $(X = \overset{+}{N}HCH_3)$. The sulphur derivative $(X = >S)$ is unprotonated and, therefore, polarographically inactive.

The rule that substitution in a position in the molecule remote from the electroactive group has little effect on the half-wave potentials (cf. Ref. 5) has been accepted as generally valid, but in the special case of steroid compounds it has been shown recently for example, Ref. 6 that effects of substituents can be transmitted from other rings and from distant positions. Even when such effects on polarographic half-wave potentials have not been stated explicitly by the authors of the original papers, they can be deduced from their published data. Thus, the polarographic reduction of a group in ring A can be affected by substituents in rings B, C, and even D, and so on. Some examples of effects based on half-wave potentials published in the literature[7-16] are given here. It can be assumed that in most cases a difference of 0.01 V in the value of a half-wave potential is significant, but, in all cases, only the values measured by one author are compared below. It has been demonstrated (e.g.,[7,9,15]) that the shifts in half-wave potential due to change in pH are parallel for all the substances compared, but the role of acid–base equilibria accompanying the electrode process[7,10] cannot be excluded, especially in the cases of α, β-unsaturated ketones and of α-halo ketones. It is possible that in these cases, the effect of proton transfer accompanying the electrode process contributes to the observed shifts in half-wave potential. Little attention has been paid so far to the form of the reduction waves, but from the published data on αn values[13,16] it seems that these values do not differ substantially for the compounds studied. Hence, the comparison of half-wave potentials should give sufficient information about the change in the relative free energy of the electrode process.

The effect of substituents in position 17 on the reduction of a carbonyl group in position 3 is demonstrated by the shifts in half-wave potential[12] for compounds (IX)–(XII) (90% ethanol, 0.05 M N(C$_4$H$_9$)$_4$Cl;

(IX) − 2.28 V   (X) − 2.37 V

(XI) − 2.38 V   (XII) − 2.44 V

In compounds (XI) and (XII) the reduction of the two groups in positions 3 and 17 and in positions 3 and 20, respectively, occurs at only slightly different potentials, so that the two waves merge and only one wave of double height is observed. Hence, strictly speaking, the measured half-wave potential of this wave, given above, does not express solely the effect of substitution.

Substituents in position 11 have much less effect on the reduction of the carbonyl in position 3, as can be seen from a comparison of the half-wave potentials[12] of (XII) and (XIII) [90% ethanol, 0.05 M N(C$_4$H$_9$)$_4$Cl].

(XII) − 2.44 V   (XIII) − 2.46 V

The available data are most numerous for substituent effects on half-wave potentials of 3-keto-α,β-unsaturated steroids. Here, the effect of substitution in several positions can be followed. The effects of substituents in position 6 can be observed for the compounds[15] (XIV) to (XVI) (in 50% ethanol, Britton–Robinson buffer pH 6.0):

(XIV) -1.385 V   (XV) -1.375 V   (XVI) -1.355 V

Differences in half-wave potential[8,9] can be observed also for substituents in position 11 [see (XVII) and (XVIII); 90% ethanol, barbital buffer pH 8.5]:

(XVII) -1.50 V   (XVIII) -1.44 V and -1.65 V

On the other hand, compounds (XIX) and (XX), studied[8] under the same conditions, and differing only in the absence of the hydroxy group in position 17, show a much less marked effect for substitution in position 11:

(XIX) -1.48 V and -1.68 V   (XX) -1.49 V and -1.64 V

The effect of a hydroxy group in position 16 also can be considerable, as is shown by a comparison of the substances[16] (XXI) and (XXII) [dimethylformamide, 0.1 M N(C$_4$H$_9$)$_4$I and 0.01 M N(C$_2$H$_5$)$_4$I]:

(XXI) -1.66 V and -2.10 V   (XXII) -1.63 V and -2.11 V

Table IX-1

Effect of Substituents in Position 17 on the Half-Wave Potential of 3-Keto Δ^4-Steroids
(50 % ethanol, NCE)

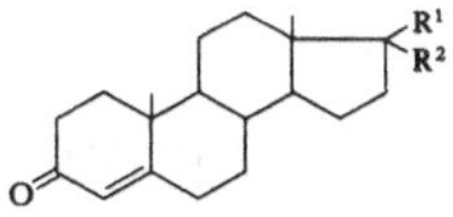

Compound	R^1	R^2	$E_{1/2}$, V 0.1 M HCl	$E_{1/2}$, V pH 10.5[a]	$dE_{1/2}/dpH$
Progesterone	$COCH_3$	H	-1.11	-1.68	0.063
Testosterone	OH	H	-1.13	-1.69	0.061
Methyl testosterone	OH	CH_3	-1.13	-1.68	0.059
Testosterone Propionate	$OOCC_2H_5$	H	-1.13	-1.68	0.060
17-Hydroxy progesterone	$COCH_3$	OH	-1.14	-1.68	0.059
11-Deoxycorticosterone	$COCH_2OH$	H	-1.12	-1.67	0.059
11-Deoxycorticosterone acetate	$COCH_2OOCCH_3$	H	-1.11	-1.67	0.060
11-Deoxy-17--hydroxycorticosterone	$COCH_2OH$	OH	-1.15	-1.69	0.059

[a] Triethylamine buffer.

Much attention has been paid to the effect of substituents in position 17 on shifts in the half-wave potentials of 3-keto Δ^4-steroids.[7-9,11,15] In 50% ethanol containing triethylamine buffer[11] at pH 10.5 the effect is small; in 0.1 M HCl it is more marked (Table IX-1). More considerable shifts were observed[8] in 90% ethanol containing a barbital buffer of pH 8.5 [(XXIII) to (XXV) and in the more complex systems (XXVI) to (XXIX)], but under these conditions the effect of proton transfer accompanying the electrode process cannot be excluded.

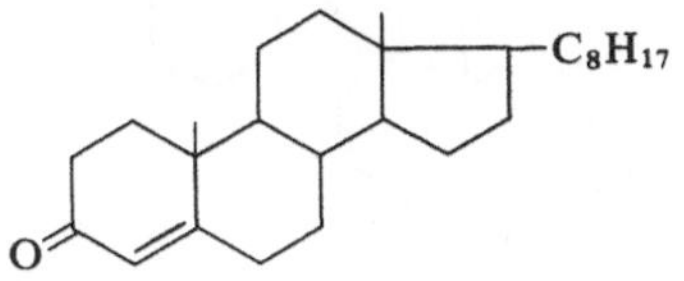

(XXIII) -1.55 V

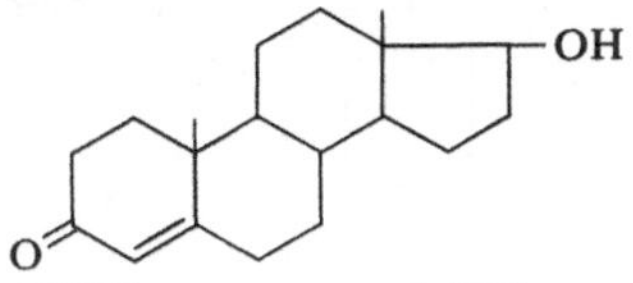

(XXIV) -1.49 V

(XXV)   − 1.45 V

(XXVI)   − 1.48 V and − 1.68 V

(XXVII)   − 1.50 V

(XXVIII)   − 1.49 V and − 1.64 V

(XXIX)   − 1.44 V and − 1.65 V

Even substituents in the side chain can affect the half-wave potentials[16] of 3-keto Δ^1-steroids [(XXX) and (XXXI)] [in dimethyl-formamide, 0.1 M N(C$_4$H$_9$)$_4$I, 0.01 M N(C$_2$H$_5$)$_4$I].

(XXX)   − 1.66 V

(XXXI)   − 1.62 V

Similarly, half-wave potentials of 3-keto $\Delta^{1,4}$- and $\Delta^{4,6}$-steroids also can be affected[15] by substituents in various positions, e.g., in position 6 [(XXXII) to (XXXV) in 50 % ethanol and Britton–Robinson buffer of pH 6.0].

(XXXII)   − 1.22 V

(XXXIII)   − 1.22 V

(XXXIII) -1.06_5 V   (XXXV) -1.09_5 V

and in position 11 [(XXXVI) and (XXXVII), in 50% ethanol].

(XXXVI)   (XXXVII)

malonate, pH 4.6: -1.21 V   malonate, pH 4.6: -1.16 V
triethylamine, pH 10.8: -1.59 V   triethylamine, pH 10.8: -1.52 V

The half-wave potentials not only of keto steroids, but also of their derivatives can be affected by remote substituents. Most available data concern the Girard derivatives, mainly trimethyl-quaternized glycyl-hydrazones.[8] For derivatives of saturated ketones such an effect can be demonstrated for compounds containing a reducible C=N bond in position 20 and varied in ring B [(XXXVIII) and (XXXIX), in 90% ethanol containing barbital buffer of pH 8.5]:

(XXXVIII) -1.51 V   (XXXIX) -1.57 V

where
$$R = NHCOCH_2\overset{+}{N}(CH_3)_3$$

Among steroids containing a double bond in the α,β-position relative to the carbonyl group, effects have been observed[8] for derivatives of 3-keto Δ^4-steroids in the presence of substituents in position 11 [comparison of (XL) with (XLI) and of (XLII) with (XLIII)] and in position 17 [comparison of (XL) with (XLII) and (XLI) with (XLIII)] [in 90% ethanol containing barbital buffer of pH 8.5;

$$R = NHCOCH_2\overset{+}{N}(CH_3)_3]:$$

(XL)   −1.35 V

(XLI)   −1.37 V
                                         −1.70 V

(XLII)   −1.42 V

(XLIII)   −1.34 V

Unlike these compounds, betainylhydrazones (quaternized glycyl-hydrazones) of compounds (XXIII) to (XXV) in 90 % ethanol at pH 8.5 all give their first reduction step at −1.34 V, even though they differ from the above compounds only in the absence of a substituent in position 11. There is no explanation available at present for the fact that the half-wave potential of the betainylhydrazones of (XXIII) to (XXV) are identical, whereas, the half-wave potentials of the parent compounds are significantly different.

In some α-hydroxy ketones the reduction of the C—O bond occurs at more positive potentials than that of the C=O group.[12] The reduction of the C—O bond in position 16 is shifted to considerably more positive values by the aromatization of ring A [compare (XLIV) and (XLV)]. The substitution of hydroxylic hydrogen for acetyl groups results in a shift in the same direction [compare (XLV) with (XLVI), in 90 % ethanol, 0.05 M N(C$_4$H$_9$)$_4$Cl]:

(XLIV)   −2.04 V

(XLV)   −1.96 V

(XLVI)   −2.09 V

A shift in the same direction was observed[12] for (XLVII) and (XLVIII) [90% ethanol, 0.05 M N(C$_4$H$_9$)Cl] for a step in which the reduction waves of C=O and C—O coalesce.

| (XLVII) | −2.38 V | (XLVIII) | −2.35 V |

The effect of remote substituents is clearly seen even from the comparison of (XLIX) and (L) [90% ethanol, 0.05 M N(C$_4$H$_9$)$_4$Cl]:

| (XLIX) | −2.43 V | (L) | −2.48 V |

(the reported values of half-wave potential correspond to the combined wave).

Further data are available for half-wave potentials of α-halo ketones. Thus for 4-bromo-3-keto steroids[13] the effects of substituents in rings C and D are demonstrated by the compounds (LI) and (LII) (80% ethanol, 2% sodium lauryl sulfate, acetate buffer of pH 6.7):

| (LI) | −0.18 V | (LII) | −0.28 V |

The effect of substituents in positions 16 and 17 on the reduction of compounds containing a fluorine group in position 6 is shown by compounds (LIII) to (LVIII) (50% ethanol, Britton–Robinson buffer of pH 6.0):

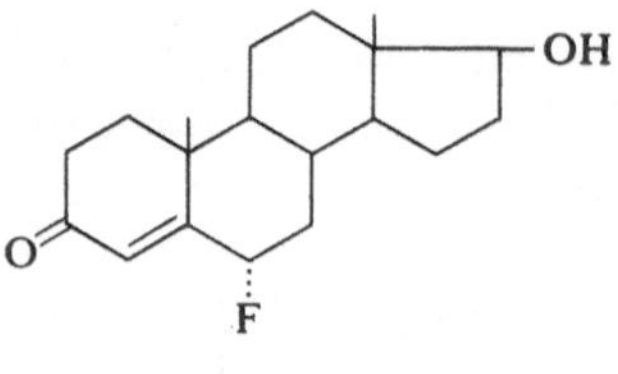

(LIII)   −1.09 V

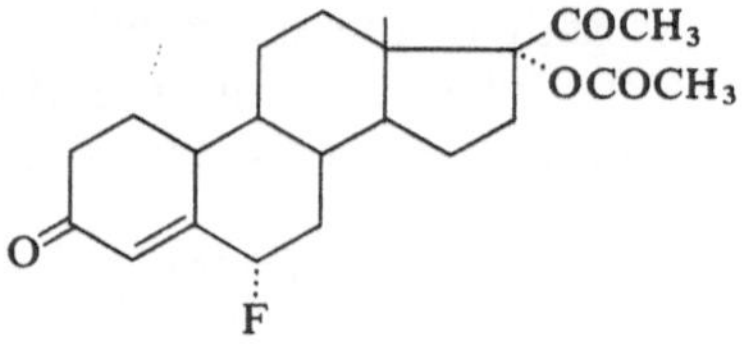

(LIV)   −1.04 V

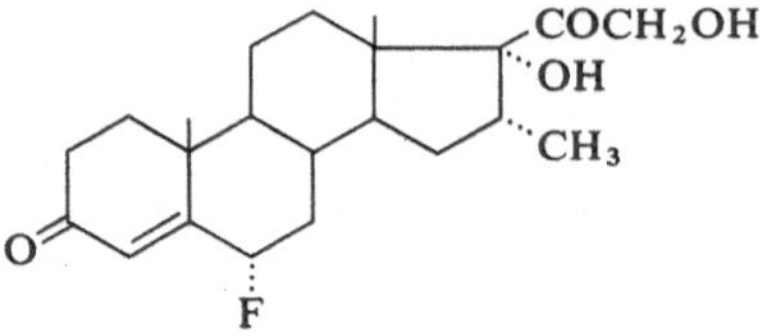

(LV)   −1.06 V

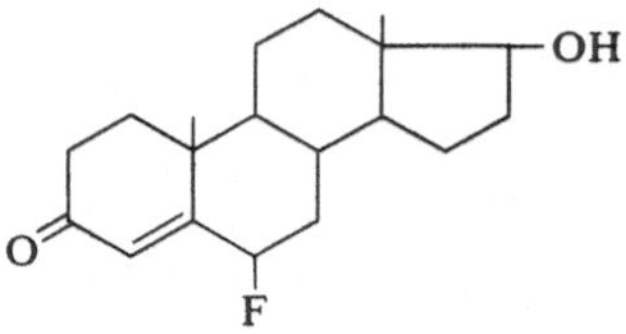

(LVI)   −0.86 V

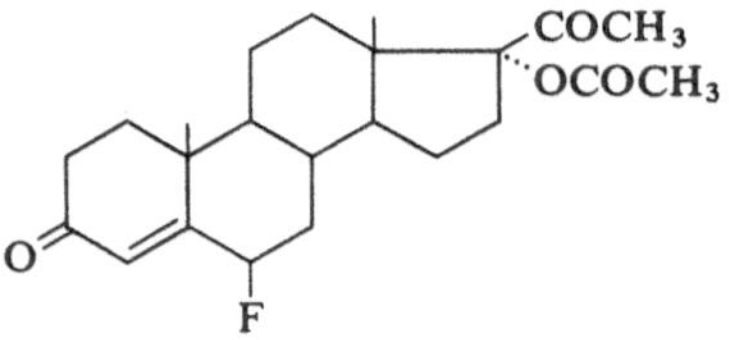

(LVII)   −0.89₅ V

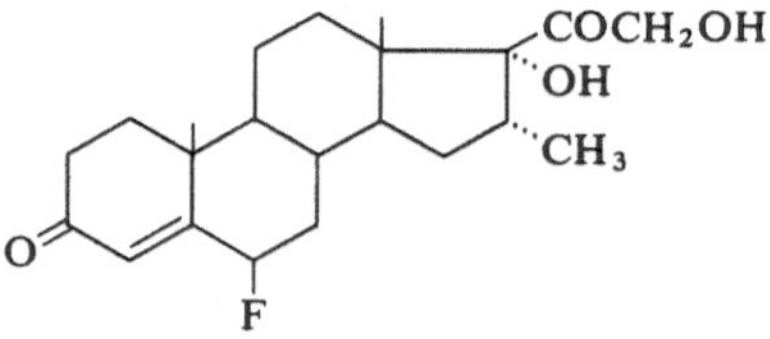

(LVIII)   −0.82 V

This type of compound is also affected[16] by substituents in position 11 [(LIX) and (LX), dimethylformamide, 0.1 M N(C$_4$H$_9$)$_4$I and 0.01 M N(C$_2$H$_5$)$_4$I],

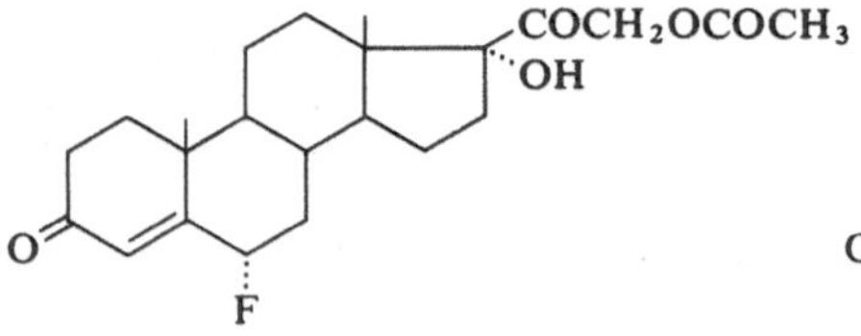

(LIX)   −1.31 V
                 −1.72 V
                 −2.03 V

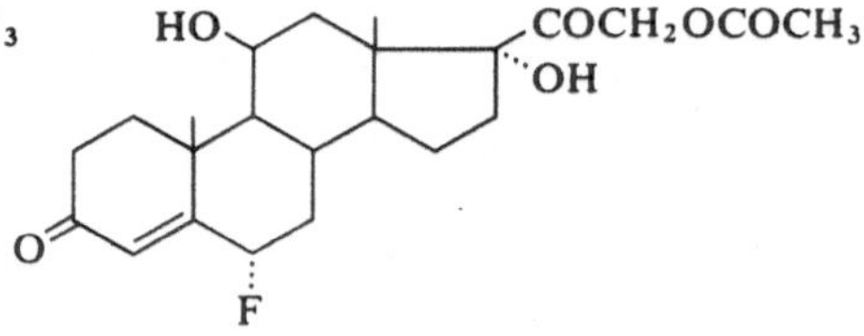

(LX)   −1.28 V
                 −1.72 V
                 −2.06 V

and by substituents in positions 9 and 11 [(LXI) and (LXII) in the same electrolyte]:

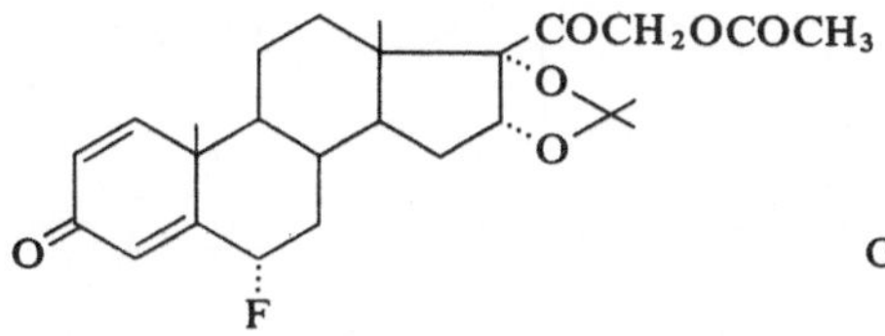

(LXI)   −1.16 V
                         −1.52 V
                         −1.77 V
                         −2.06 V

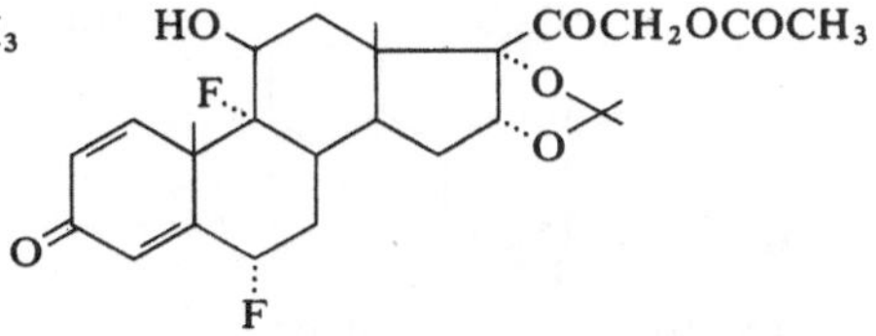

(LXII)   −1.20 V
                         −1.43 V
                         −1.76 V
                         −2.06 V

The half-wave potentials of 17-bromo derivatives are affected by substituents in positions 11 and 3 [(LXIII) to (LXV), 80% ethanol, 2% sodium lauryl sulfate, toluenesulfonic acid, "pH" 2.1)]:

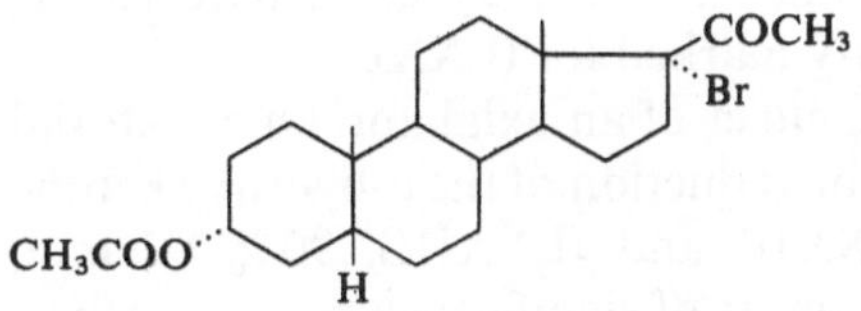

(LXIII)   −0.09 V

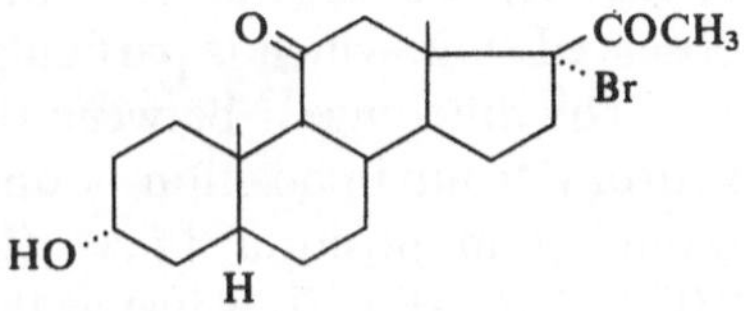

(LXIV)   +0.01 V

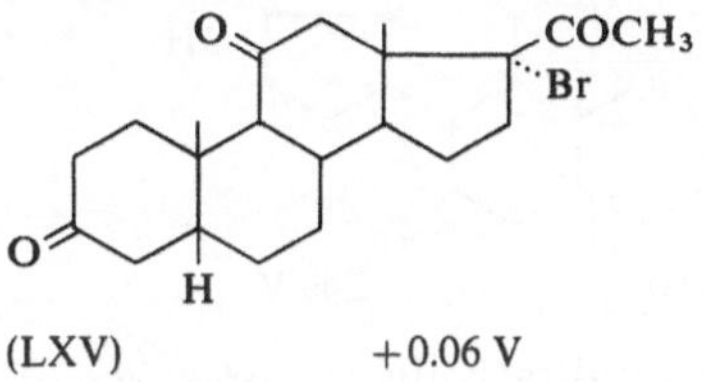

(LXV)   +0.06 V

Even the reduction of bromine in position 21 is affected by substituents in positions 3 and 11 [(LXVI) to (LXXI), 80% ethanol, 2% solution lauryl sulfate, toluenesulfonic acid "pH" 2.1]:

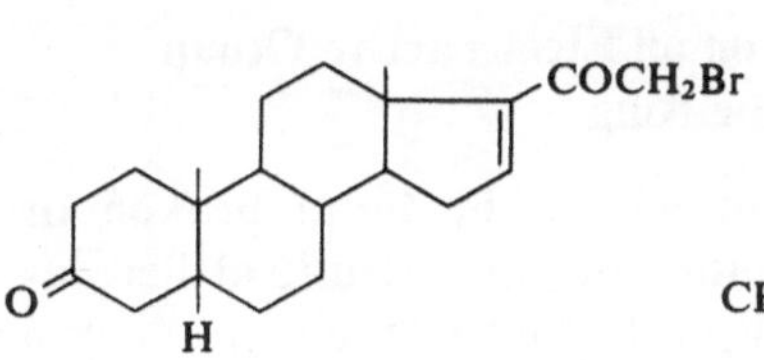

(LXVI)   −0.23 V

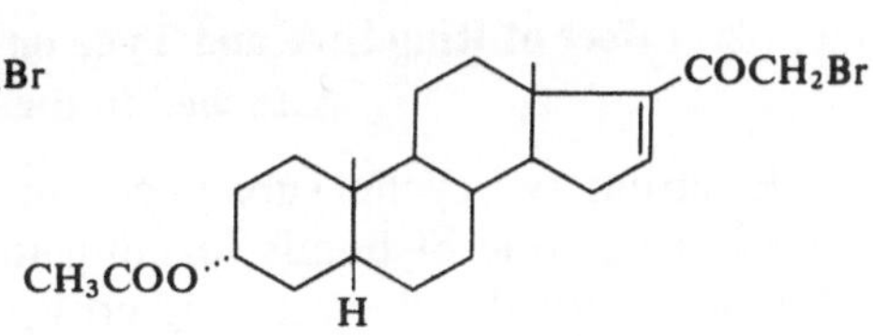

(LXVII)   −0.27 V

(LXVIII)   −0.30 V   (LXIX)   −0.24 V

(LXX)   −0.41 V   (LXXI)   −0.26 V

In the substance (LXX) the hydroxy group in position 16, and in (LXXI) the unsaturation of ring A, also may contribute to the observed shift, which is particularly marked for (LXX).

The difference[12] between the effect of an axial and an equatorial hydroxy group in position 11 on the reduction of the α-hydroxy ketone grouping in position 17,20 [(LXXII) and (LXXIII), 90% ethanol, 0.05 M $N(C_4H_9)_4Cl$] is just at the limits of significance.

(LXXII)   −2.48 V   (LXXIII)   −2.49 V

The examples cited allow us to infer, qualitatively, the presence of long-range effects. More systematic treatment, based on the study of a larger group of substituents for each combination of the positions of the electroactive group and the substituent, seems necessary before any further conclusions can be drawn.

2. Effect of Ring Size and Type on an Electroactive Group Attached to the Ring

Electroactive cyclic substances in which the bond broken or formed is not one of bonds forming the ring are described here as substances with the electroactive group attached to the ring. Such a definition includes cycloalkyl halides and saturated cycloalkanones and their derivatives, including 2-alkylcycloalkanones and similar compounds. The half-wave potentials of these substances can vary according to the ring size and conformation, and for unsymmetrical systems it can also vary according to the position of the electroactive group in the ring. These problems are discussed in the following paragraphs. Because of the multitude of possibilities and the variety of the data, which are not very consistent, the comparison of half-wave potentials corresponding to a given electroactive group in various positions of the steroid ring is discussed separately. For steroid compounds, sufficient related data were available to allow various electroactive groups to be compared.

(a) Effect of Ring Size. When a single electroactive group is attached to an alicyclic ring (as in cycloalkyl bromides†), two types of quantitative treatments are possible.

First, we can consider the ring as a special substituent attached to the electroactive substance. Because the ring-group is directly attached and because no mesomeric interaction is to be expected, the Taft polar substituent constants σ_X^* and equation (29), $\Delta E_{1/2} = \rho_{\pi,R}^* \sigma_X^*$, can be applied. With noncyclic substituents the value of the reaction constant $\rho_{\pi,R}^*$ can be determined first. The half-wave potentials for cyclopentyl and cyclohexyl derivatives (for which σ_X^* values are known) are next compared with the $E_{1/2}$–σ_X^* relation. Departures from a linear plot are to be explained either by steric effects or by a change in the mechanism of the electrode process. For larger rings, it is possible to calculate values of σ_X^* from $\Delta E_{1/2}$ and $\rho_{\pi,R}^*$ and to discuss the magnitude of substituent constants obtained in this way.

Second, the half-wave potentials of a given reaction series can be correlated with either the half-wave potentials of another reaction series (always compounds of the same ring size being compared) or with rate constants (or equilibrium constants) of a homogeneous reaction which compounds of the same reaction series undergo.

Finally, for those cases in which neither quantitative treatment is possible the effects of ring size can be discussed in qualitative terms. It is also possible to compare the shape of the plot of the half-wave potential as a function of the number of ring carbons.

† Another reaction series allowing this kind of treatment would be 2-(piperidinomethyl)-cycloalkanones, but for these substances the half-wave potentials of the corresponding 1-(piperidinomethyl)-2-alkanones ($C_5H_{11}N$-CH_2CO-R, where R = alkyl), are unknown.

For cycloalkyl bromides the first type of treatment involves the complications indicated in Fig. V-5. Whereas, the value of the half-wave potential of cyclohexyl bromide satisfactorily fits the linear relation between the half-wave potentials and the substituent constants σ^* obtained for open-chain compounds, the value for cyclopentyl bromide deviates. A deviation in a similar direction has been observed for isopropyl and t-butyl bromide. It was deduced (p. 184) that a mechanism can participate in the reduction of the last three substances which is different from that operating in the reduction of straight-chain bromides and cyclohexyl bromide.

Because nothing is known about the participation of one or other mechanism in the case of other cycloalkyl bromides, all further deductions involve uncertainty to a high degree. Moreover, the large value of the reaction constant ($\rho_{\pi,Br}^* = 4.5$ V has been accepted as the best value) makes the determination of σ^* values from polarographic measurements insufficiently sensitive to substituent effects. This means that a large shift in half-wave potential results in a relatively small change in the calculated value of the substituent constant σ_X^*.

Bringing the numerous experimental data[17–21] to the same scale and using the value $\rho_{\pi,Br}^* = 4.5$ V, we obtained the values in Table IX-2. These values are of little use in the discussion of the effect of ring size and stereochemistry.

The other possibility, i.e., the correlation of half-wave potentials with logarithms of the rate constants of reactions of cycloalkyl bromides, presents another difficulty. Whereas, for the nucleophilic reactions of branched alkyl halides attention has been paid to the participation of a continuously changing transition state or to various contributions of a canonical structure with a positive charge on carbon to the resonance hybrid in the transition state (commonly referred to as the parallel participation of varying contributions of mechanisms S_N1 and S_N2), no such detailed studies are known for cycloalkyl halides. For these compounds reaction with a given reagent has been assumed to follow either an S_N1 or an S_N2 mechanism (according to the reagent used) for all the cycloalkyl halides studied.

Little correlation has been detected between the half-wave potentials of cycloalkyl bromides and the logarithms of the rate constants of reactions for which an S_N1 mechanism was to be expected, such as the solvolysis of methylcycloalkyl chlorides,[22,23] of cycloalkyl p-toluenesulfonates,[24] and of cycloalkyl chlorides.[24] The greatest deviations were observed for the cyclobutyl derivatives.

Table IX-2

Approximate Values of Polar Substituent Constants for Cyclic Substituents [computed from half-wave potentials[17-21] of cycloalkyl bromides using equation (29) and $\rho_{\pi,Br} = 4.5$ V]

Substituent	σ^*
Cyclopropyl	−0.15
Cyclobutyl	−0.15
Cyclopentyl	−0.20[a]
Cyclohexyl	−0.15[a]
trans-t-Butylcyclohexyl	−0.17
cis-t-Butylcyclohexyl	−0.14
Cycloheptyl	−0.13
Cyclooctyl	−0.13
Cyclononyl	−0.12
Cyclodecyl	−0.13
Cycloundecyl	−0.13
Cyclododecyl	−0.13
Bicyclo[2.2.2]oct-1-yl	−0.18
Adamantyl	−0.16
endo-Bornyl	−0.17
exo-Bornyl	−0.15

[a] Values given by Taft.

On the other hand, half-wave potentials of cycloalkyl bromides have shown relatively good correlation with the rate constants of those reactions to which an S_N2 mechanism has been attributed, such as the exchange reaction of cycloalkyl iodides with radioiodine[25] or the reaction of cycloalkyl bromides with potassium[24,26] or lithium iodide (Fig. IX-2).[27]

Since the participation of various mechanisms cannot be excluded for homogenous reactions, it is not safe to deduce that all cycloalkyl bromides are reduced by an S_N2-like mechanism. It has been shown in fact (p. 184 and cf. Fig. V-5) that the contributions of particular mechanisms differ for cyclopentyl and cyclohexyl derivatives, and it is therefore preferable to state that the participation of the particular mechanisms in the polarographic reduction of cycloalkyl bromides is similar to that found in the homogeneous reactions of these compounds with iodides.

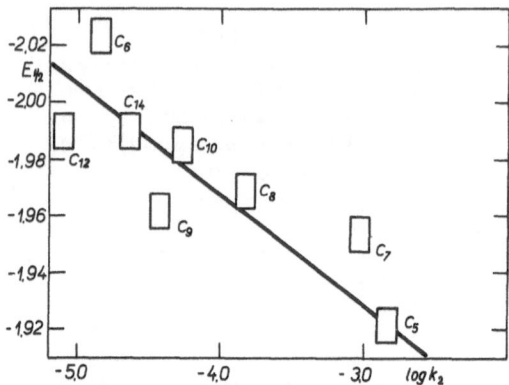

Fig. IX-2. Relation of half-wave potentials of cycloalkyl bromides to logarithms of the rate constants of their reaction with potassium iodide in acetone. Half-wave potentials in 99.97% dimethylformamide containing 0.1 M N(C$_2$H$_5$)$_4$Br from Ref. 18, rate constants at 60°C from Ref. 26.

For some compounds the possibility that a homolytic mechanism is involved cannot be ruled out.[20] On the other hand, it seems improbable that this mechanism predominates throughout the reaction series. First, it seems unlikely that at negative potentials, at which the electrical double layer is negatively charged, the cycloalkyl bromide will be oriented with the negative end of its dipole toward the negatively charged surface of the electrode, as would be expected according to the halogen-bridge theory. Second, no organometallic compounds with mercury are formed[28] during the electrolysis of alkyl halides except in the case of benzyl and allyl halides. Finally, no correlation of the half-wave potentials of cycloalkyl bromides with kinetic data on reactions involving the homolytic cleavage of the C—Br bond has been found.[18]

There is also the possibility of correlating the half-wave potentials of cycloalkyl bromides with the half-wave potentials of other alicyclic compounds using equation (49). Satisfactory correlation has been found for cycloalkanones,[12] for their betainylhydrazones,[29] and for 2-(piperidinomethyl)cycloalkanones[30] (Fig. IX-3). For cycloalkanones the cyclopropanone, and for betainylhydrazones, the cyclopentanone derivative deviated. For 2-(piperidinomethyl)cycloalkanones no correlation was found for cyclooctanone, cyclononanone, and cyclodecanone derivatives, for which the shape of the waves[30] indicate a change in the mechanism of the reduction process. For cycloalkyl bromides, as well for cycloalkanones, the cyclopentyl derivative is reduced at more positive potentials than the cyclohexyl derivative.

On the contrary, for cycloalkanone betainylhydrazones and for Mannich bases the reverse relation is observed, namely, the reduction of the cyclohexyl derivative occurs at more positive potentials than that of the cyclopentyl derivative. For this fact, which is expressed by the different slope of the plot 1 from that of plots 2 and 3, no explanation is known at present.

A very good correlation has been observed (Fig. IX-4) between half-wave potentials of 2-(piperidinomethyl)cycloalkanones and the

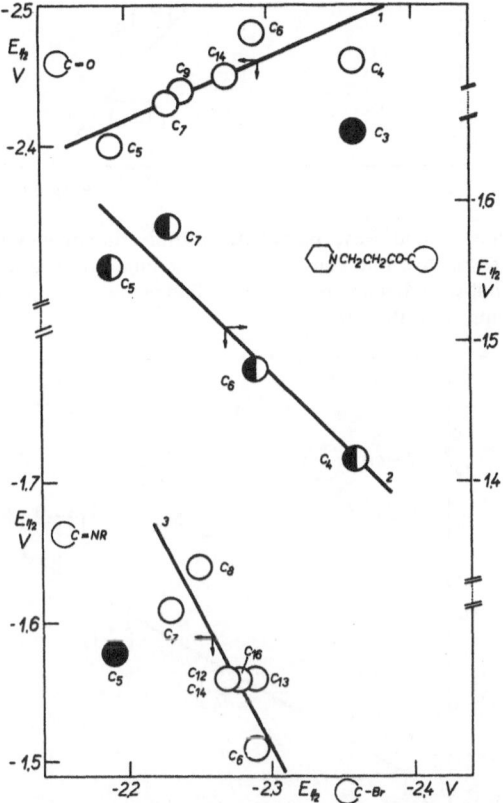

Fig. IX-3. Relation of half-wave potentials of (1) cycloalkanones, (3) their betainylhydrazones, and (2) 2-(piperidinomethyl)cycloalkanones to half-wave potentials of cycloalkyl bromides. Half-wave potentials of cycloalkanones (O) from Ref. 12, of 2-(piperidinomethyl)cycloalkanones (◑) from Ref. 30, of betainylhydrazones of cycloalkanones (O) from Ref. 29, and of cycloalkyl bromides from Ref. 17, 18. Full points deviate.

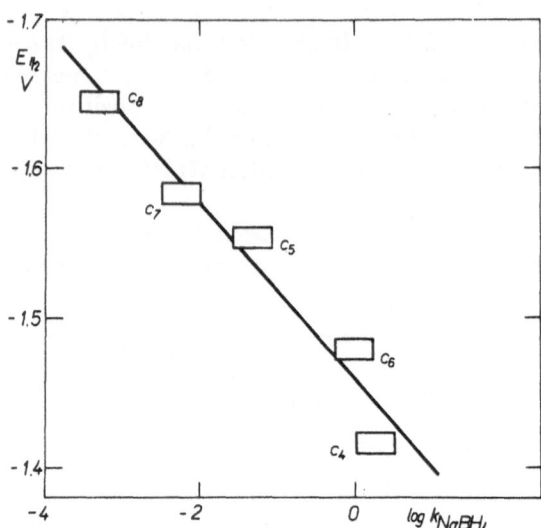

Fig. IX-4. Relation of half-wave potentials of 2-(piperidinomethyl)cycloalkanones to logarithms of the relative rate constants of the sodium borohydride reduction of cycloalkanones. Half-wave potentials of 2-(piperidinomethyl)cycloalkanones at pH 8, and also rate constants, from Ref. 30.

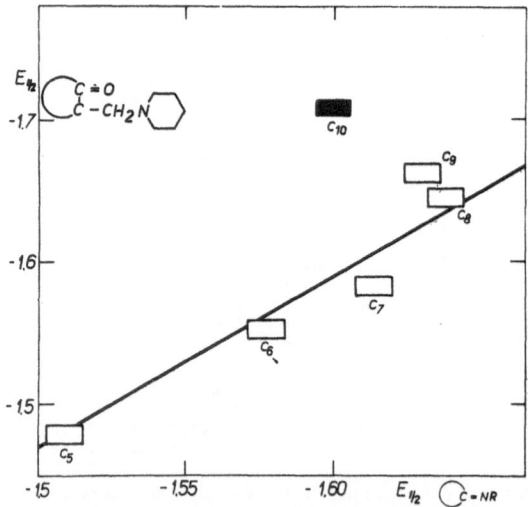

Fig. IX-5. Relation of half-wave potentials of 2-(piperidinomethyl)cycloalkanones to half-wave potentials of cycloalkanone betainylhydrazones. Half-wave potentials of 2-(piperidinomethyl)cycloalkanones at pH 8 from Ref. 30, those of betainylhydrazones from Ref. 29.

logarithms of the rate constants of sodium borohydride reductions[30] of the corresponding unsubstituted cycloalkanones. This is evidence that nucleophilic attack on the carbonyl group occurs in both cases. On the contrary, attempts to correlate the half-wave potentials with logarithms of rate constants of elimination reactions of these Mannich bases failed. This can be taken as a proof that in the elimination process not the electron density on the carbonyl group, but the proton transfer from the α-carbon to nitrogen is essential in the transition state.

The possibility of correlating half-wave potentials of 2-(piperidinomethyl)cycloalkanones[30] with half-wave potentials of cycloalkyl bromides was mentioned earlier (Fig. IX-3). Similarly, equation (49)

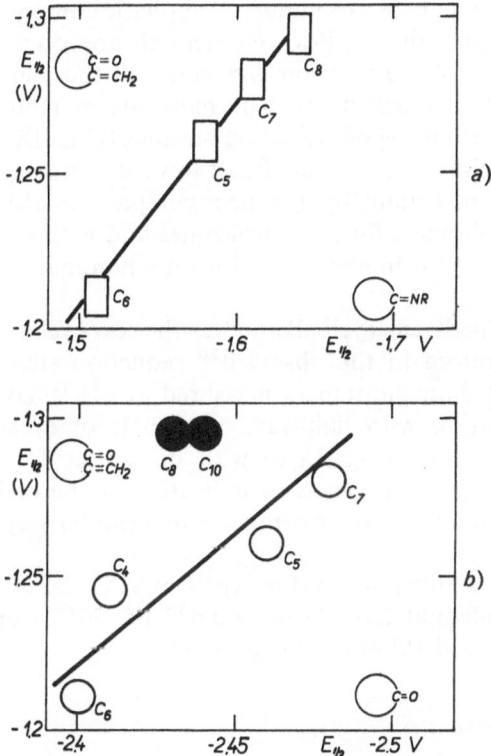

Fig. IX-6. Relation of half-wave potentials of 2-methylenecycloalkanones to half-wave potentials of a) cycloalkanone betainylhydrazones, and b) cycloalkanones. Half-wave potentials of 2-methylenecycloalkanones at pH 9 from Ref. 30, those of betainylhydrazones from Ref. 29, and those of cycloalkanones from Ref. 12. Full points deviate.

can be used for the correlation of half-wave potentials of 2-(piperidinomethyl)cycloalkanones with half-wave potentials of betainylhydrazones of cycloalkanones[29] (Fig. IX-5). Good correlation is observed even for cyclooctane and cyclononane Mannich bases, which deviate in polarographic behavior from the lower members of the series. Only for the cyclodecane derivative is this change in reduction mechanism reflected by deviation from the linear relation.

A similar situation exists for cycloalkanones reduced[12] at about -2.4 V in 90% ethanol containing 0.05 M $N(C_4H_9)_4Cl$. The correlation of their half-wave potentials with those of cycloalkyl bromides is given in Fig. IX-3.

The situation is similar again for betainylhydrazones of cycloalkanones,[29] and here the number of aliphatic compounds studied and the differences in their half-wave potentials are too small to allow the determination of the reaction constant and the computation of apparent substituent constants. The half-wave potentials have shown correlation with those of cycloalkyl bromides (Fig. IX-3) and Mannich bases (Fig. IX-5). The shift of 70 mV toward more negative values for cyclopentanone betainylhydrazone is comparable with the values[31] of 80–120 mV obtained for semicarbazones and with the shifts of 20–50 mV observed[32] for imines derived from ammonia, methylamine, and glycine.

For 2-methylenecycloalkanones the exocyclic C=C bond is probably reduced in the observed[30] reduction step. The half-wave potentials of these substances measured at pH 9 have shown rather good correlation with half-wave potentials of cycloalkanones and betainylhydrazones (Fig. IX-6) with deviations for eight-and-nine-membered rings. This seems to indicate that the carbonyl group in cycloalkanones is affected by ring size in a similar way to an exocyclic C=C bond.

For cyclic diketones of the type (LXXIV) the following shifts in half-wave potential have been found[33] (in 50% isopropyl alcohol, 0.1 M acetic acid, 0.1 M sodium acetate)

	n	$E_{1/2}$, V
	$n = 2$	-0.75
	3	-1.16
	4	not reducible
	14	-1.25

and compared with values for aliphatic diketones:

$$RCOCOR \qquad R=CH_3 \qquad -0.81$$
$$C_3H_7 \qquad -0.83$$
$$CH(CH_3)_2 \quad -0.87$$
$$C(CH_3)_3 \quad -1.27$$

The reduction of the cyclic diketones is assumed to take place in the *cis* form, the nonreducibility of 3,3,8,8-tetramethyl-1,2-cyclo-octanedione being a result of steric hindrance to the coplanarity of the COCO grouping and perhaps other effects of the α-methyl groups. For $n = 2$ the reduction occurs in a similar way to that of sterically unhindered aliphatic diketones, but with increasing n steric effects analogous to those operating in 2,5-dimethyl-3,4-hexanedione play an increasingly important role.

(b) Position of the Reducible Group in the Steroid Ring. Numerous examples of the effect of the position of the electroactive grouping on half-wave potential have been reported for α-halo-keto steroids.[13,14] Some examples are summarized in Tables IX-3–5. More numerous data would be necessary for each structure for general conclusions to be possible.

The mutual positions[12] of two [(LXXV) to (LXXVII)]

| (LXXV) | -2.38 V | (LXXVI) | -2.44 V |

(LXXVII) -2.46 V

or three [(LXXVIII) and (LXXIX)] simultaneously reducible

(LXXVIII)   −2.37 V   (LXXIX)   −2.41 V

carbonyl groups also affect the half-wave potential.

Also, the reducibility of an aldehydic group depends on its position in the steroid ring. Thus, substances of the type (LXXX), such as cymarin or convallotoxin are reducible[34] at −2.2 to −2.3 V, which is similar to the behavior of saturated terpene aldehydes.[35] On the other hand, compounds of the holarrhimine type (LXXXI) are reduced[36] in one pH-independent wave at −1.65 V. As regards the chemical inactivity of (LXXXI), it is assumed that even the hydration of the aldehydic group is sterically hindered. The more positive reduction process has been attributed[36] to the reduction of the unhydrated carbonyl group.

(LXXX)   (LXXXI)

There is also a difference between the reduction[3] of 8-oxocamphor (LXXXII) and 10-oxocamphor (LXXXIII) (pH 6.8)

(LXXXII)   −1.73 V   (LXXXIII)   −1.70 V

(c) Comparison of Halogens. In practically all cases reported so far (Table IX-6), the half-wave potentials of α-halo ketones are increasingly negative in the sequence I < Br < Cl < F. This sequence would be in agreement both with a mechanism involving nucleophilic attack and with change in deformability, which would explain the change in reactivity in terms of a bridge mechanism. Hence, the observation of this sequence cannot be used as an argument for or against one of these possibilities.

Table IX-3
Comparison of Half-Wave Potentials α-Bromo-Keto-Steroids

Substance, $E_{1/2}$ (V)	Br	CO
-0.06 V[14]	2α(e)	3
-0.14 V[13]		
-0.28 V[13]		
-0.04 V[13]	4α(e)	3
-0.33 V[14]	6α(e)	7
-0.05 V[14]		3

Table IX-3 (*continued*)

Substance, $E_{1/2}$ (V)	Br	CO
-0.08 V[14]	$11\alpha(e)$	12
-0.16 V[14]	$16\alpha(b)$	17
-0.08 V[14]	$17\alpha(p\text{-}a)$	20
-0.09 V[13]		
-0.21 V[14]	21	20

Table IX-4
Comparison of Half-Wave Potentials of β-Bromo-Keto Steroids

Substance, $E_{1/2}$ (V)	Br	CO
COCH₂OCOCH₃ / OH −0.45 V[13]	2β(a)	3
COCH₂OCOCH₃ / OH −0.19 V[13]	4β(e)	3
C₈H₁₇ −0.14 V[14]	6β(a)	7
−0.02 V[14]		3
COOCH₃ −0.07 V[14]	11β(a)	12
−0.28 V[14]	16β(b)	17

Table IX-4 (*continued*)

Substance, $E_{1/2}$ (V)	Br	CO
−0.21 V[14]	21	20
−0.24 V[13]		
−0.23 V[13]		

Table IX-5
Comparison of Half-Wave Potentials of α-Fluoro and α-Chloro-Keto Steroids

Substance, $E_{1/2}$ V	$E_{1/2}$ V	Halogen	CO	Ref.
	−2.23	2α-F(*e*)	3	14
	−1.46[a]	6α-F(*e*)	3	14

[a] Probably reduction of the COCH=CH system.

Table IX-5 (*continued*)

Substance, $E_{1/2}$ V	$E_{1/2}$ V	Halogen	CO	Ref.
COCH$_3$	-2.15	17α-F(p-a)	20	14
C$_8$H$_{17}$	-0.94	2α-Cl(e)	3	14
COCH$_3$ OCOCH$_3$	-0.94	6α-Cl(e)	3	15
COCH$_3$ OCOCH$_3$	-0.96_5	6α-Cl(e)	3	15
COCH$_3$ Cl	-0.93	17α-Cl(p-a)	20	14
COCH$_2$Cl	-1.25	21	20	14

Table IX-6
Comparison of the Reducibilities of Various Halogens

Type	Substituent	X	$E_{1/2}$		Ref
	$2\alpha(e)$	I	−0.02 V	−0.26	14
		Br	−0.06 V		
		Cl	−0.94 V		
		F	−2.23 V		
	$6\alpha(e)$ $R^1, R^2 = O$ $R^1 = COCH_3 \quad R^2 = OH$ $R^1 = COCH_3 \quad R^2 = H$	Br	−0.05 V		14
		Cl	−1.12 V		
		F	−1.46 V		
	$6\alpha(e)$	Cl	−0.94 V		15
		F	−1.04 V		
	$6\beta(a)$ $R^1, R^2 = O$ $R^1 = COCH_3 \quad R^2 = OH$ $R^1 = COCH_3 \quad R^2 = H$	Br	−0.02 V		14
		Cl	−0.74 V		
		F	−1.18 V	−1.48 V[a]	

	X			
15	Cl	> -0.7 V		$6\beta(a)$
	F	-0.895 V		
14	Br	-1.18 V	-1.48 V[a]	4
	Cl	-1.19 V	-1.50 V[a]	
16	Cl	-1.59 V	-1.72 V	$16\beta(b)$
	Br	-1.23 V	-1.71 V	
14	Br	-0.08 V[b]		$17\alpha(p\text{-}a)$
	Cl	-0.93 V[c]		
	F	-2.15 V[d]		

Structures (left to right): COCH$_3$ / OCOCH$_3$; COCH$_2$OCOCH$_3$ / OH; (ketone with X); COCH$_3$ / X with CH$_3$COO.

[a] The possibility of the reduction of the system COCH=CH was not excluded.
[b] Without the 11-carbonyl group.
[c] With a β-hydroxy group in position 3 instead of α-acetoxy.
[d] With an α-hydroxy group in position 3 instead of α-acetoxy.

Table IX-6
Comparison of the Reducibilities of Various Halogens

Type	Substituent	X	$E_{1/2}$	Ref.
	21	I Br Cl	−0.22 V −0.21 V −1.25 V	14

(d) Stereochemical Factors. In addition to the effect of ring size and the position of the electroactive group in the steroid ring, both of which involve stereochemical factors even when these are not always clearly understood, some other stereochemical factors affecting half-wave potentials will be discussed in this chapter.

Reduction of α-halo ketones will be discussed first because the most extensive material is available for this group of compounds.

Among simpler compounds, 2-halo-4-*t*-butylcyclohexanones have been studied in some detail.[37] For these substances it is assumed that the bulky *t*-butyl group takes up an equatorial position almost exclusively. Moreover, it is taken for granted that *cis* and *trans* isomers are almost conformationally homogeneous, *cis* halogen being equatorial (*e*) and *trans* halogen axial (*a*):

(*e*)   (*a*)

In both pairs of 2-halo-4-*t*-butylcyclohexanones compared (Table IX-7), the thermodynamically more stable equatorial halogen is reduced at more negative potentials than axial halogen.

Similarly, in the rigid system of the B and C rings of steroids equatorial halogen is again reduced at more negative potentials than axial halogen (Table IX-8). The difference in the half-wave potentials of equatorial and axial halogen in α-halo ketones varies according to the position of the halogen in the ring system and to the position of the vicinal carbonyl (Tables IX-7 and -8). The differences depend also on the nature of the halogen involved, increasing in the sequence Br < Cl < F (Tables IX-7 and -8). This sequence parallels the observation made for benzene derivatives, viz., that the more negative the reduction potential the greater the susceptibility to substituent effects.

The observed differences in the half-wave potentials of equatorial and axial halogen have been used also in the discussion of the reduction mechanism of vicinal halo ketones. Whereas, for alkyl halides nucleophilic substitution and bridging mechanisms have been principally considered (p. 184), for α-halo ketones an elimination process has been taken into consideration[13,37] in which an enolate is formed in the electrode process (this is subsequently transformed into the ketone in a reaction consecutive to the electrode process proper).

Table IX-7
Reduction of 2-Halocyclohexanones[37] in Dimethylformamide with
0.1 M N(C$_4$H$_9$)$_4$Br

Substituent	$E_{1/2}$, V			$\Delta(e - a)$
	cis (*e*)	Flexible[a]	*trans* (*a*)	
2-Chloro-4-*t*-butyl	−1.57		−1.43	−0.14
2-Chloro		−1.40		
trans-2-Chloro-5-methyl		−1.45$_5$		
2-Fluoro-4-*t*-butyl	−2.08		−1.85	−0.23
2-Fluoro		−2.00		

[a] The position in this column should indicate whether the conformation approached was axial or equatorial.

An important experimental result which contributed to the proof of the formation of an enolate as intermediate (considered also in other reductions of carbonyl compounds[38,39]) was obtained in a comparison of the wave shape with the results of controlled-potential electrolysis of 2-halo-3-keto Δ^4-steroids.[13] For these compounds electrolysis at the limiting current of the first, two-electron, step yielded the 3-keto Δ^4-steroid. It was confirmed that an equivalent amount of halide ions was generated simultaneously. On the other hand, on polarographic curves of 2-bromo-3-keto Δ^4-steroids only the wave corresponding to the reduction of the C—Br bond was observed, and there was no wave due to the corresponding 3-keto Δ^4-steroid. This occurred even in media in which the wave for the 3-keto Δ^4-steroid was well developed when a sample of the particular steroid was added to the supporting electrolyte used. This can be explained by the formation of an intermediate at the dropping mercury electrode which is subsequently transformed into the 3-keto Δ^4-steroid. An enolate structure is attributed to this intermediate, and its formation can be interpreted principally in two ways: as an attack of the electron on the carbon of

$$\text{(I)}$$

$$\text{C=O} \quad \cdots\text{C}-\text{O}^{(-)} \quad \text{C}-\text{O}^{(-)}$$
$$\quad\quad R \quad\quad\quad\quad R \quad\quad\quad\quad R$$
$$\backslash \text{CH}_2-\text{X} \quad \backslash \text{CH}_2-\text{X} \quad \text{CH}_2 + \text{X}^{(-)}$$

(II)

$$\underset{\text{CH}_2}{\overset{R}{\text{C}-\text{OH}}} \rightleftharpoons \text{H}^{(+)} + \underset{\text{CH}_2}{\overset{R}{\text{C}-\text{O}^{(-)}}} \longleftrightarrow \underset{\text{CH}_2}{\overset{R}{\text{C}=\text{O}}} + \text{H}^{(+)} \rightleftharpoons \underset{\text{CH}_3}{\overset{R}{\text{C}=\text{O}}}$$

the C—Br bond (I), or as an attack of the electron on the carbonyl carbon (II). In the first case, it must be assumed that the carbanion formed in the electrode process can accept hydrogen more readily on its carbonyl oxygen than on the carbon from which the halogen left.

To follow mechanism (I) the carbon atom carrying bromine would have to pass through a trigonal form. It is assumed that for vicinal halo keto steroids with the halogen in ring B or C the rigidity and shape of the molecule prevent attack from the rear.

Hence, mechanism (II) is preferred. Both C—X and C=O are expected[13] to be oriented in the electric field of the electrode in the way depicted in Fig. IX-7 with the double bond of the carbonyl group parallel with the surface of the electrode. Simultaneously, the ring attains a pseudochair form. For axial halogen such deformation is only slight, whereas, for equatorial halogen the carbon carrying the halogen is separated by an axial hydrogen from the surface of the electrode. To attain the most preferable configuration relative to the electrode, additional energy is necessary. This explains why equatorially bound bromine is reduced at more negative potentials than axial bromine (Tables IX-7 and -8).

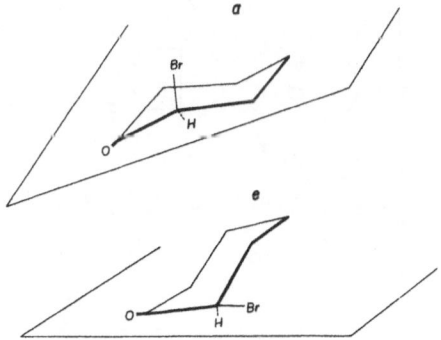

Fig. IX-7. Scheme of the orientation of 2-bromo 3-keto steroids at the electrode surface.

Table IX-8
Comparison of Vicinal Halo Keto Steroids with Axial and Equatorial Halogens

Compound	$E_{1/2}$, V			X	Position of CO	Ref.
	$\alpha(e)$	$\beta(a)$	$\Delta(\alpha - \beta)$			
	−0.33	−0.14	−0.19	6-Br	7	14
	−0.08	−0.07	−0.01	11-Br	12	14
	−0.05	−0.02	−0.03	6-Br	3	14

Structure					
-1.12	-0.74	-0.38	6-Cl	3	14
-1.46	-1.18	-0.28	6-F	3	14
-1.09	-0.86	-0.23	6-F	3	15
-1.04	-0.89_5	-0.14_5	6-F	3	15

Table IX-8 (*continued*)

Compound	$E_{1/2}$, V			X	Position of CO	Ref.
	$\alpha(e)$	$\beta(a)$	$\Delta(\alpha - \beta)$			
	−1.06	−0.82	−0.24	6-F	3	15
	−1.31[a]	−1.18[a]	−0.13	6-F	3	16
	$\alpha(b)$	$\beta(b)$	$\Delta(\alpha - \beta)$			
	−0.16	−0.28	+0.08	16-Br	17	14

[a] First wave.

For flexible molecules, such as cyclohexyl derivatives without a
t-butyl group, an equilibrium of the type $(e) \rightleftharpoons (a)$ is to be considered

If the equilibrium $(e) \rightleftharpoons (a)$ were mobile in comparison with the
rate of the electrode process, only one wave would be observed on
polarographic curves at potentials at which the reduction of the more
positively reducible form (a) occurs.

If the rate of the establishment of the equilibrium $(e) \rightleftharpoons (a)$ were
comparable with or slower than the rate of the electrode process, two
waves of kinetic (or diffusion) character would be observed. One of
these would lie at potentials characteristic for the axial bond in (a), and
the other, the more negative wave, would lie at potentials characteristic
for the equatorial bond in (e). The observation of two waves was not
reported.[37] This means that either the equilibrium is established
rapidly in comparison with the electrode process and the observed
wave corresponds to the reduction of the axial bond, or the equilibrium
is not established rapidly and the impossibility of distinguishing two
waves arises from the fact that these two waves are not sufficiently
separated. The differences in half-wave potentials of -0.14 V for
chlorine and -0.23 V for fluorine (Table IX-7) would be sufficient for
the development of well separated waves if it were not for the high
irreversibility of the waves measured ($RT/\alpha n$F ranges from 0.12 to
0.23 V). With such drawn-out waves the separation of two waves
differing by less than 0.3 or 0.4 V cannot be expected.

In such a case of two merged waves, the measured effective "half-
wave potential" would be affected by the ratio of the concentrations
of the two forms. Because the observed values for flexible molecules
(Table IX-7) are nearer either to the observed values for the reduction
of an equatorial bond or to those for the reduction of an axial bond, it
can be assumed that for these flexible compounds equilibrium is not
established rapidly, and the two waves merge. Quantitative deductions
concerning the position of the equilibrium will be possible only when
the role of the polar effect of the alkyl (t-butyl, methyl) group introduced can be expressed quantitatively. At present, it is impossible to
distinguish which part of the observed shift (when one of the t-butyl

conformers is compared with the 2-halocyclohexanone) is due to the polar effect of the $C(CH_3)_3$ group and which to the equilibrium involved.

It is now possible to summarize the role that polarography can play in the distinguishing of epimers.

For truly rigid systems, such as the B and C rings of the steroid system, it seems possible from polarographic data to distinguish between the epimers of an α-halo ketone if both epimers are available and their half-wave potentials can be measured. In this respect, polarography is superior to standard methods, using ultraviolet spectra or optical rotatory dispersion for axial and infrared spectra for equatorial halogen. Each of these methods determines either axial or equatorial halogen—not both, as polarography does. Moreover, the above-mentioned methods depend principally on the effect exerted by the halogen on the carbonyl group, whereas, the polarographic method depends primarily on the halogen itself.[14]

Moreover, polarography enables even the bisectional bonds existing in 16α- and 16β-halogen derivatives to be distinguished.[14] Spectral methods do not, for the wavelengths of the two isomers are identical.

The data given in Table IX-8 show that the regions of potentials overlap at which α-halo ketones with the groups in various positions in the steroid ring are reduced. Hence, the possibility of the determination of conformation when only one epimer is available seems questionable. The data available are too limited to allow a final judgment.

For flexible molecules the situation is more complicated. Here, quantitative data about the equilibria of the two conformers can be expected only from measurements of the ratio of wave heights in cases of sluggishly established equilibria for which two separate waves are observed. To draw at least qualitative deductions from the position of the effective half-wave potential of the merged wave, it would be necessary to possess information about other factors affecting the half-wave potential in the transition from the flexible systems studied to the chosen rigid model system. For example, for rigid systems containing t-butylcyclohexyl, it would be necessary first to determine the effect of other alkyl groups in order to separate the contribution due to the polar effect of the t-butyl group from the observed shift for t-butyl-cyclohexyl.

For α-dichlorocyclohexanones[40] the half-wave potentials in water–dioxane mixtures are in the sequence:

$$gem < cis < trans$$

$$ae \quad aa \rightleftharpoons ee \quad ae$$

In view of the results obtained for monohalo ketones, the more positive reduction of the *cis* form would be more readily explained by pre-dominance of (*aa*) than of (*ee*). In dimethylformamide the waves of the *cis* and *trans* forms do not differ significantly.

In general, the reduction of a bond at position 16α occurs at a more positive potential than that of a bond at position 16β. Examples are the 16-C—Br bonds in Table IX-8 and the 16-C—O bonds in (LXXXIV) and (LXXXV) [(in 80% ethanol containing[12] 0.05 M N $(C_4H_9)_4Cl)$].

(LXXXIV) α: −1.99 V

(LXXXV) β: −2.04 V

The relative positions of the A and B rings are of considerable importance for the influence on half-wave potentials (Table IX-9).

The reduction of some bridgehead bromides, such as 1-bromo-bicyclo[2.2.2]octane, *endo*-1-bromobicyclo[3.2.1]octane, and 1-bromo-adamantane (Table IX-10) occurs[19-21] at potentials of about 0.2 V to 0.1 V (according to the medium used) more negative than the reduction of straight-chain bromides. This difference has been inter-preted as a result of (*i*) change in mechanism, or (*ii*) steric strain. The bridgehead molecules are assumed not to be accessible to a back-side attack on carbon. This type of attack, characteristic for an S_N2-like reduction mechanism is thought to predominate for straight-chain bromides (*cf.* Ref. 21 and p. 184). A front-side attack on bromine[20] or an S_N1-like mechanism[21] is suggested.

The more positive reduction of *exo*-norbornyl bromide than that of *endo*-norbornyl bromide[21] is quoted in support of the S_N1-like mech-anism. It is considered that ionization in the field of the electrode with formation of carbonium ions will occur more readily in *exo* norbornyl bromide than in *endo*-norbornyl bromide because of anchimeric aid in the *exo* bromide. Further confirmation of the S_N-like mechanism of the reduction of bridgehead bromides may be provided by the fact that the reduction of *cis*-1-bromo-4-*t*-butylcyclohexane occurs at

Table IX-9
Effect of the Relative Positions of Rings A and B in Steroids on Half-Wave Potentials

Substance	$\Delta(V)$	Ref.
-1.34 V[a]		
-1.44 V[a]	0.10	8
-0.04 V $\quad \alpha n = 0.52$[c]		
-0.19 V $\quad \alpha n = 0.35$[c]	0.15	13
-0.14 V $\quad \alpha n = 0.54$[c]		
-0.45 V $\quad \alpha n = 0.27$[c]	0.31[b]	13

[a] $R = NHCOCH_2N(CH_3)_3Cl^-$.

[b] This value will involve also contributions from shifts resulting from other changes in the molecule.

[c] The difference in the values of αn is rather too great for a direct comparison of half-wave potentials to be made.

Table IX-10
Reduction of Bridgehead Bromides

Type	95% dimethylformamide; 0.03 M N(CH$_3$)$_4$Br[19]			Dimethylformamide; 0.01 M N(C$_2$H$_5$)$_4$Br[21]		Dimethylformamide; 0.03 M N(C$_2$H$_5$)$_4$Br[20]	
	Substance	$\frac{2.3RT}{anF}$	$E_{1/2}$[a] vs. SCE	Substance	$E_{1/2}$ vs. SCE	Substance	$E_{1/2}$ vs. Ag/AgBr
Aliphatic	Butyl bromide	0.254	−1.90	Butyl bromide	−2.23	Butyl bromide	−1.77
	Isopropyl bromide	0.200	−1.91	Isopropyl bromide	−2.26	Octyl bromide	−1.81
Cyclic	Cyclohexyl bromide	0.231	−1.91	trans-4-bromo-t-cyclohexane	−2.45	—	
	Cyclodecyl bromide	0.225	−1.92	cis-4-bromo-t-cyclohexane	−2.32	—	
Bridgehead	1-Bromobicyclo[2.2.2]octane	0.150	−1.99	1-Bromobicyclo[2.2.2]-octane	−2.48	1-Bromobicyclo[2.2.2]octane	−1.79
	endo-1-bromobicyclo[3.2.1]octane	0.187	−2.15	endo-norbornyl bromide	−2.43	1-Bromobicyclo[2.2.1]heptane	−2.17
	endo-1-bromobicyclo[3.2.1]oct-2-ene	0.183	−2.16	exo-norbornyl bromide	−2.34		
	1-Bromoadamantane	0.210	−2.02	1-Bromoadamantane	−2.38		

[a] Corrected against half-wave potential of potassium, $E_{1/2} = -1.86$ V.

more positive potentials than that of its *trans* isomer. Because of the steric conditions, the back side of the carbon attached to the equatorial bromine in *trans*-1-bromo-4-*t*-butylcyclohexane is almost as hindered as in a bridgehead bromide. Hence, an S_N2-like mechanism cannot occur, and the negative reduction wave is ascribed predominantly to an S_N1-like mechanism. The far less hindered *cis*-1-bromo-4-*t*-butyl-cyclohexane can be attacked from the rear. A predominantly S_N2-like mechanism is in agreement with the more positive potential, which is almost in the range of the reduction potentials of aliphatic bromides.[21]

That steric strain is the main reason for the more negative potential of 1-bromobicyclo [2.2.2] octane was corroborated by comparison with 1-bromoadamantane. The reduction of the latter occurs at potentials which are about 0.1 V more positive† than those of the bicyclooctane derivative. It is assumed that in the transition state adamantyl is less strained than bicyclooctyl.[20]

The fact that the half-wave potential of 1-bromobicyclo [2.2.2] octane (see Table IX-10 and Ref. 20) also can be in the region in which the reduction of alicyclic bromides has been observed (Table IX-10) makes the evidence given above less conclusive. A compensation of various factors cannot be ruled out. It is clear that further studies are needed in this field, not only on more critically selected model compounds, but also on the basis of more detailed experimental data (dependence of half-wave potentials and wave shapes on the concentrations of the depolarizer and the supporting electrolyte, on the composition of the solvent, on the kind of the cation used, on the temperature, etc.) so that we may reach more dependable conclusions.

3. Effect of Ring Size and Type on Electroactive Groups that Form Part of the Cyclic System

In this section, systems will be discussed in which during the electrode process either a bond in the ring system is reductively hydrogenated or a new bond is formed in the ring system. Both of these processes can be affected by the ring size, by the position of the electroactive group or the group that activates it in the ring system, by the kind and extent of the conjugation present, and by some stereochemical factors.

---

† Závada, Sicher and Krupička,[19] however, found a more negative value for 1-bromoadamantane.

Table IX-11
Effect of Ring Size on Reversible Oxidation–Reduction Potentials[41,42]
(10% acetic acid, 40% ethanol)

	n	$E_{1/2}$, V	
	9	−0.029	—
	10	+0.026	+0.081
	11	+0.086	—
	12	+0.117	+0.178
	13	+0.139	+0.200
	14	+0.143	+0.192
	15	+0.160	+0.211
	16	+0.151	+0.205
	17	+0.158	+0.202
	18	+0.144	+0.190
	19	+0.144	+0.200

(a) Effect of Ring Size. The effect of ring size on reversible oxidation-reduction systems has been studied systematically using the reduction of 2,6-polymethylene-*p*-benzoquinones and the oxidation of 2,6-polymethylene-4-aminophenols[41,42] (Table IX-11). It has been shown that a decrease in ring size below $n = 14$ causes a shift in the oxidation-reduction potential towards negative values, when compared with the value for dialkyl derivatives. For smaller rings, therefore, the stability of the quinone relative to the hydroquinone is greater. Two quinonoid systems in a 16-membered ring show no mutual interaction, which was demonstrated by the coincidence of the half-wave potential of 2,2′:6,6′ bispentamethylene di-*p*-benzoquinone (LXXXVI) with that of higher 2,6-polymethylene-*p*-benzoquinones.[43]

(LXXXVI)

Table IX-12

Half-Wave Potentials of Naphthylalkenes[2,44] [75 % dioxane–water, 0.1 M N(C₄H₉)₄I]

Compound	$n E_{1/2}(V$ vs. SCE)				
	1	2	Δ^a	3	Δ
(CH₂)ₙ naphthalenyl cyclopentenyl	−2.27	−2.46	−0.19	−2.41	−0.14
(CH₂)ₙ CH₃ naphthalenyl cyclopentenyl	(−2.42)	(−2.46)	(−0.04)	—	—
(CH₂)ₙ CH₃ naphthalenyl cyclopentenyl	−2.36	(−2.42)	(−0.06)	—	—
naphthalenyl (CH₂)ₙ cyclopentenyl	−2.30	−2.38	−0.08	−2.35	−0.05
CH₃ naphthalenyl (CH₂)ₙ cyclopentenyl	−2.37	−2.42	−0.05	—	—

^{a} Δ = Difference relative to cyclopentenyl derivative.

Among irreversible systems, unsaturated hydrocarbons will be discussed first. The half-wave potentials of naphthylalkenes[2,44] (Table IX-12) roughly follow the sequence found for the relative values of the conjugative power C_t as determined on the basis of ultraviolet absorption spectra and of the rates of maleic anhydride addition in the Diels–

Alder reaction. It is assumed[44] that the half-wave potential tends to be a function of the angle of the twist in the moment of the electrode process (ϑ_p) rather than of the angle of twist in the original molecule (ϑ). The electric field of the electrode will, if possible, orient the depolarizer preferably flatwise in the double layer, so that $\vartheta_p < \vartheta$. The inspection of molecular models[44] indicates little or no hindrance to the attainment of coplanarity in naphthalenes bearing the substituent in position 2. For these compounds it is assumed that in the course of the electrode process they became virtually coplanar, i.e., that $\vartheta_p \approx 0°$.

For (1-cyclopenten-1-yl)naphthalenes there is only a small difference between the half-wave potentials of 1- and 2-substituted compounds. The direction of the observed shift toward more negative values for 2-naphthyl derivatives is in accordance with the prediction from C_t values. Hence, it is assumed that, like 2-naphthyl derivatives, the molecule of 1-(1-cyclopenten-1-yl)naphthalene is able to achieve an almost planar arrangement during the electrode process ($\vartheta_p \approx 0°$). For 1-(1-cyclohexen-1-yl) and 1-(1-cyclohepten-1-yl) derivatives the steric hindrance to the rotation of the alkenyl group into coplanarity is no longer negligible and results in shifts in the half-wave potentials of 1-naphthyl derivatives toward more negative values.[44]

Further evidence for this kind of deduction can be drawn from the behavior of methyl derivatives. The shifts caused by the introduction of the methyl group into the cycloalkenyl group (-0.07 V or -0.04 V, respectively) can give us a rough idea of the magnitude of the polar effect resulting from methyl substitution in the cyclopentenyl ring. For 1-(1-cyclopenten-1-yl)naphthalene the observed shift (-0.17 V) can be attributed partly to steric hindrance to coplanarity. For 1-(1-cyclohexen-1-yl)naphthalene the introduction of the methyl group is practically without effect, showing that in the unsubstituted 1-(1-cyclohexen-1-yl) derivative there is a negligible contribution from the conjugative interaction of the two rings. Similarly, only a slight effect is produced by the introduction of a methyl group into position 8 of the naphthalene ring in 1-(1-cyclohexen-1-yl)naphthalene. The greater shift (and toward more negative potentials) of -0.09 V caused by an 8-methyl group in 1-(1-cyclopenten-1-yl)naphthalene indicates that this shift is, at least, partly due to steric hindrance to coplanarity, even though such substitution is less effective than the substitution of the methyl group in the cyclopentenyl ring.

Whereas, in the hydrocarbons discussed the reduction of the double bond occurs at more positive potentials in the cyclopentenyl

ring than in the cyclohexenyl ring, the reduction of the system COCH=
CH (to $COCH_2CH_2$) follows the reverse order. Both in α,β-unsaturated
ketones,[16] such as (LXXXVII) and (LXXXVIII) [in dimethylformamide
with 0.1 M $N(C_4H_9)_4I$ and 0.01 M $N(C_2H_5)_4I$], and in α,β-unsaturated
lactones[45] (in which the reduction of the double bond occurs at
potentials about 0.4 V more positive than in the corresponding cardiac
glycosides with an unsaturated five-membered lactone ring) the reduction of the six-membered compound occurs at more positive
potentials than that of the five-membered compound.

(LXXXVII)   − 1.66 V   (LXXXVIII)   − 1.85 V

For monocyclic terpene ketones,[46,47] the difference between the
five- and six-membered compound is rather small [(LXXXIX) to
(XCII) in 80 % ethanol with 0.1 M $N(C_2H_5)_4I$)]; no direct comparison
can be made, however, because there are two methyl groups on the
double bond in isothujone (XCII), but only one in (LXXXIX) and (XC).
That the introduction of the second alkyl group can have a considerable
effect on the half-wave potential is demonstrated by a comparison of

(LXXXIX)   (XC)   (XCI)   (XCII)
−1.77 V   −1.80 V   −1.81 V   −1.81 V

the half-wave potentials[47] of jasmone (XCIII) and isodihydrojasmone
(XCIV):

CH_2CH = CHCH_2CH_3   CH_2CH_2CH_2CH_2CH_2CH_3

(XCIII)   − 2.00 V   (XCIV)   − 1.87 V

As examples of electrode processes which result in the formation
of a new bond, we shall now discuss the reduction of vicinal dibromides.
In the main, these compounds are reduced[48] in a one, two-electron

Table IX-13
Half-Wave Potentials of Vicinal Dibromides[18]

No.	Substance	cis			trans		
		$I.10^3$	$E_{1/2}$, V	$\dfrac{2.3RT}{\alpha nF}$	$I.10^3$	$E_{1/2}$, V	$\dfrac{2.3RT}{\alpha nF}$
1	1-t-Butyl-trans-3-cis-4-dibromocyclohexane	—	—	—	2.30	-0.86ᵃ	0.359
2	1-t-Butyl-cis-3-trans-4-dibromocyclohexane	2.89	-1.67ᵃ	0.180	—	—	—
3	2β,3α-Dibromo(4aα-H,8aβ-H)decalin	—	—	—	3.27	-0.82ᵃ	0.374
4	exo-cis-2,3-Dibromobicyclo[2.2.1]heptane	3.32	-1.53	0.150	—	—	—
5	endo-cis-2,3-Dibromobicyclo[2.2.1]heptane	3.78	-1.21	0.200	—	—	—
6	trans-2,3-Dibromobicyclo[2.2.1]heptane	—	—	—	4.10	-1.56	0.216
7, 8	2,3-Dibromobicyclo[2.2.2]octane	4.05	-1.28	0.186	2.82	-1.33₅	0.189
9	1,2-Dibromocyclopentane	—	—	—	3.60	-0.94	0.215
10, 11	1,2-Dibromocyclohexane	2.62	-1.64	0.182	2.98	-1.04	0.247
12	1,2-Dibromocycloheptane	—	—	—	3.25	-1.00	0.288
13	1,2-Dibromocyclooctane	—	—	—	2.08	-1.05	0.346
14	5,5,8,8-Tetramethyl-trans-1,2-dibromocyclodecane	—	—	—	2.67	-1.15	0.313
15, 16	1,2-Dibromocyclodecane	2.13	-1.51	0.230	2.53	-1.44	0.182
17	1,2-Dibromocyclotetradecane	2.21	-1.33	0.312	—	—	—
18	1,2-Dibromocyclohexadecane	2.06	-1.21	0.440	—	—	—
19	Ethylene dibromide	3.28	-1.23ᵃ	0.185	—	—	—
20	erythro-5,6-Dibromodecane	3.01	-1.14	0.386	—	—	—
21	threo-5,6-Dibromodecane	—	—	—	2.71	-1.24	0.340
22, 23	1,6-Dibromocyclodecane	3.16	-1.78	0.210	2.52	-1.73	0.178

ᵃ The inclusion of the value in this column is deliberate.

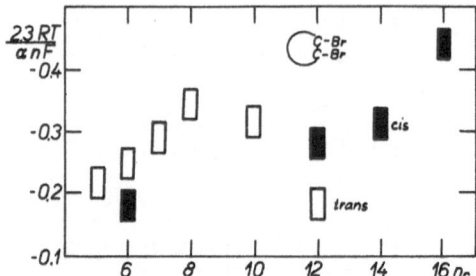

Fig. IX-8. Relation of values of 2.3 $RT/\alpha nF$ for the reduction waves of cyclic vicinal dibromides to the number of ring atoms. Values of 2.3 $RT/\alpha nF$ from Ref. 19 for trans ☐ and cis ■ isomers.

step. Since the main product of this step has been shown to be the unsaturated hydrocarbon, and since no other wave is apparent on the polarographic curve which could correspond to the reduction of the monobromo derivative, it may be assumed that an elimination process takes place with participation of the second bromine atom in the formation of the transition state.

To what extent this assumption is valid can be verified by an inspection of Table IX-13. For monobromo derivatives, for which the two-electron process can be taken for granted, the value of the limiting current constant I varied under identical experimental conditions between $I = 2.2$ and 2.9. All values of I in Table IX-13 that are greater than about $I = 3.2$ seem to indicate that in these cases, apart from the two-electron elimination process, a four-electron substitution can participate in the electrode process. An indication that this type of process may occur is given in the statement[48] that controlled-potential electrolysis of ethylene dibromide at a mercury pool electrode gives rise to some ethane, in addition to the main product that is ethylene. Controlled-potential electrolysis and microcoulometry using the mercury dropping electrode would be of importance for checking this, especially in the case of compounds No. 5, 6, 7, 9, and 12 (Table IX-13). The two-electron reduction of 1,6-dibromocyclodecanes (No. 22 and 23) seems to indicate transannular cyclization to decalin.

In homologous series and for those reaction series in which the properties and shape of the molecule do not vary over too wide a range it has been repeatedly proved that the value of the parameter 2.3 $RT/\alpha nF$ varies relatively little. For vicinal dibromides, on the other hand, the variation of this parameter is quite considerable (from 0.15 to 0.44;

Table IX-13). Since there is no correlation between this value and the value of I, a change in the overall number of electrons n transferred does not seem to be the essential reason for this change. Nevertheless, a change in mechanism involving a change in the number of electrons transferred in the potential-determining step does not seem very probable, though it cannot be completely excluded. A change in the value of the transfer coefficient α seems to be the deciding factor. It can be assumed that this change is connected with the orientation of the electroactive substance or with its adsorbability. As no information is available on the relation between the adsorbability and chemical structure of compounds of the type discussed, the assumption must be regarded as only tentative.

The values of $2.3\,RT/\alpha nF$ form two groups—one with $2.3\,RT/\alpha nF = 0.18 \pm 0.04$, and the second with $2.3\,RT/\alpha nF = 0.36 \pm 0.07$. The dibromides of the first group can in most, but not all, cases attain an *anti*-periplanar arrangement of the bromine atoms with relative ease, whereas, for those belonging to the second group such an arrangement can be attained either not at all or only with difficulty.[18] The relation of the value of $2.3RT/\alpha nF$ to the number of ring carbons is given in Fig. IX-8.

It has been stressed several times in previous chapters that quantitative comparisons of half-wave potentials can be made only in reaction series in which the value of αn remains constant. As this condition is not fulfilled for vicinal dibromides, the discussion of structural effects on half-wave potentials (Table IX-13) given below should be regarded as a first approximation.

Compounds No. 1–8 (Table IX-13) with a rigid structure, in which the torsion angle between the C—Br bonds is known, are considered first. The graph showing the dependence of half-wave potentials on torsion angle φ (Fig. IX-9A) resembles the Karplus curve.[49] The most positive waves were observed for $\varphi \approx 180°$ (*anti*-periplanar arrangement) and $0°$ (*syn*-periplanar arrangement), and the most negative were observed for $\varphi \approx 60-120°$. Only the value for compound No. 4 does not follow a smooth curve. An analogous trend also is shown by the relation of $RT/\alpha nF$ to the torsion angle (Fig. IX-9B).

Flexible dibromides, which exist as equilibria of two or more forms include compounds which are known[18] to be capable of readily attaining an *anti*-periplanar arrangement of the C—Br bond (No. 9, 12–14, Table IX-13), which is reduced at relatively positive potentials. They would correspond to rigid compounds with torsion angles of

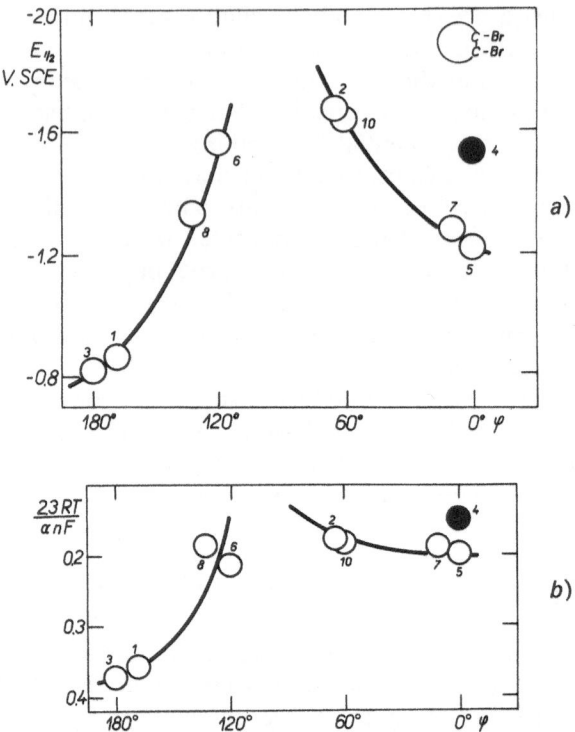

Fig. IX-9. Relation of *a*) half-wave potentials, and *b*) values of 2.3 $RT/\alpha n$F, for the reduction waves of cyclic vicinal dibromides to estimated values of the torsion angle (φ). Half-wave potentials, values of 2.3 $RT/\alpha n$F, and estimated values of the torsion angle from Ref. 19; numbering of compounds according to Table IX-13, full points deviate.

$\varphi > 150°$. On the contrary, those dibromides (No. 11, 15, 16) which attain an *anti*-periplanar arrangement only with a great expenditure of energy,[18] or not at all, are reduced at more negative potentials, which would correspond to rigid dibromides with $\varphi \approx 60°$.

The half-wave potentials of monocyclic, flexible vicinal dibromides, as well as of some rigid dibromides, are approximately linear functions of logarithms of rates of homogeneous elimination reactions.[50] The more positive reduction of *erythro*-5,6-dibromo decane (No. 20, Table IX-13) than that of the *threo* isomer (No. 21) is also in agreement with results obtained with bimolecular homogeneous reactions.

(b) Effects of Position and Conjugation. The discussion in this section is restricted to the reduction of α,β-unsaturated keto steroids and their derivatives.

Whereas, for keto steroids considerable differences in half-wave potential have been observed[8] according to the position of the $COCH=CH$ system [(XCV) to (C) in 90% ethanol, barbital buffer, pH 8.5], little difference has been found for the corresponding betainylhydrazones[8] [(CI) and (CII)].

(XCV) -1.55 V	(XCVI)	no wave
(XCVII) -1.18 V	(XCVIII)	-1.24 V
(XCIX) -1.16 V	(C)	-1.10 V

$R = NHCOCH_2\overset{+}{N}(CH_3)_3Cl^-$

(CI) -1.34 V	(CII)	-1.34 V

The compounds (XCIX) and (C) (in toluenesulfonate, pH 2.1, 80% ethanol[13]) differ in more than the position of the double bond, but the effect of other changes probably does not predominate.

Increasing conjugation shifts the half-wave potentials toward more positive values (Table IX-14). $\Delta^{1,4}$ and $\Delta^{4,6}$ 3-keto steroids are reduced at more positive potentials than their Δ^4 analogs by about

Table IX-14

Effect of Conjugation on the Reduction of Keto Steroids

Δ^4	$\Delta^{1,4}$	$\Delta^{4,6}$	Ref.
C₈H₁₇ structure, -1.55 V	—	C₈H₁₇ structure, -1.18 V $(0.37$ V$)$	8
OH structure, -1.38_5 V	OH structure, -1.23_5 V $(0.15$ V$)$	—	15
COCH₃/OCOCH₃ structure, -1.37_5 V	COCH₃/OCOCH₃ structure, -1.22 V $(0.15_5$ V$)$	COCH₃/OCOCH₃ structure, -1.06_5 V $(0.31$ V$)$	15
COCH₃/OCOCH₃ structure, -1.38 V	COCH₃/OCOCH₃ structure, -1.22 V $(0.16$ V$)$	COCH₃/OCOCH₃ structure, -1.09_5 V $(0.29$ V$)$	15

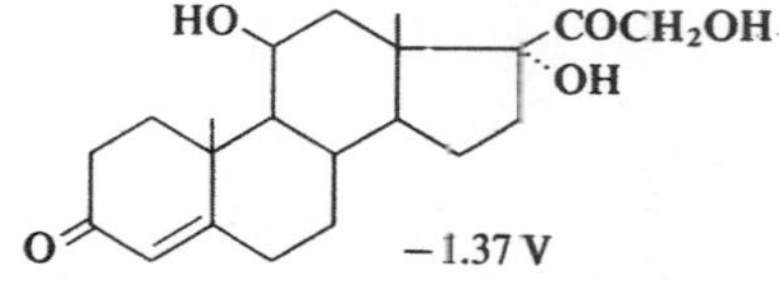

9

9

16

16

8

COCH$_2$OH
OH
HO
−1.21 V
(0.16 V)

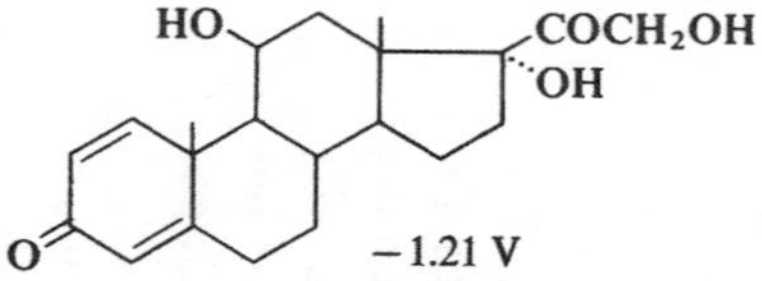

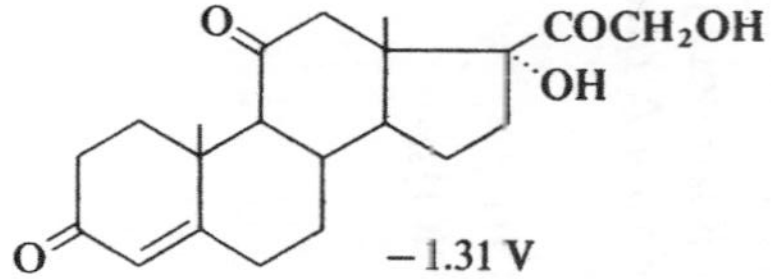

COCH$_2$OH
OH
HO
−1.37 V

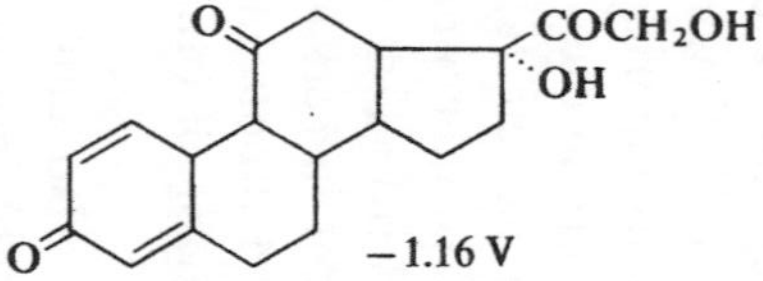

COCH$_2$OH
OH
O
−1.16 V
(0.15 V)

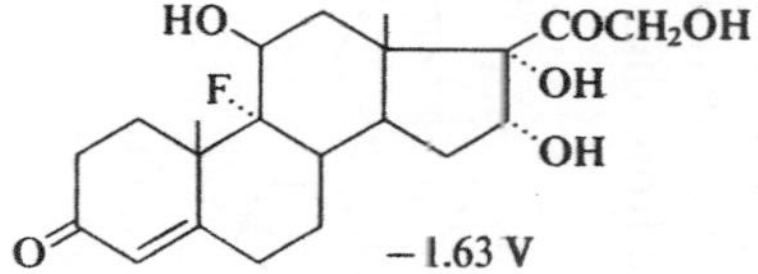

COCH$_2$OH
OH
O
−1.31 V

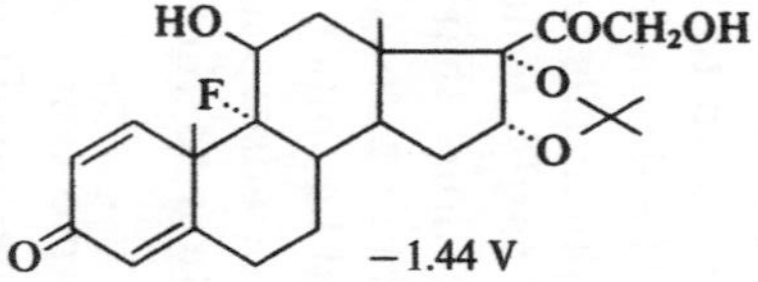

COCH$_2$OH
HO
F
−1.44 V
(0.19 V)

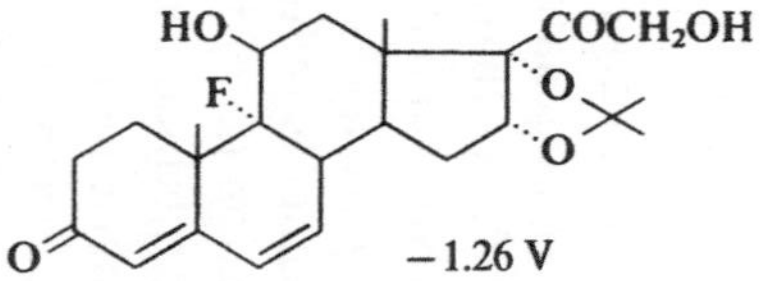

COCH$_2$OH
OH
OH
HO
F
−1.63 V

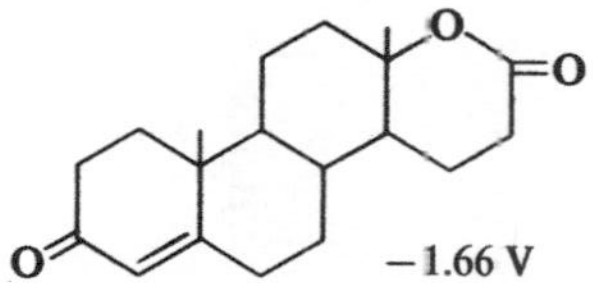

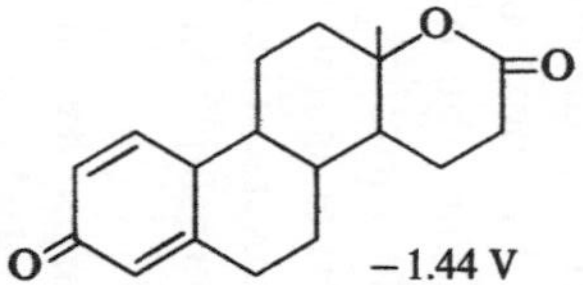

O
O
−1.44 V
(0.22 V)

O
O
−1.66 V

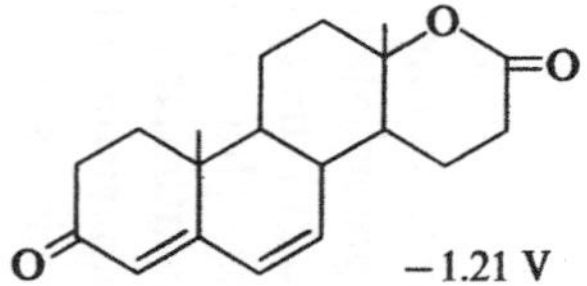

COCH$_2$OH
O   O
HO
F
−1.26 V
(0.37 V)

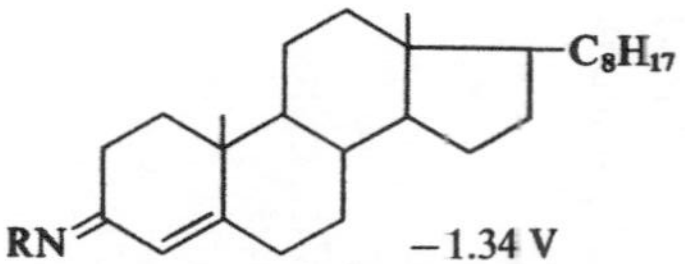

O
O
−1.21 V
(0.45 V)

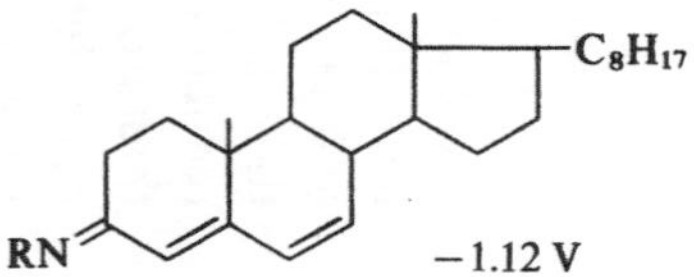

C$_8$H$_{17}$
RN
−1.12 V
(0.22 V)

C$_8$H$_{17}$
RN
−1.34 V

R = NHCOCH$_2$N(CH$_3$)$_3$Cl$^-$

0.16 and 0.30 V, respectively. Crossed conjugation is less effective than linear. The effect in betainylhydrazones seems slightly less marked.

(c) Steric Factors. A few comments can be added concerning the polarographic reduction of polyhalogenated compounds.

Inspection of Table IX-13 shows that, of the available pairs of *cis* and *trans* isomers, in one case (No. 7 and 8) the *cis* isomer is reduced at more positive potentials than the *trans* isomer, but in two cases (No. 10 and 11; 15 and 16) the *trans* isomer is reduced at more positive potentials. It has been shown in Section (*a*) of this chapter that the ease of reductive elimination is better explained on the basis of torsion angles φ than from comparisons of isomeric pairs. We, therefore, refrain from a more detailed discussion of, for example, the results given in Ref. 51. Nevertheless, the particular case of hexachlorocyclohexane should be mentioned. This substance is reduced to benzene[51-55] with the consumption of six electrons and the liberation of six chloride ions. The gamma isomer is reduced at potentials which are about 0.6 V more positive than the next most easily reducible isomer (alpha). This behavior parallels the observations that this isomer possesses the greatest dipole moment, is more strongly adsorbed at a mercury surface than other isomers, and has much greater insecticidal activity than other isomers. The role of adsorption and dipole moment has also been discussed[55] in relation to the behavior of other halogen derivatives.

The adsorptivity of organic compounds also can be studied polarographically by an examination of capacity phenomena and even the suppression of maxima.[36] When the 3-amino group in hollarhimine (CIII) is in the α-position, the compound is much more easily adsorbed at the surface of the mercury electrode than the β-isomer. It has been deduced that adsorption occurs from the rear of ring A of the steroid molecule.

(CIII)

4. Conclusions

If quantitative treatments similar to those used in other groups of compounds are applied in the future to alicyclic systems, it will be necessary to carry out even more extensive polarographic research in this field than in other groups of compounds, because of the great variability of alicyclic systems. In this chapter we have included more qualitative information than in others: this was done with the aim of demonstrating the possibilities of the application of polarography in the solution of problems in this field and of exposing the gaps that exist in our knowledge.

References

[1] Krupička, J., *Collection Czech. Chem. Commun.* **24**: 2324 (1959).

[2] L. H. Klemm, W. Hodes and W. B. Schaap, *J. Org. Chem.* **19**: 451 (1954).

[3] Y. Asahi, *J. Pharm. Soc. Japan* **77**: 128, 132, 136, 140, 145 (1957).

[4] P. Zuman and V. Horák, unpublished results.

[5] Ch. Prévost and P. Souchay, *Chim. Anal.* **37**: 3 (1955).

[6] D. H. R. Barton and A. J. Head, *J. Chem. Soc.* **1956**: 932.

[7] P. Zuman, J. Tenygl, and M. Březina, *Chem. Listy* **47**: 1152 (1953); *Collection Czech. Chem. Commun.* **19**: 46 (1954).

[8] D. M. Robertson, *Biochem. J.* **61**: 681 (1955).

[9] P. Kabasakalian and J. McGlotten, *J. Am. Chem. Soc.* **78**: 5032 (1956).

[10] P. Zuman, *J. Electrochem. Soc.* **105**: 758 (1958).

[11] P. Kabasakalian and J. McGlotten, *J. Electrochem. Soc.* **105**: 261 (1958).

[12] P. Kabasakalian and J. McGlotten, *Anal. Chem.* **31**: 1091 (1959).

[13] V. Delaroff, M. Bolla, and M. Legrand, *Bull. Soc. Chim. France* **1961**: 1912.

[14] P. Kabasakalian and J. McGlotten, *Anal. Chem.* **34**: 1440 (1962).

[15] O. Hrdý, *Collection Czech. Chem. Commun.* **27**: 2447 (1962).

[16] A. I. Cohen, *Anal. Chem.* **35**: 128 (1963).

[17] F. L. Lambert and K. Kobayashi, *J. Am. Chem. Soc.* **82**: 5326 (1960).

[18] J. Krupička, J. Závada, and J. Sicher, *Collection Czech. Chem. Commun.* (1966).

[19] J. Závada, J. Krupička, and J. Sicher, *Collection Czech. Chem. Commun.* **28**: 1664 (1963).

[20] J. W. Sease, P. Chang, and J. L. Groth, *J. Am. Chem. Soc.* **86**: 3154 (1964).

[21] F. L. Lambert, A. H. Albert, and J. P. Hardy, *J. Am. Chem. Soc.* **86**: 3154 (1964).

[22] H. C. Brown, R. S. Fletcher, and R. B. Johannsen, *J. Am. Chem. Soc.* **73**: 212 (1951).

[23] H. C. Brown and M. Borkowski, *J. Am. Chem. Soc.* **74**: 1894 (1952).

[24] J. D. Roberts and V. C. Chambers, *J. Am. Chem. Soc.* **73**: 5034 (1951).

[25] S. F. Van Straten, R. V. V. Nichols, and C. A. Winckler, *Can. J. Research* **29**: 372 (1951).

[26] L. Schotsmans, P. J. C. Fierens, and T. Verlie, *Bull. Soc. Chim. Belg.* **68**: 580 (1959).

[27] P. J. Fierens and P. Verschelden, *Bull. Soc. Chim. Belg.* **61**: 427, 609 (1952).

[28] N. S. Hush and K. B. Oldham, *J. Electroanal. Chem.* **6**: 34 (1963).

[29] V. Prelog and O. Häfliger, *Helv. Chim. Acta* **32**: 2088 (1949).

[30] M. Mühlstädt and R. Herzschuh, *J. Prakt. Chem.* (4) **20**: 20 (1963).

[31] P. Souchay and M. Graizon, *Chim. Anal.* **36**: 85 (1954).

[32] M. Březina and P. Zuman, *Chem. Listy* **47**: 975 (1953).

[33] N. J. Leonard, H. A. Laitinen, and E. H. Mottus, *J. Am. Chem. Soc.* **75**: 3300 (1953).

[34] P. Zuman and F. Šantavý, *Chem. Listy* **46**: 393 (1952); *Collection Czech. Chem. Commun.* **18**: 28 (1953).

[35] K. Schwabe, G. Ohloff, and H. Berg, *Z. Elektrochem.* **51**: 293 (1953).

[36] P. Zuman and V. Černý, *Chem. Listy* **52**: 1468 (1958).

[37] A. M. Wilson and N. L. Alinger, *J. Am. Chem. Soc.* **83**: 1999 (1961).

[38] P. Zuman and J. Michl, *Nature* **192**: 655 (1961).

[39] P. Zuman and S. Tang, *Collection Czech. Chem. Commun.* **28**: 829 (1963).

[40] Q. Q. Dang, Thesis, Faculté des Sciences, Paris, 1961.

[41] V. Prelog, O. Häfliger, and K. Wiesner, *Helv. Chim. Acta* **31**: 877 (1948).

[42] V. Prelog, K. Wiesner, W. Ingold, and O. Häfliger, *Helv. Chim. Acta* **31**: 1328 (1948).

[43] V. Prelog, K. Wiesner, and O. Häfliger, *Collection Czech. Chem. Commun.* **15**: 900 (1950).

[44] L. H. Klemm, C. D. Lind, and J. T. Spence, *J. Org. Chem.* **25**: 611 (1960).

[45] F. Šantavý, O. Čapka, and J. Malinský, *Collection Czech. Chem. Commun.* **15**: 953 (1950).

[46] K. Schwabe and H. Berg, *Z. Elektrochem.* **56**: 961 (1952).

[47] K. Schwabe, G. Ohloff, and H. Berg, *Z. Elektrochem.* **57**: 34 (1953).

[48] M. v. Stackelberg and W. Stracke, *Z. Elektrochem.* **53**: 118 (1949).

[49] M. Karplus, *J. Chem. Phys.* **30**: 11 (1959).

[50] J. Závada, private communication.

[51] M. Nakazima, Y. Katumura, and T. Okubo, "Proceedings of the First International Polarography Congress," Vol. I, p. 173, Prague, 1951.

[52] G. B. Ingram and H. K. Southern, *Nature* **161**: 437 (1948).

[53] K. Schwabe and H. Frind, *Z. Physik. Chem.* **196**: 342 (1951).

[54] W. Kemula and A. Cisak, *Roczniki Chem.* **28**: 275 (1954).

[55] A. Cisak, *Roczniki Chem.* **37**: 1025 (1963).

[56] P. Zuman, *Talanta* **12**: 1137 (1965).

Author Index

(Pages on which a full reference is quoted are given in italics)

Subject Index

375